15TH
INTERNATIONAL
ANAMED ANNUAL
SYMPOSIUM

Winds of Change: Environment and Society in Anatolia

Koç University Research Center for Anatolian Civilizations (ANAMED) Series

KOÇ UNIVERSITY SUNA KIRAÇ LIBRARY CATALOGING-IN-PUBLICATION DATA
Koç University Research Center for Anatolian Civilizations (ANAMED) Series
International ANAMED Annual Symposium, (15th : 2021 : Istanbul, Turkey)
15th International ANAMED Annual Symposium: Winds of change: environment and society in Anatolia that was held at the Koç Üniversity Research Center for Anatolian Civilizations on 29-30 April 2021 / Bülent Arıkan [and twenty four others] ; editors Christopher H. Roosevelt, John Haldon ; copy editor Tara Ingman ; production editor Alican Kutlay ; book and cover design Burak Şuşut ; cover images Alican Kutluay, Charlotte Pearson ; symposium coordinator Naz Uğurlu ; project manager Buket Coşkuner ; ANAMED director Chris Roosevelt. Istanbul : Koç University, 2021.
308 pages ; 19,5 × 25 cm.
ISBN 978-605-7685-70-4
1. Turkey--Antiquities. 2. Human ecology--Turkey--History. 3. Social change--Turkey--History. 4. Excavations (Archaeology)--Turkey. 5. Turkey--Civilization. I. Arıkan, Bülent. II. Haldon, John. III. Ingman, Tara. IV. Roosevelt, Christopher H. V. Kutlay, Alican. VI. Şuşut, Burak. VII. Uğurlu, Naz. VIII. Coşkuner, Buket. IX. Title.
DR431.I58 2021

EDITORS
Christopher H. Roosevelt
John Haldon

AUTHORS
Bülent Arıkan
Eleni Asouti
Semih Çelik
Mustafa Doğan
Uğur Doğan
Warren J. Eastwood
Hugh Elton
Josh England
Patrick Free
Anneley Hadland
John Haldon
Adam Izdebski
Ceren Kabukcu
Ahmet Köse
Fikri Kulakoğlu
Christina Luke
Lee Mordechai
James Newhard
Aziz Ören
Charlotte Pearson
Neil Roberts
Christopher H. Roosevelt
Çetin Şenkul
Tevfik Emre Şerifoğlu
Madelynn von Baeyer

COPY EDITOR
Tara Ingman

PRODUCTION EDITOR
Alican Kutlay

BOOK AND COVER DESIGN
Burak Şuşut, Fika

COVER IMAGES
Alican Kutlay
Charlotte Pearson

PRE-PRESS PRODUCTION
Fika

SYMPOSIUM COORDINATOR
Naz Uğurlu

ANAMED MANAGER
Buket Coşkuner

ANAMED DIRECTOR
Chris Roosevelt

PRINT
12. Matbaa
İbrahim Karaoğlanoğlu Cad. No. 35
34418 Kağıthane, Istanbul
T. +90 212 281 25 80
info@onikincimatbaa.com
Certificate No: 46618

PUBLISHER
KOÇ UNIVERSITY PRESS
Rumelifeneri Yolu, 34450
Sarıyer, Istanbul/Turkey
T. +90 212 338 10 00
kup.ku.edu.tr
kup@ku.edu.tr
Certificate No: 18318

FIRST EDITION
Istanbul, November 2021

ISBN
978-605-7685 70-4

© 2021. This book has been published on the occasion of the 15th International ANAMED Annual Symposium "Winds of Change: Environment and Society in Anatolia" that was held at the Koç University Research Center for Anatolian Civilizations (ANAMED) on 29-30 April 2021.

WINDS OF CHANGE

Environment and Society in Anatolia

EDITED BY
CHRISTOPHER H. ROOSEVELT AND JOHN HALDON

ANAMED
KOÇ UNIVERSITY
RESEARCH CENTER FOR
ANATOLIAN CIVILIZATIONS

KUP

Table of Contents

Preface

The papers in this volume bring together a selection of research presented at the 15th International ANAMED Annual Symposium (IAAS), organized by John Haldon of Princeton University and its Climate Change and History Research Initiative (or CCHRI) and Christopher H. Roosevelt of Koç University and its Research Center for Anatolian Civilizations (or ANAMED) and bearing the same name as this volume: "Winds of Change: Environment and Society in Anatolia." The CCHRI was established in 2013 in order to promote and better integrate collaborative work between paleoenvironmental and paleoclimatic science on the one hand and historical and archaeological study on the other. ANAMED was established in 2005 in order to support research on the archaeology, art, heritage, and history of Anatolia and to make it publicly accessible via fellowships, exhibitions, and library resources, as well as symposia and publications. In organizing the "Winds of Change" symposium and assembling this publication, we aimed to bring together the foci of these institutions to highlight the range and depth of recent research exploring the varied and dynamic interactions between environment and society in the specific geography of Anatolia.

Over the last decade or so, interest has accelerated significantly in relationships between environmental and climate factors, on the one hand, and the history, or histories, of human societies, on the other. Anatolia is one of the great "cradles of civilization" in the Old World, along with the fertile crescent, the Indus valley, and the Yellow River plain in China. Its complex and highly differentiated landscapes lie at the intersection of three climate systems converging from Asia, Europe, and Africa, which means that Anatolian geographies offer numerous opportunities for exploring human-environment relationships over their long history of human activity.

Understanding precisely how environmental, especially climatic, disruptions affect human societies is of course not just the work of archaeologists and historians, just as understanding how human activities affect local environments (and more recently global climates) is not just the task of (paleo-)climate and environmental scientists.

The range of correlations and casual relationships between climatic, environmental, and societal phenomena is non-linear and complex, requiring collaborative approaches to bring diverse evidence to bear. These subjects have begun to attract a great deal of attention from the public and policy makers, as well, with the results of research into past human-environment dynamics ever more likely to influence present and future planning efforts. It is all the more crucial, then, that such understandings be based on the results of holistic, collaborative work among researchers looking across the divides that typically separate their fields. Only by pulling together approaches and results from disparate disciplines can we hope to come to terms with some of the complexity that confronts us and to avoid simplistic, deterministic, or overly dramatic answers.

As we hope the case studies in this volume demonstrate, collaborations that integrate archaeological, written, environmental, and climate records allow exploration of the causal mechanisms underlying societal change across millennia and highlight the social origins of significant landscape and vegetation changes. Some chapters offer regional, microregional, or site-based studies that throw significant light on questions that until recently could not be answered, much less addressed, satisfactorily. Others challenge assumptions about the impact of climatic factors on socio-political organization and medium-term historical change, further highlighting the importance of collaboration between archaeologists, historians, and environmental and climate scientists. Despite their diversity and whether based on fieldwork results, laboratory or computer analyses, or archival research, all chapters are united by the shared challenges presented by the geography of Anatolia and a common goal of improving understandings of entangled environmental and social systems and the range of complex relationships between them.

For allowing us to include their work as examples of the types of approaches we think are required for such research, we are very grateful to the contributors to this volume. All have been extremely patient and understanding with our process—from the postponement of the original symposium owing to pandemic precautions, to its reincarnation as an online event, to our perpetual pestering to adhere to a rapid publication schedule—and we remain in their debt. We are similarly thankful to those who participated in the symposium yet were unable to contribute to the volume, including Sena Akçer-Ön, Peter Biehl, Namık Çağatay, Jean-Pascal Dumoulin, Yılmaz Selim Erdal, Sevinç Günel, Ali Gürel, Catherine Kuzucuoğlu, Tom Maltas, Zeki Bora Ön, Mehmet Sakınç, Özlem Sert, Mehmet Somel, Vasıf Şahoğlu, Ralf Vandam, Nükhet Varlık, Patrick T. Willett, Elena Xoplaki. We would like to thank Neil Roberts, in particular, for setting the tone of the symposium and this volume alike, with his keynote lecture and Prologue chapter, respectively; his approach exemplifies the holistic panoply of methods we believe will continue to be the mainstay of this research area for some time to come.

Many others deserve our thanks for financial, administrative, logistical, and personal support of both the symposium and this volume, beginning with the Vehbi

Koç Foundation (VKV) and Koç University (KU), which provide ANAMED's institutional framework, along with Ömer M. Koç, VKV President Cihan Özsönmez, KU President Umran S. İnan, KU Vice President for Research and Development İrşadi Aksun, and College of Social Sciences and Humanities Dean Aylin Küntay. All ANAMED activities result from the enthusiastic work of a dedicated team, and it is a pleasure to thank them here: Buket Coşkuner, Duygu Tarkan, Yağmur Uyanık, Murat Türk, Mukaddes Gül, Vasia Mole, İrem Ünal, and Defne Gier. Among this ANAMED team, we give special thanks to Naz Uğurlu, for so professionally transforming the symposium from an in-person event to an online format, and to Alican Kutlay, for his always smooth and dedicated coordination of publication processes. For copyediting, translation, proofreading, design, personal support, and other tangible and intangible efforts that improved this work substantively, we offer further thanks to Tara Ingman, Burak Şuşut, Christina Luke, Dalila Alberghina, Tunç Kaner, and Orhan Efe Yavuz.

As a final note to this brief preface, we would like to make clear that the title of the symposium and volume was selected for its appropriateness to the subject, rather than in reference to either the Scorpions' 1990 rock ballad "Winds of Change" or Patrick Radden Keefe's very entertaining 2020 podcast on Cold War intrigues about that song, about which we learned later. Nonetheless, we hope you may find the volume as enjoyable, if not as melodic or mysterious, as both!

Chris Roosevelt & John Haldon

Abbreviations

aDNA	Ancient DNA
ANAMED	Koç University Research Center for Anatolian Civilizations
AP	Arboreal pollen
ARIT	American Research Institute in Turkey
BIAA	British Institute at Ankara
BOP	Beyşehir Occupation Phase
BP	Before present
CCHRI	Climate Change and History Research Initiative
EBA	Early Bronze Age
EBZ	Early Byzantine
EIA	Early Iron Age
ERO	Early Roman
FEDD	Fruits of Eurasia, Domestication and Dispersal
GCM	Global Circulation Model
GI	Lateglacial Interstadial
GIS	Geographic Information Systems
GS	Greenland Stadial
IntCal20	Northern Hemisphere International Atmospheric Calibration Curve, version 2020
IRD	Ice-rafted debris
JP	Justinianic Plague
KU	Koç University
LALIA	Late Antique Little Ice Age
LBA	Late Bronze Age
LCh	Late Chalcolithic
LM	Late Minoan
LRO	Late Roman
Masl	Meters above sea level
MBA	Middle Bronze Age
MBZ	Middle Byzantine
MCM	Macrophysical Climate Modeling
MRO	Middle Roman
MTK	Medieval Turkish
N	Number
NAO	North Atlantic Oscillation
OPDA	Ottoman Public Debt Administration
OTT	Ottoman
PMO	Prime Ministry Ottoman Archives
PVS	Istanbul Municipality Atatürk Library, Pertevniyal Valide Sultan Collection
SPD	Summed Probability Density
SDM	Species distribution modeling
SNP	Single nucleotide polymorphism
SUNY	State University of New York
TÜBİTAK	Scientific and Technological Research Council of Turkey
UCL	University College London
VERA	Vienna Environmental Research Accelerator
VKV	Vehbi Koç Foundation

PAPERS

PROLOGUE

Boon or Curse? The Role of Climate Change in the Rise and Demise of Anatolian Civilizations

Neil Roberts

Introduction

There is a paradox at the heart of the climate-population nexus. According to what might be termed the Standard Relationship, a favorable climate should increase food supply and hence allow population to grow. Similarly, adverse climatic conditions, such as drought, should be associated with demographic overshoot and even societal collapse. On the other hand, pre-industrial population levels rose substantially despite a long-term "deterioration" in climate after the mid-Holocene optimum (6000–4000 BCE) in most of the Old World. How can we explain this dilemma? In order to resolve this paradox, I argue here that climate change played an essential role in the process of societal transformation, notably following periods of crisis when population levels and prevailing climatic conditions were "out of sync." In the Standard Relationship, any demographic overshoot would be corrected by population reduction to the previous base level. In reality, the archaeological-historical record tells us that some societies responded to climate-related crises by innovating, for example via new irrigation technologies or via food storage and redistribution systems. Once these innovations were adopted by other societies, they shifted the population-climate nexus upwards to a new baseline state. This Inverted Relationship allowed population to rise over the *longue durée* even while climatic conditions became overall less favorable.

In this chapter, a conceptual cyclical model is set out that links climate dynamics and human demography via the Theory of Adaptive Change (Holling and Gunderson 2002). This model allows the Standard and Inverted Relationships to be reconciled because different causal mechanisms prevailed at different stages in the cycle. The model is illustrated and tested with reference to trends in climate, population, and societal complexity in southern Anatolia and northern Greece between the Neolithic and the early Iron Age. The discussion is then expanded to consider later time periods, including a "natural experiment" involving Ottoman responses to major drought events.

Initially, I briefly review the relevant literature linking climate and history and then describe the empirical evidence available for reconstructing past climatic change and demography, deriving from paleo-science and archaeology, respectively.

Context

The idea that changes in climate have acted as a stimulus for events in human history is a long-standing one. More than a century ago in his book *The Pulse of Asia*, Ellsworth Huntington (1907) hypothesized that periods of extended drought and severe winters risked decimating the livestock herds on which pastoral nomads depended, prompting waves of population migration out of central Asia. Climatic cycles, he argued, pushed mobile horse-riding "barbarians" outwards from dryland steppes to the rim of Eurasia, notably during the "Migration Period" (375–568 CE) and again in Medieval times (1205–1405 CE). During these key time intervals, the warlike invasions of the Huns, Alans, Mongols, and other nomads helped to destroy, disturb, and transform the sedentary civilizations of Europe, the Middle East, India, and China, and so changed the course of history. Although widely criticized at the time as over-deterministic, Huntington's ideas have parallels in more recent research, particularly in the work of Harvey Weiss who, like Huntington, is based at Yale University. Weiss (2017) has argued that some major discontinuities in the archaeological and historical record can be linked to episodes of abrupt climatic change. His best-known example is the "collapse" of the Akkadian empire at the end of the Early Bronze Age, which he linked to a period of severe drought around 2200 BCE, the "4.2 ka event" (Weiss et al. 1993 but see Lawrence, Palmisano, and de Gruchy 2021). There are some interesting similarities here with Huntington's explanatory model, with Weiss envisaging nomadic pastoralists from the desert steppe over-running the sedentary agrarian populations of the Fertile Crescent, notwithstanding Akkadian attempts to stop them by building a Trump-like great wall.

There are comparable intellectual continuities for other times of historical crisis, such as the Greek Dark Ages that separated the Bronze Age world remembered by Homer from that of Archaic Greece (from ca. 1150 to ca. 750 BCE). Drawing on the work of climatologist Reid Bryson (e.g., Bryson, Lamb, and Donley 1974), classicist Rhys Carpenter explained this—and also the post-antique—Dark Age as the result of adverse climatic changes in his *Discontinuity in Greek Civilization* (1966). This explanation is mirrored in more recent research, which frames the "collapse" of the Late Bronze Age World System across the eastern Mediterranean as a consequence of an abrupt decline in rainfall at the end of the second millennium BCE, the so-called "3.2 ka event," notably in the work of David Kaniewski (Kaniewski, Guiot, and Van Campo 2015) and others (e.g., Drake 2012). Some higher-order polities, such as the Hittite Empire, experienced significant political instability at this time, and archaeologically inferred population levels fell sharply in regions such as southern Greece (Weiberg et al. 2019). Some centuries later, the fall of the Neo-Assyrian Empire in the seventh century BCE has also been linked to drought conditions (Sinha et al. 2019). Recent research has similarly related the demise of Imperial Rome and the end of antiquity to climatic and environmental crises, for example in Kyle Harper's *The Fate of Rome* (2017). This highlighted the disruptive impact of both disease pandemics, such as the sixth century onset of the Justinianic Plague, and synchronous climatic cooling, notably during the so-called Late Antique Little Ice Age (536–660 CE) (Büntgen et al. 2016). The later main Little Ice Age has also been invoked by historians such as Geoffrey Parker (2013) as an underpinning cause of the general crisis of the seventeenth century CE in different regions of the world, from Europe to China via the Ottoman Empire.

This wide body of literature has argued for a close relationship between climatic variations and changes in complex societies at key transformative moments during the second half of the Holocene. Much of this work sees the relationship as a deterministic one, in which climatic adversity prompted societal decline or collapse, often inferred from archaeological evidence of regional site abandonment. But whether determinist or possibilist in character, the relationship between climate and society has been envisaged as one in which periods of favorable climate would expand the food supply, notably via higher crop yields, and hence allow human populations to grow. By the same logic, adverse climatic conditions, such as major droughts, have been linked to societal and demographic crises, as the food supply shrank and human populations exceeded the available resources. On the other hand, human populations have grown greatly since the emergence of farming economies in Neolithic times, while the climate in almost all of the Old World has become less favorable for food production over the same time period. Between 8000 and 3000 BCE, empirical evidence indicates a global thermal optimum followed by a gradual, but oscillating, cooling trend (Marcott et al. 2013; Kaufman et al. 2020). This downward temperature trend would have had its greatest impact at high latitudes but also affected high altitude and continental interior regions such as upland eastern Anatolia. Even more significant were the changes in hydrology caused by climatic fluctuations (Jones et al. 2019). Patterns of drought and flood are more spatially heterogenous than those of temperature (Seager et al. 2007), but, even so, it is clear that large areas of south and east Asia, along with northern Africa, experienced a long-term trajectory towards climatic desiccation after 4000 BCE (e.g., Ponton et al. 2012; Shanahan et al. 2015). This was linked to the weakening of the Indian Ocean and west Pacific monsoon circulation systems, as orbitally forced summer heating over northern land masses reduced; changes that led to the end of the Green Sahara and Arabia. A similar mid-Holocene drying trend affected most circum-Mediterranean lands (Finné et al. 2019).

According to the Standard Relationship, the mid-Holocene drying and cooling of the climate should have reduced food supply, lowered population carrying capacities, and led to a reversal in the course of human cultural development. Instead, the centuries between 6000 and 4000 BCE saw the rise of the first Old World civilizations.

In the previously "green" northeast Sahara, some human communities responded to mid-Holocene climatic desiccation by relocating into well-watered lands along the river Nile (Hassan 2002; Kuper and Kröpelin 2006). Here they used flood irrigation to provide the economic basis for pre-Dynastic Nilotic civilization that, within a millennium, would lead to the construction of the Pyramids of Giza. It is therefore clear that complex, urban, literate societies emerged in places like Egypt in the face of an unfavorable long-term trajectory of climate, rather than the reverse. Dan Lawrence and colleagues (2016) have recognized the same paradox for northern Mesopotamia, where demographic increase and the rise of urban populations coincided with a long-term aridification trend. Lawrence and colleagues argue that the development of urban settlements, complex polities, and later imperial powers controlling large territories allowed Near Eastern civilizations to become effectively decoupled from climate from the end of the Early Bronze Age. The evidence linking the demise of Late Bronze Age imperial powers such as the Hittites to late second millennium BCE mega-droughts or the role that climatic deterioration played in the disintegration of the Roman politico-economic world system would suggest otherwise. The emergence of complex societies during and after the Bronze Age inevitably changed societal relationships with climate, but increasing freedom from climatic constraints does not provide a satisfactory explanation for the climate-population paradox. Rather, it lies in the fact that climatic crises are capable of prompting societal innovation as well as, or instead of, economic and demographic decline, in what we could describe as an Inverted Model. Arlene Rosen (2007, ch. 7) provides an example of this process for the Levant during the Early Bronze Age, where the 4.2 ka mega-drought led to some cultural groups developing new irrigation and other hydro-technologies along with food storage and redistribution systems that allowed them not only to survive but subsequently to flourish, while other groups diminished in importance. Of course, innovations can potentially occur at any time, but their exploitation and dissemination take place faster and wider at times of necessity, just as most military innovations emerge during war rather than peace time.

Past Climate Changes: The Evidence

Climatic variations during the Holocene are recorded in a wide array of proxy data, although some of these are more useful than others if we want to relate them to long-term societal change. Ideally, paleoclimate indicators should be continuous through time, well-dated, and able to register short-term variations as well as long-term trends. From an archaeological point of view, they should also be geographically close to areas of human occupation. Using these criteria, some of the most useful paleoclimate proxies include tree rings (e.g., Cook et al. 2016), cave speleothems (e.g., Göktürk et al. 2011; Burstyn et al. 2019), and lake records. The last of these can reveal past climatic conditions via fluctuations in water levels, salinity, and isotope hydrology (e.g., Roberts et al. 2008). Ice cores and deep-sea sediment cores, although geographically remote, also provide key evidence for past climate change. Over multi-decadal climate changes, the advance and retreat of valley glaciers can be used to reconstruct changes in mass balance caused by winter snow accumulation and summer ablation. Additionally, documentary sources can record exceptional weather conditions, such as rivers freezing or flooding, that can be usefully compared against longer and more continuous records of climate change (e.g., McCormick et al. 2012; Haldon et al. 2014). These primary proxies can be distinguished from secondary and/or response variables, which are influenced not only by climate but also by human impact, along with ecological or geomorphological dynamics. This distinction is particularly significant in regions such as the eastern Mediterranean that have a longstanding human footprint. Secondary proxies include pollen evidence (reflecting past vegetation and land use; e.g., Gaillard et al. 2010), microcharcoals (indicating wildfire history; e.g., Turner, Roberts, and Jones 2008), onsite archaeological evidence (Fiorentino et al. 2008), and geomorphological data (e.g., alluvial histories; see Walsh et al. 2019).

Empirical paleoclimatic data are often presented stratigraphically against depth or time, but it is important to note these are only proxies for climate, rather than climate variables per se. To convert proxy data into true climate variables, such as mean annual temperature, requires a further step of climate calibration, which is possible with some, but not all, datasets. Most commonly, it involves statistical calibration of the chosen proxy against an historical time series of instrumental measurements. This is the approach used by dendro-climatologists to convert annual tree ring widths into growing season temperatures. For proxy data that are not annually resolved, an alternative—although less robust—approach involves calibration in space rather than in time. This uses modern "training sets" that have been sampled in different locations on a climate gradient and compared against measured climate variables from nearby weather stations. Using a statistical transfer function, this relationship can then be reconstructed for past biotic assemblages. In this way, Samartin and colleagues (2017) used the head capsules of chironomid midges preserved in lake sediments to reconstruct summer temperatures in northern Italy and to show that between 9000 and 5000 BP these were, on average, 1–2°C warmer than in recent, pre-industrial times.

Using multiple different types of archive in combination makes it possible to create a synthetic record of past climatic changes, although comparing them is not always a straightforward task (Haldon et al. 2014). One way to do this is to put proxy climate data into a standard measurement unit of variance, such as z-scores. Even though these do not represent climate variables per se, they create a "common currency" for different proxies that allows them to be combined in order to reconstruct regional patterns of past climate variability. In this way, Labuhn and colleagues (2018) and Finné and colleagues (2019) used z-scores to identify areas within the Mediterranean that had similar—or contrasting—climate trajectories during different periods of the Holocene.

Computer-based simulations of past climate offer an alternative and complementary approach to that based on empirical evidence. In addition to modeled reconstructions at low spatial resolution for defined time windows such as the mid-Holocene, there are now down-scaled regional climate reconstructions (e.g., Brayshaw, Rambeau, and Smith 2011), transient models, and others coupled interactively to ocean and vegetation models (Harrison, Bartlein, and Prentice 2016). Even so, it should be borne in mind that standard Global Circulation Model (GCM) experiments simulate climate changes that are averaged over multi-centennial time periods, rather than the shorter-term variations experienced by specific past human societies. Despite claims to the contrary, notably in the "Macrophysical Climate Model" devised by Bryson and Bryson (2000) that has been employed by some archaeologists (e.g., Arıkan 2015 and this volume), it is not possible to obtain reliable climate reconstructions that are place- and time-specific using conventional computer models, even using high-resolution simulations with small grid cells (i.e., smaller than $1° \times 1°$) (Xoplaki et al. 2018).

An important feature of most climate reconstructions based on proxy data is that they display fluctuations over time rather than linear trajectories. These fluctuations are typically quasi-cyclical in nature, and the question of whether these cycles have a regular periodicity has been the subject of extensive debate among paleo-climatologists. Statistical methods such as spectral and wavelet analyses have been employed in an attempt to identify the dominant frequencies in proxy climate time series (e.g., Jones et al. 2006). One of the most influential cyclical patterns was proposed by Gerard Bond and colleagues (1997) based on sand layers in North Atlantic deep-sea cores that had been carried in icebergs from Greenland. The ice-rafted debris (IRD) layers reached a peak every 1000–1500 years during the Holocene. These so-called Bond cycles have, in turn, been related to solar variations (Bond et al. 2001), and this link offers a potential controlling mechanism that exhibits regular cyclical behavior. Proxy evidence for solar variability prior to observational records has been derived from short-term fluctuations in the ^{14}C content of the atmosphere, as well as from isotopes of beryllium. In a comprehensive global analysis, Heinz Wanner and colleagues (2008) confirmed that that there is evidence for "Bond events" in Northern Hemisphere proxy climate datasets for the last 6000 years, even if these do not seem to represent true, regular cycles, partly because of the complex teleconnections between solar forcing and climate response.

Reconstructing Long-Term Population Dynamics

Human population trends prior to the availability of historical documents, such as census data, can be derived from the results of archaeological excavations and site surveys (Bintliff and Sbonias 1999). Another method involving aggregated "big data" that is increasingly being used involves compilations of radiocarbon (^{14}C) dates from archaeological site excavations (Williams 2012; McCoy 2017). These two approaches based on big data sets have the potential to provide continuous demographic trajectories for different regions from the advent of Neolithic farming through Medieval times for the former and until the Iron Age for the latter.

The "dates as data" approach utilizes statistical measures based on calibrated radiocarbon date Summed Probability Density (SPDs) distributions through time. It assumes that if there is a sufficiently big sample, then it will exhibit highs and lows in date frequency in step with varying intensity of human activity, and hence can be used as a chronologically precise proxy for long-term population dynamics. Because human populations have grown over the long term, peaks and troughs lying outside a 95% confidence envelope of a theoretical null model can be used to highlight periods that depart significantly from an expected upward trend. In some cases, it is possible that the number of ^{14}C dates reflects intensity of archaeological research as much as the intensity of past human activity, although the potential bias of oversampling individual site-phases can be reduced by aggregating multiple uncalibrated dates from the same site into time bins (Timpson et al. 2014). Even so, inferred demographic peaks need to be treated with caution in some instances. There are over 600 ^{14}C dates from the single Neolithic-early Chalcolithic site of Çatalhöyük, for example, which might give the false impression of a huge demographic peak if taken at face value. As with all numerical methods, it is therefore important to evaluate ^{14}C SPD results critically in case of bias or statistically spurious outcomes; this can be done, for example, by cross-comparing them against results based on other methods (see below). While radiocarbon dating can be used for dating archaeological sites of Classical or Medieval age, other methods of dating (e.g., ceramic typology) are more commonly used, so that ^{14}C date distributions cannot be relied on as a demographic proxy in the eastern Mediterranean after about 500 BCE and in some regions, including Anatolia, earlier than that in time.

A second popular archaeo-demographic proxy involves systematic synthesis of past settlement data from archaeological surface survey projects. Archaeological sites have been mapped, dated, and counted for many years (Wilkinson 2003), but larger datasets deriving from multiple sources have become available in recent years, both from intensive systematic surveys and more extensive reconnaissance or site gazetteers. The basic information collected in this manner comprises settlement numbers, locations, and sizes, with each site attributed to particular cultural periods based on the typology of ceramics and lithics found at the surface. These cultural periods, which vary in duration, can be assigned absolute ages based on the chronology of stratified finds from excavated archaeological sites in the same region. As with ^{14}C date SPD analysis, there are methodological limitations to transforming regional archaeological site survey data into past settlement or population density (Cassis et al. 2018; Bevan and Crema 2021). As well as raw counts, site occupation data can be expressed via aoristic analysis (Crema 2012; Bevan et al. 2013), which derives the summed probabilities across defined time-blocks to give results directly comparable to the SPDs. In general, site survey results have a lower temporal precision than ^{14}C SPD data, especially for earlier archaeological phases such as the Neolithic, where defined cultural periods typically last many centuries (Palmisano et al. 2021). On the other hand, site survey data span historical as well as prehistoric times, and therefore extend the archaeo-demographic time range forward to the Medieval period, which then allows overlap with documentary data from historical demography. Site survey data have, for example, been used to assess the settlement history in different regions of Anatolia and the Levant during the transition between antiquity and medieval times (Izdebski et al. 2016).

A database of over 16,000 radiocarbon dates from excavated archaeological sites in more than 20 different Mediterranean regions was compiled by Alessio Palmisano, Andrew Bevan, Stephen Shennan, and colleagues (Palmisano et al. 2018, 2021; Katsianis et al. 2020). This database begins around 10,000 years ago and ends (in terms of usable data) during the first half of the first millennium BCE; it therefore spans the Neolithic, Chalcolithic, and Bronze Ages, along with the Early Iron Age. Alongside radiocarbon dates, Palmisano and colleagues compiled archaeological survey data from several regions, which altogether includes over 30,000 sites from pre-Neolithic to Medieval times. These systematic regional survey data sets were then analyzed using a combination of GIS, multivariate generalization, Monte Carlo simulation, and probability-based statistical techniques. The methods used to analyze and compare settlement survey and radiocarbon dates have been set out using west-central Italy as a case study (Palmisano, Bevan, and Shennan 2017, 2018). Importantly, archaeo-demographic reconstructions were possible using both site survey and ^{14}C SPD methods for west-central Italy, as well as southern France and the southern Levant, so here it is possible to compare results. In all three regions, there is a good match between the two approaches for the period of overlap, giving strong confidence that they provide a trustworthy index of past population change (Roberts et al. 2019a, fig. 3).

Given the good correspondence between the two approaches, a combination of SPDs and settlement numbers from site surveys should provide reliable demographic trends through time in different regions. Figure 1 shows the ^{14}C SPD index for pre-Classical times in different Mediterranean regions. In between the initial Neolithic horizon and Classical times, there have been a series of long-term cycles of regional population growth and decline. This phenomenon of demographic "boom and bust" has been recognized in many previous studies of long-term settlement history (e.g., Shennan et al. 2013). Karl Butzer, for example, saw Mediterranean environmental history in terms of alternating periods of intensification and dis-intensification, with the former linked to systemic growth in agrarian production and human population, and the latter to demographic decline and land abandonment (Butzer 2005, 1776 ff.). In dryland zones such as central Anatolia and Jordan, Mark Whittow (2003) similarly described alternating cycles, which he termed intensification and abatement.

The evidence for long-term demographic and climatic cycles in regions such as the Mediterranean is thus persuasive and ubiquitous, even if the cycle length was often uneven rather than regular.

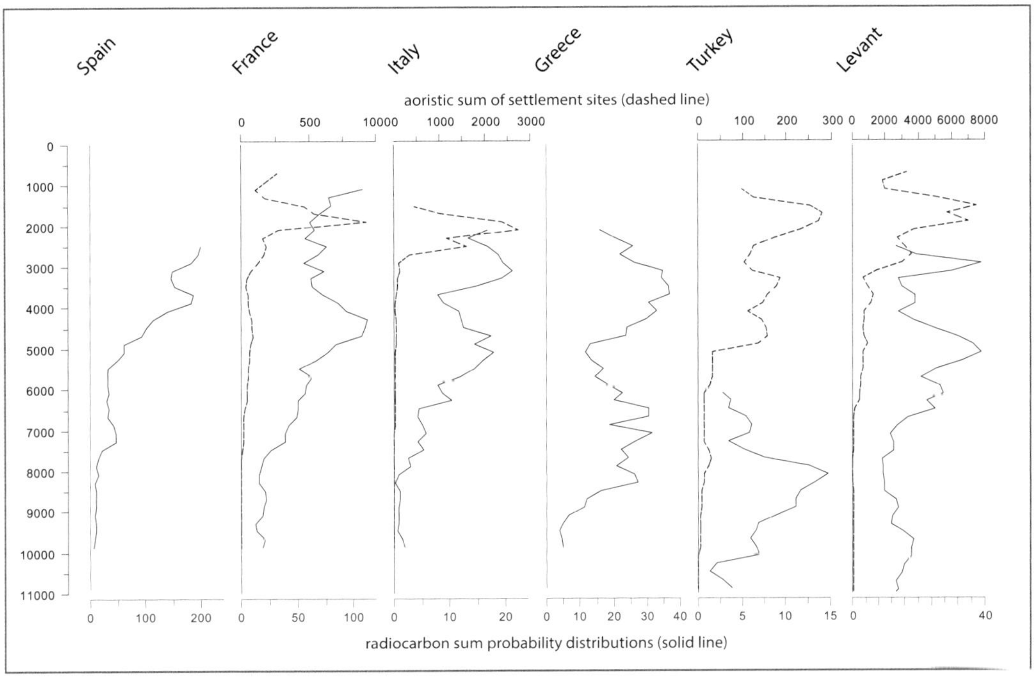

Fig. 1. ^{14}C SPD and aorist sum curves as archaeo-demographic proxies for selected Mediterranean regions (eastern Spain, southern France, central Italy, Greece, southern Anatolia, and southern Levant) for the last 11,000 years (modified from Roberts et al. 2019a). Note that the two scales are independent.

Modeling the Climate-Population Nexus

Figure 2 shows a simple model relating fluctuations in climate to levels of human population, with the former acting as an independent variable and the latter as the dependent one. In an agrarian system such as rain-fed cereal farming, more favorable climatic conditions for crop growth would lead to higher and more reliable yields and allow for rural population growth, following Malthusian logic. In the eastern Mediterranean, this involves climatic conditions that were wetter and, for upland areas, warmer. However, and crucially, population would have lagged behind carrying capacity and hence also behind changes in climate. The length of the lag depends on the timescale of analysis, so that for multi-decadal climatic fluctuations, the time lag in demographic response would have occurred over a decade or two. For longer (e.g., multi-centennial) climate cycles, human demography would have responded over a more extended timescale, as needed to allow a physical expansion of settlement into lands that had previously been climatically marginal for farming, such as the semi-arid zone of central Anatolia. Figure 2 (A and B) uses an arbitrary climatic cycle length of 2,500 years, comparable to the multi-centennial Bond cycles of Holocene climate, and a demographic time lag of 200 years. In this case, a linear population response to climate would produce the same cycle length, but with a phase of demographic overshoot midway through the declining limb of the climate curve, when population and resources (i.e., climate) became out of sync and carrying capacity was exceeded. In practice, population levels are unlikely to have declined gradually in the face of a deteriorating climate and resource scarcity, but instead would have fallen rapidly or even catastrophically. This is partly because of the inertia that exists in human decision-making systems, given that societies are reluctant to change unless forced to do so—which means that responses usually take place after the event. Furthermore, food scarcity or famine bring with them a raft of other problems that amplify the crisis, such as breakdown in social order and vulnerability to disease (Turchin 2003; Tainter 2006). This rapid decline represents a "Seneca cliff," named after the Roman philosopher who observed that growth is slow, but ruin is rapid. In consequence, a more realistic population trajectory during the period of overshoot is shown in a modified version of the model (Fig. 2C), with a gradual demographic rise during the period of improving climate but a more abrupt fall during the phase of climatic deterioration. This asymmetric trajectory reduces both the size and duration of the gap between population and climatic resources. It is a pattern seen in archaeologically and historically reconstructed long-term population time series which display gradual growth followed by rapid decline (e.g., Weiberg et al. 2019 for the Peloponnese).

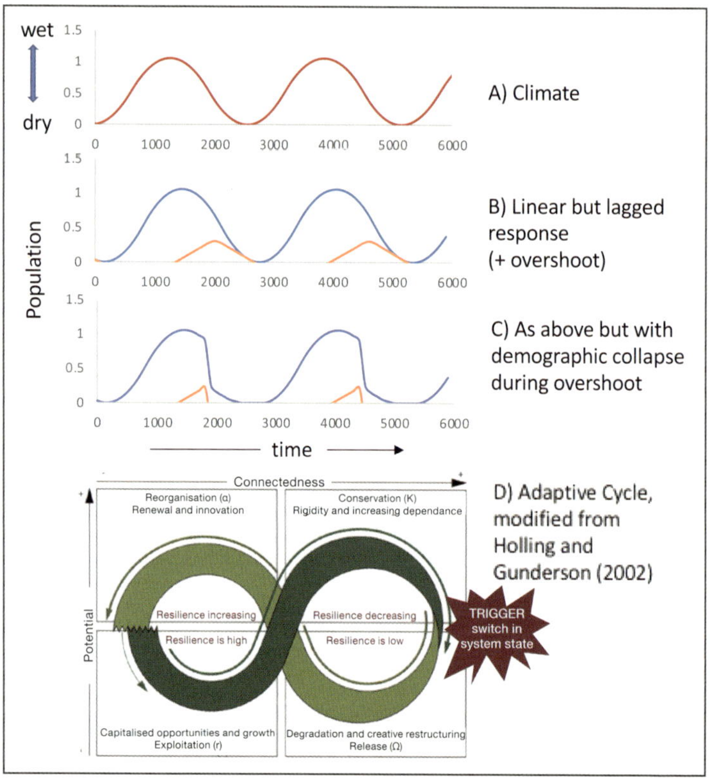

Fig. 2. Schematic model relating fluctuations in climate (A) to levels of human population (B–C), along with the Adaptive Cycle (D), modified from Holling and Gunderson (2002). Demographic overshoot represents any excess of population above the climatic carrying capacity at a point in time.

This sequence of changes matches well with the Adaptive Cycle (or Resiliency) model that was originally developed for natural ecosystems but which has subsequently been applied also to socio-environmental systems linking people, climate, and resources (Holling and Gunderson 2002). This model envisages systems as growing gradually (the r phase) to reach a peak in system inter-connectedness coupled to rigidity (K phase), before they over-reach and collapse (Ω). They then have to restructure themselves and start again (α phase) (Fig. 2D). Consequently, cyclical change need not be a smooth process but can include periods of crisis leading to an abrupt shift to a new, simplified system state. Samantha Allcock (2017) showed how the Adaptive Cycle model could be applied to changes in rural settlement using archaeological site survey data from Cappadocia; here she identified four major long-term cycles since Neolithic times, each characterized by demographic growth and decline.

If a passive "Standard" response was the only way that societies reacted to climatic change, then each demographic correction would return the system to a point set by the prevailing climate. In the simple scenario shown in Figure 2 where the climate cycles are identical, the population level would return to its previous starting point and there would be no long-term demographic increase. Evidently this was not the case, partly because climatic cycles varied in intensity and duration, but more importantly because societies actively adapted in the face of a changing environment. Moreover, they did so primarily during times of necessity, when previously successful solutions no longer sufficed. This argument has been made for more recent time periods by Ugo Bardi (2020), who has attempted to show that societies can rebound after collapse to be more successful than previously. It was also integral to the model of Sing Chew (2007), who saw recurring Dark Ages as formative in human history. In demographic terms, new innovation meant that the baseline shifted upwards so that the next cycle had a higher carrying capacity. The overall population was thus able to rise over multiple cycles, in line with trends reconstructed from archaeological data. Figure 3B builds an adaptive element into the population-climate model that results in a shift to a higher baseline state with each new cycle. Significantly, it reveals that the relationship between climate and population was not constant over time but shifted during different phases of each adaptive cycle. Specifically, a positive correspondence (the Standard Relationship) can be observed between climatic favorability and population during the growth (r) phase, but this breaks down during the over-extension (K) phase. During transition from the release (Ω) to the succeeding growth phase (r), the relationship is no longer "normal" but inverted, with climatic adversity prompting innovation that allowed population to stabilize rather than continue a downward decline. These two contrasting relationships can be highlighted by using a moving-window statistical correlation between climate and population. As shown in Figure 3B, they display a clear positive correlation for most of the time (i.e., the Standard Relationship), but periodically switch to a negative correlation between climate and population, notably at the start of each new cycle.

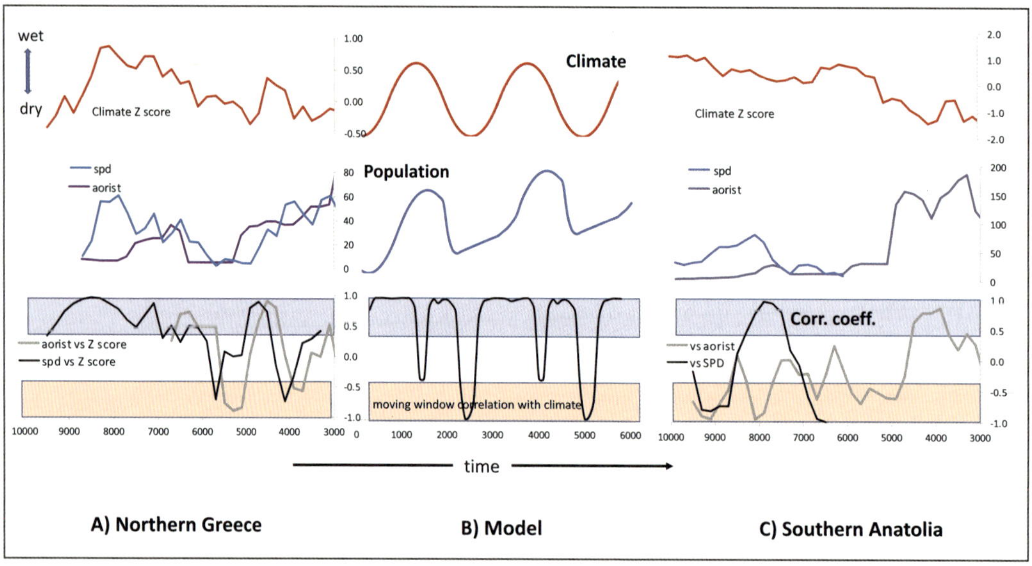

Fig. 3. An adaptive version of the climate-population model (B) compared to empirical data from (A) northern Greece and (C) southern Anatolia, 10,000–3000 BP (8000–1000 BCE); data from Woodbridge and colleagues (2019); Finné and colleagues (2019); Weiberg and colleagues (2019).

Case Studies: Southern Anatolia and Northern Greece, 8000–1000 BCE

While remaining a greatly simplified representation of reality, this schematic model has the advantage of being testable against proxy data from paleoclimatology and archaeodemography. To illustrate and test this model, two regional case studies are used, covering the Neolithic and Bronze Age socio-ecological cycles that led to the rise of some of the first eastern Mediterranean civilizations. The first of these is from southern Anatolia and draws on data presented by Woodbridge and colleagues (2019), while the second derives from northern Greece, based on the regional synthesis of Weiberg and colleagues (2019) (Fig. 4).

As described above, climate z-scores reflect regional variations, in this case between wetter (positive value) and drier (negative value) conditions. The regional z-score mean for the Balkans, as reported by Finné and colleagues (2019), has been used here for northern Greece. For southern Anatolia, a sub-set of the regional z-score data for Anatolia is employed here, drawing only on those sites located inside the study region, namely Eski Acıgöl (Roberts et al. 2001), Nar Lake (Dean et al. 2015), and Gölhisar (Eastwood et al. 2007). Because of uncertainties in chronological precision (e.g., calibrated ^{14}C dates), z-scores have been calculated for 200-year time intervals (e.g., 1200–1000 BCE), and so reflect mainly multi-centennial changes rather than shorter-lived climate events. The global climate event at 8.2 ka BP, for example, is not reflected in these curves, although that around 4.2 ka BP is clearly identifiable. Data “binning” also tends to dampen cyclical trends that are present in individual climate records.

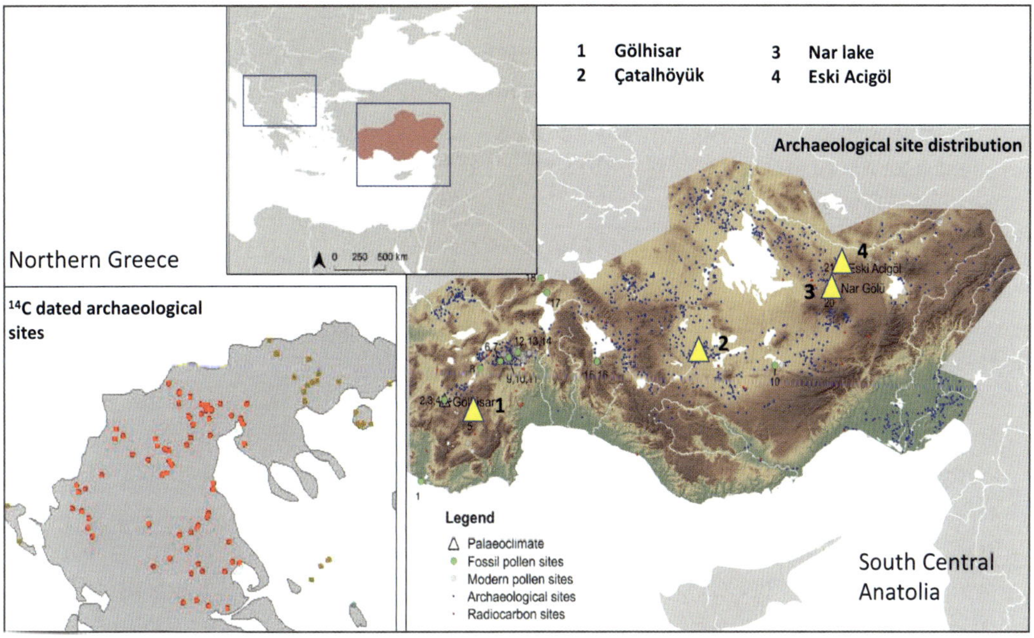

Fig. 4. Maps showing study areas and key sites discussed in the text.

Regional demographic changes are based firstly on SPD functions for the number of ^{14}C dates from excavated archaeological sites and, secondly, on the aorist sum of site numbers for each archaeological phase from regional survey data (in the Greek case, for Macedonia only). For southern Anatolia, there are too few radiocarbon dates to yield statistically significant results for the Bronze Age, so the SPD curve covers only the Neolithic-Chalcolithic demographic cycle, prior to ca. 4000 BCE.

Figure 3 shows climate z-scores, reconstructed archaeo-demographic trends (showing SPD and aorist sum curves separately), and 1000-year moving-window statistical correlation between climate and population for each of the two regions at 200-year time steps, compared to model results. It is evident that the synthetic trends in climate are less cyclical than those shown in the model, with its regular repeated oscillations, and instead are dominated by the long-term trajectory from a wetter early Holocene climate to drier conditions by the Late Holocene. Nonetheless, there are some striking similarities between the model and the data, especially for northern Greece. There were demographic peaks in northern Greece during the early Neolithic (ca. 6000 BCE) and again in the Middle/Late Bronze Age (second millennium BCE), separated by an intervening trough during the fourth millennium BCE. Population rose following periods of wetter climate and declined during a phase of climatic desiccation, but with a ca. 200-year time lag. The ^{14}C SPD data are matched by a similar trend in population inferred from archaeological site survey data, although the latter indicates a higher overall population level during the later Bronze Age than in the Neolithic. In addition, the moving-window correlation between proxies for climate and population from northern Greece shows a pattern much like that seen in the model. There is a strong positive correlation during periods of population growth, followed by shorter-lived periods of negative correlation at the end of the Neolithic and Early Bronze Age demographic cycles.

In southern Anatolia, as in northern Greece, two major demographic cycles are identifiable between 8000 and 1000 BCE, with the first of them seeing a population peak during later Neolithic to early Chalcolithic times. The SPD data for this region may be partly distorted by the large number of ^{14}C dates deriving from the ceramic Neolithic mega-settlement at Çatalhöyük East (ca. 6500 BCE). Site survey data, by contrast, indicate that the number and total area of sites was greater during the subsequent early Chalcolithic period (6000-5500 BCE). The regional climate was relatively humid throughout this first demographic cycle, but it was interrupted by an abrupt mega-drought around 6200 BCE lasting up to 160 years, the so-called 8.2 ka climate event. This was ultimately triggered by catastrophic melting of the residual Laurentide ice sheet over Hudson Bay, with the resulting meltwater altering ocean circulation in the Atlantic (Alley and Ágústsdóttir 2005; Daley et al. 2011). The 8.2 ka climate event is not easily identified in the graphs shown in Figure 3, which are based on the synthesis of Finné and colleagues (2019). As noted above, this is because the hydro-climate datasets they used were put into the 200-year long time windows, although the event is present in many of the original datasets. At Çatalhöyük, for example, Roffet-Salque and colleagues (2018) used hydrogen isotopes on lipid biomarkers preserved in pottery vessels to show that climatically driven variations in δ^2H values coincided with important changes in animal husbandry and consumption practices at ca. 6200 BCE. Prior to this, population in the Konya basis had been nucleated in this single large settlement site, whereas afterwards it became dispersed into a series of medium-sized settlements, such as Çatalhöyük West and Can Hasan I. There is a strong positive correlation between ^{14}C SPDs and climate z-cores in southern Anatolia during the period of population growth and subsequent initial decline (Fig. 3C), followed by an extended period of negative correlation between the two variables after the end of the Neolithic-Early Chalcolithic demographic cycle, 5000-3000 BCE. Site survey data shows a rapid rise in inferred population at the start of the third millennium BCE, marking the onset of the Bronze Age demographic cycle. ^{14}C SPD data from elsewhere in Anatolia suggest that this cycle probably began earlier than 3000 BCE, during Late Chalcolithic times (Palmisano et al. 2021). The correlation between archaeologically inferred population and climate became positive again during Early and Middle Bronze Age times.

Complex Societies, Complex Responses

The two regional examples described above therefore give some empirical support to the conceptual model for the changing climate-population nexus, at least on a heuristic basis. They are also supported by Palmisano and colleagues' (2021) analysis of Pearson's correlations (r) between the climate record of Jeita Cave (Lebanon) and archaeo-demographic ^{14}C SPDs for both Cyprus and the Levant. They show periods of statistically significant positive correlation (e.g., early aceramic Neolithic) alternating with phases of inverse correlation (e.g., ceramic Neolithic), as in the model presented here. A comparable set of relations between climate and demography are likely to have operated for later (i.e., post-Bronze Age) socio-ecological cycles. The gradual rise in population during the course of the first millennium BCE, for example, followed a climatic amelioration after a series of major eastern Mediterranean droughts between 1200 and 600 BCE (Finné et al. 2019). On the other hand, climatic changes did not have the same effect on complex, socially stratified societies as they did on subsistence farmers. The latter typically rely on a relatively narrow range of crops and domestic livestock, and they become vulnerable if these staples are threatened by drought, frost, pests, or diseases. Perhaps for this reason, a positive correlation is commonly seen between population increase/decline and favorable/adverse climatic conditions in Neolithic societies in many different parts of the world (e.g., Bevan et al. 2017). Sedentary village life also allows for food storage, and this becomes even more important for complex societies, especially for urban populations. Food storage and redistribution became essential to feeding the population of great empires of the Mediterranean and the Near East. Andrew Bevan (2020) has tracked the increasing scale and complexity of grain storage needed to support the growth in population during the first millennium BCE, from pit stores to above-ground storage structures and eventually to Roman *Horrea*. By the end of the millennium during the Roman Warm Period, the largest granary-warehouse in imperial Rome—the *Horrea Galbae*—covered an area of over 20,000 m^2 and could hold enough food to supply the city's million-strong population for seven years.

The increasing spatial scale in societal organization and economic activity from village to city-state to empire therefore acted to mediate human relationships with changes in climate. Neolithic communities had a relationship that was experienced relatively directly via crop yields and livestock numbers. In contrast, the population of sixth-century CE Constantinople drew on foodstuffs and other resources from throughout the Mediterranean basin and, indeed, beyond it. Its economic catchment consequently became larger than the spatial scale over which coherent climatic changes took place, at least for droughts and floods. A wide variety of climate archives have shown that there has been no Mediterranean-wide pattern of precipitation variations within the current interglacial, regardless of the timescale of analysis, except during abrupt events such as at 8.2 ka BP. Several patterns can be identified for the period of instrumental observations, including a Mediterranean Oscillation, linked to the North Atlantic Oscillation (NAO), in which southeastern regions have been wet while northwestern ones have been dry, and vice-versa (Dünkeloh and Jacobeit 2003; Türkeş and Erlat 2003; Xoplaki et al. 2004). A similar southeast-northwest Mediterranean climate see-saw is also evident when the timescale is extended back to the last millennium (Cook et al. 2016) or the last ten millennia (Finné et al. 2019). As a consequence of this, early Byzantine Constantinople never suffered from food shortages caused by pan-imperial drought events; instead, times of feast or famine would have been regionalized across the empire (e.g., Anatolia, Egypt, or Sicily). This would have allowed regions of grain surplus to aid those in deficit, so long as inter-regional transport and redistribution mechanisms operated efficiently. On the downside, in the interconnected world of empires, the consequences of climate change could be felt indirectly over long distances via movements of people, as Ellsworth Huntington identified, or by the spread of disease pathogens, such as plague during the Black Death. Indirect climatic knock-on effects may have operated as early as the end of the second millennium BCE, when the eastern Mediterranean Late Bronze Age world system came under threat from large-scale population migrations, notably by the various "Sea Peoples" described in the victory inscriptions of Rameses III and other texts (Cline 2015).

Climatic changes were not the only factor that altered human carrying capacities. Another significant factor was availability of natural resources, both renewable (e.g., timber and water) and non-renewable (e.g., minerals and—in practical terms—soil), that limited the production of things for human use or consumption. The former could be degraded if over-exploited, while the latter were intrinsically finite. As early as Neolithic Çatalhöyük, there is evidence that a rising population led to over-exploitation of natural resources lying close to this large, nucleated settlement. The record of wood charcoals at the site, for example, shows an initial reliance on local riparian trees such as Ulmaceae and Salicaceae (elm and willow families) which, once used up, gave way to woody taxa that grew on drier and more distant terrain, including oak and juniper (Kabukcu 2017). Further afield, the annually layered (varved) sediments from Nar Lake in Cappadocia include a series of soil inwash layers caused by erosion of the surrounding watershed between 7200 and 6200 BCE; that is, coincident with Çatalhöyük East (Roberts et al. 2019b). Nar Lake lies close to Nenezi Dağı, one of the two main obsidian sources exploited by the inhabitants of Neolithic Çatalhöyük for making stone tools, and human activity around the lake seems to have triggered soil inwash. Depletion of natural resources such as timber and soil would have reduced the local-regional carrying capacity and increased the likelihood of a mismatch between population and resources, especially during the K phase of the Adaptive Cycle.

If human impact on resource availability was modest during the Neolithic, it became widespread during Bronze and Iron Age times in the eastern Mediterranean (Roberts 2018; Roberts et al. 2019a; Walsh et al. 2019). The climate-population baseline "carrying capacity" was reduced most significantly during Classical (Hellenistic-Roman-Byzantine) times as a consequence of degradation of resources such as water, vegetation, and—above all—soil. The impact of environmental degradation was spatially variable, creating both losers and winners, but its overall effect was to make system failure abrupt rather than gradual when it occurred (Chew 2007). As a counterbalance to self-inflicted reduction in environmental carrying capacity, human ingenuity and innovation were capable of reducing or reversing the threat of resource depletion, for example, via the development of hillslope terracing to prevent soil erosion. Thus, neither the climate nor the underlying resource base was fixed but instead could change over time.

An Ottoman Natural Experiment in Climate History

These complexities meant that the consequences of climatic change became increasingly contingent on antecedent conditions, both environmental and societal. A good illustration is provided by a series of major droughts that affected central Anatolia at different times during the Ottoman period and whose societal and demographic consequences strongly diverged. Ottoman rule in central Anatolia began in the mid-fourteenth century, but the largest area, the Beylik of Karamanoğlu (including Konya), came under its control between 1420 and 1487, while Kayseri fell to the Ottoman Empire as late as 1515. Ottoman rule ended, as in the rest of Anatolia, with the declaration of the Turkish Republic in 1923. The Ottoman empire coincided almost exactly in time with the period of the Little Ice Age. Within this, a continuous high-resolution record of changes in climatic moisture balance across the central Anatolian plateau is available from stable isotope analysis of varved sediments from Nar Lake in Cappadocia (Fig. 5). This provides a decadal mean value for regional water balance across a large area of central Anatolia with annual dating precision (see Yiğitbaşıoğlu et al. 2015). While this does not reveal the year-to-year fluctuations in weather that trees rings do, the smoothing out of high-frequency "noise" creates a record that makes it well-suited for comparison with the historical *longue durée*. The Nar record indicates that the most extreme Ottoman drought events occurred during the early fifteenth century (with minimum values ca. 1430), at the end of the sixteenth century (1594–1604), and during the mid-late nineteenth century (notably 1857–1874), with less severe dry phases at 1455–1478 and 1724–1725. Although the three major droughts were similar in intensity and duration (ca. 15, ca. 13, and ca. 22 years, respectively, in the Nar record), their societal impact appears to have been quite different, primarily because of antecedent circumstances. In terms of hydrology, the first and third drought phases might have been expected to have had the greatest effect on water resource availability, as they were preceded and followed by other drying trends or dry episodes. The drought at the turn of the seventeenth century, by contrast, was book-ended by periods of relatively favorable water balance.

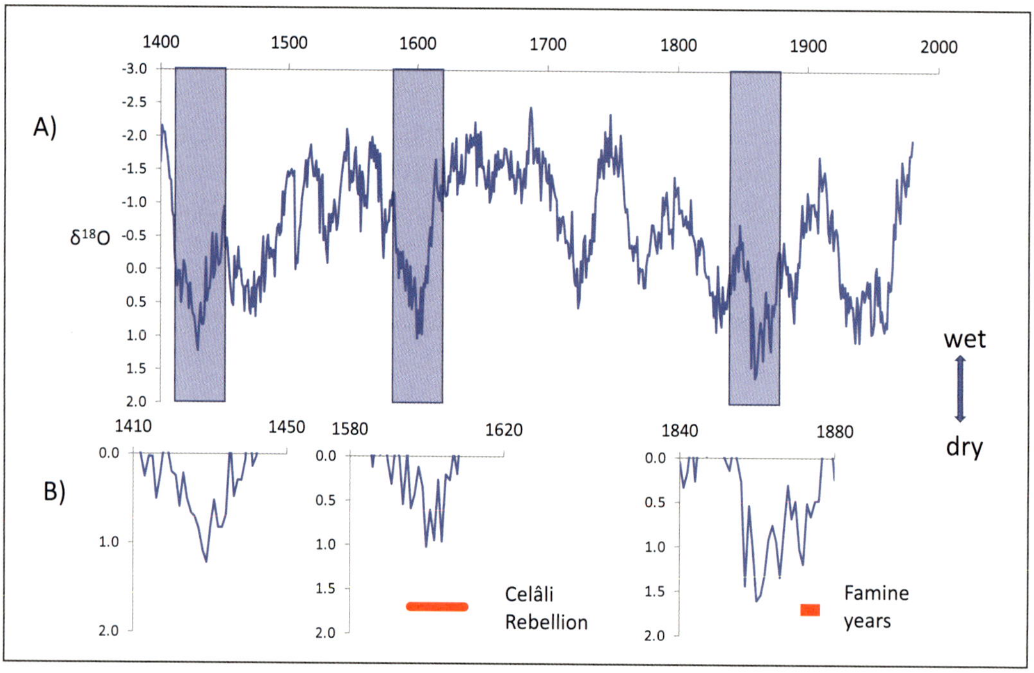

Fig. 5. A) Ottoman drought events in central Anatolia recorded in stable isotopes from Nar Lake (based on data in Jones et al. 2006; Yiğitbaşıoğlu et al. 2015). B) The three main drought episodes in detail.

The wider impact of the early fifteenth-century dry period is not well recorded historically, in part because central Anatolia was still in the process of being incorporated into the Ottoman empire. Documentary sources such as the state *defter* tax register system for recording crop yields, population, etc., were not yet in place (the oldest available *defter* in Turkish archives is dated 1431–32 [Coşgel 2004]). Ottoman rule had been disrupted following military defeat by Timur at the battle of Ankara (1402), and Constantinople was still in Byzantine hands until 1453. Nonetheless, it is evident that regions such as the Konya plain were well-populated and prosperous at this time; Bertrandon de la Brocquière (1807), for example, describes a "fine plain ... well furnished with villages." He crossed central Anatolia in 1433, stopping at Ereğli, Karaman, etc., but makes no mention of rural distress or food shortage, although his journey occurred soon after a time of important drought, according to the Nar record. The Konya plain lies at the lower limit for rain-fed cereal cultivation, but water was available from rivers flowing into the basin (Roberts 1990). So long as a system of irrigation and drainage canals and a parallel system for administering water rights were in place, the effects of drought could be offset. The limited evidence available therefore points to the early fifteenth-century episode of adverse climate change having had only limited societal consequences, plausibly due to the resilience of the socio-ecological system during this "r" growth phase of the Adaptive Cycle.

By the end of the following century, Ottoman rule was well-established in central Anatolia, and *defter* and other documentary sources for the period are relatively numerous. Settlements were abundant in regions like the Konya plain, with prosperous agriculture based on complex irrigation works (Beldiceanu and Beldiceanu-Steinherr 1968). In his map for 1580, Hütteroth (1968) shows around 30 villages on the alluvial fan of the Çarşamba River alone. There is evidence of rural demographic increase during the sixteenth century, with population almost doubling in Karaman province between the 1520s and 1570s (Barkan 1970, cited in İnalcık and Faroqhi 1997), although whether this amounted to "population pressure" is contested (Özel 2004; cf. Cook 1972). The late sixteenth century drought had major negative societal consequences for communities in central Anatolia, and indeed for the whole Ottoman state, a story that has been reconstructed in detail by Sam White in *The Climate of Rebellion* (2011). This drought, which is also recorded in Anatolian tree-ring records (Kuniholm 1990; Touchan et al. 2007), occurred at a time when Ottoman resources were stretched to a maximum, primarily to supply its armies engaged in the Long War with the Habsburg Empire in central Europe (1593–1607). Central Anatolian farmers were required to provide sheep and grain for these military supply lines, even though drought and severe winters reduced crop yields and zoonotic disease killed many livestock. During the drought, food prices doubled, famine conditions prevailed in some parts of Anatolia, and tax records suggest that there was as much as 50% mortality in parts of Anatolia between the 1580s and 1640s (Özel 2004). The resulting social discontent and lack of security allowed rural resentment to develop into a full-scale uprising, the Celâli Rebellion (1596–1610). White (2011) has argued that this revolt represents a key tipping point in Ottoman history, after which it never regained the momentum of the preceding century and instead entered a long, slow decline. In terms of the Adaptive Cycle, the Ottoman Empire at the end of the sixteenth century represented the K-phase of maximum inter-connectedness coupled to a risk of over-extension that would have limited the system's resilience to stress, in this case climatic and demographic.

The third major dry phase occurred during the nineteenth century, with drought affecting the Ottoman Empire in seventeen of the forty-two years between 1838 and 1880, which caused severe shortage of grain (Ayalon 2015, 199). By this time, the Ottoman Empire was approaching its nadir, despite still controlling vast territories in southeast Europe, southwest Asia, and north Africa. The Konya basin irrigation system was no longer operating, and only four villages remained on the Çarşamba alluvial fan by 1870 (Hütteroth 1968); instead, its flat plains were seasonally home to Türkmen nomads and their herds of sheep and goats (Roberts 1990). The drought conditions in central Turkey, coupled to an extremely cold winter in 1874–75, are reported to have killed up to 250,000 people and at least 100,000 head of livestock (40% of all herds) (Naumann 1893, cited in Kuniholm 1990). In one village in Konya vilayet alone, the famine reduced the number of cattle, sheep, and goats by more than 90% (Duffy 2020). Although triggered by drought, the famines of the 1870s contributed to and were exacerbated by political-economic conditions prevailing in Late Ottoman society. These included the Russo-Ottoman war of 1877–78 for which taxes continued to be levied (Ertem 2017), and which formed part of the "Great Eastern Crisis" (1874–78). The reduced tax income from agriculture coupled with rising military expenditure led the Ottoman state to default on its sovereign debt in 1875. Adverse climatic conditions thus hastened the transformation of the late Ottoman empire during the late nineteenth century, but this impact was contingent on wider factors, including state weakness both internally and internationally. By this stage, the Ottoman socio-ecological system was evidently approaching the release (Ω) phase of the Resiliency cycle.

The three most important central Anatolian drought phases to have affected the Ottomans occurred at different stages in the evolution of its imperial power; the first (fifteenth century) during its growth phase, the second (end of the sixteenth century) close to its apogee, and the third (late nineteenth century) near its end. In principle, for the latter two periods, the fact that the state could draw on resources from provinces with different, often contrasting, climatic histories should have allowed food redistribution within the empire, as was the case in Imperial Rome and Early Byzantine Constantinople. In practice, this seems not to have been the case, at least in part because Ottoman wars meant that military provisioning took precedence over famine relief. Wars stretched resources to full capacity and left no spare capacity—or redundancy—in the system of food production and procurement. In contrast, in the fifteenth century, the Ottomans inherited a relatively flourishing rural economy in central Anatolia, mainly based on agro-pastoralism (England et al. 2008) but also including irrigation, that proved resilient in the face of adverse climatic conditions. Interestingly, a fourth dry climatic period occurred during the early years of the Turkish Republic (Figure 5). Meteorological records from the continental central Anatolian region show notably low annual precipitation during the mid-1930s, with 1932 and 1934 as two of the driest years of the period of instrumental monitoring (Türkeş 1996). The lack of water and grain led to many deaths in places such as Karapınar in the Konya plain during 1933 (Naim Aydınbelge, pers. comm., to Catherine Kuzucuoğlu). Nonetheless, these dry years did not derail the modernization of the state-run rural economy, and they plausibly represent the re-organization (α) phase of the Adaptive Cycle, during which system resilience should be rising.

This natural experiment in how a major, long-lasting polity responded to climate emergencies at different stages in its ca. 600-year history indicates that the impact on rural Anatolia was ultimately contingent on antecedent socio-environmental conditions as much as on drought severity and duration. It also shows how sub-millennial cycles of climate change and adaptive response were nested within longer ones. While the available evidence makes them harder to identify for earlier time periods, it seems likely that similar nested cycles existed within earlier empires, such as the successive

Hellenistic-Roman-Byzantine polities. Importantly these different temporal scales would have had the potential to interact to create complex responses in the wider socio-environmental system.

Conclusion

In summary, by using a schematic cyclical model linked to the Theory of Adaptive Change, it is possible to reconcile the two contrasting societal reactions to climatic change, namely, the "business-as-usual" Standard Relationship response to climatic opportunity or adversity versus the Inverted Relationship involving societal adaptation and innovation under changed conditions. A key element that allows these two opposing relationships to be reconciled is that the model is cyclical in nature, with different causal mechanisms prevailing at different stages in the cycle. While almost everyone did well during extended periods of "normal" relations between climate and population when social and environmental conditions were improving, the challenge posed at times of crisis and over-reach resulted in a range of different societal responses, some of them more successful than others. Overall population levels often declined, economies simplified, and communities re-localized at these times, with a shortening of supply chains at the expense of long-distance trade in specialist products. At the same time, some societies responded to climate-related crises by innovating, for example, via the development of secondary animal products such as wool (McCorriston 1997) or the domestication of new crops like olive trees (Langgut et al. 2019). These, along with new organizational adaptive strategies, allowed the subsequent cycle of transformation to start from a new footing. In some instances, climate changes in more distant regions may have prompted a response that impacted on a wider geographical space, an obvious example being via out-migration, as Huntington (1907) proposed with his "Pulse of Asia" model.

Periodic environmental crises thus helped shift the population-climate nexus upwards to a new, higher baseline state. In this way, changes in climate may have played an essential role in long-term societal transformation, with environmental crises acting not only as a stressor, but also as a stimulus to social evolution. Even if climate-induced food scarcity or famine would have had negative consequences for human populations in the short term, it is striking that over the long arc of history—and even longer arc of prehistory—this did not impede social evolution, because climatic challenges also prompted technological and societal innovation. Over the *longue durée*, societal resilience to climate shocks may have been less important than the capacity for re-invention following periods of socio-environmental rupture, a lesson that may prove to be equally relevant to the unfolding climate crisis of the early third millennium CE.

Acknowledgments

I am grateful to Alessio Palmisano for providing the archaeo-demographic data used in this chapter and to other colleagues involved in the Leverhulme-funded "Changing the Face of the Mediterranean" project. Alessio, Dan Lawrence, and Sam White kindly provided helpful comments on an earlier draft of the chapter. My thanks also go to Catherine Kuzucuoğlu for information about the 1930s drought and to Chris Roosevelt and John Haldon as convenors of the "Winds of Change" ANAMED symposium.

Works Cited

Allcock, Samantha Lee. 2017. "Long-Term Socio-Environmental Dynamics and Adaptive Cycles in Cappadocia, Turkey during the Holocene." *Quaternary International* 446: 66–82.

Alley, Richard B., and Anna Maria Ágústsdóttir. 2005. "The 8k Event: Cause and Consequences of a Major Holocene Abrupt Climate Change." *Quaternary Science Reviews* 24, 10–11: 1123–49.

Arıkan, Bülent. 2015. "Modeling the Paleoclimate (ca. 6000-3200 cal BP) in Eastern Anatolia: The Method of Macrophysical Climate Model and Comparisons with Proxy Data." *Journal of Archaeological Science* 57: 158–67.

Ayalon, Yaron 2015. *Natural Disasters in the Ottoman Empire: Plagues, Famines, and Other Misfortunes*. New York: Cambridge University Press.

Bardi, Ugo. 2020. *Before the Collapse*. Cham: Springer International Publishing.

Barkan, Ömer Lutfi. 1970. *İstanbul Vakıfları Tahrir Defteri: 953 (1546) Tarihli*. Vol. 61. Baha Matbaası.

Beldiceanu, Nicoară, and Irène Beldiceanu-Steinherr. 1968. *Recherches sur la province de Qaraman au XVI siècle*. Leiden: Brill Archive.

Bevan, Andrew. 2020. "A Stored-Products Revolution in the 1st Millennium BC." *Archaeology International* 22: 127–44.

Bevan, Andrew, Sue Colledge, Dorian Fuller, Ralph Fyfe, Stephen Shennan, and Chris Stevens. 2017. "Holocene Fluctuations in Human Population Demonstrate Repeated Links to Food Production and Climate." *Proceedings of the National Academy of Sciences* 114, no. 49: E10524–31.

Bevan, Andrew, and Enrico Crema. 2021. Modifiable reporting unit problems and time series of long-term human activity. *Phil. Trans. R. Soc. B* 376: 20190726.

Bevan, Andrew, Enrico Crema, Xiuzhen Li, and Alessio Palmisano. 2013. "Intensities, Interactions and Uncertainties: Some New Approaches to Archaeological Distributions." In *Computational Approaches to Archaeological Spaces*, edited by Andrew Bevan and Mark Lake, 27–52. Walnut Creek: Left Coast Press.

Bintliff, John, and Kostas Sbonias, eds. 1999. *Reconstructing Past Population Trends in Mediterranean Europe (3000 BC-AD 1800)*. Vol. 1. Oxford: Oxford University Press.

Bond, Gerard, Bernd Kromer, Juerg Beer, Raimund Muscheler, Michael N. Evans, William Showers, et al. 2001. "Persistent Solar Influence on North Atlantic Climate During the Holocene." *Science* 294, no. 5549: 2130–36.

Bond, Gerard, William Showers, Maziet Cheseby, Rusty Lotti, Peter Almasi, Peter DeMenocal, et al. 1997. "A Pervasive Millennial-Scale Cycle in North Atlantic Holocene and Glacial Climates." *Science* 278, no. 5341: 1257–66.

Brayshaw, David J., Claire MC Rambeau, and Samuel J. Smith. 2011. "Changes in Mediterranean Climate During the Holocene: Insights from Global and Regional Climate Modelling." *The Holocene* 21, no. 1: 15–31.

de La Brocquière, Bertrandon. 1807. *The Travels of Bertrandon de La Brocquière, to Palestine: And His Return from Jerusalem Overland to France, During the Years 1432 & 1433. Extracted and Put into Modern French from a Manuscript in the National Library at Paris*. Translated by T. Johnes. London: I B Tauris.

Bryson, Reid A., and Robert U. Bryson. 2000. "Site-Specific High-Resolution Models of the Monsoon for Africa and Asia." *Global and Planetary Change* 26, no. 1-3: 77–84.

Bryson, Reid A., H. H. Lamb, and David L. Donley. 1974. "Drought and the Decline of Mycenae." *Antiquity* 48, no. 189: 46–50.

Büntgen, Ulf, Vladimir S. Myglan, Fredrik Charpentier Ljungqvist, Michael McCormick, Nicola Di Cosmo, Michael Sigl, Johann Jungclaus, et al. 2016. "Cooling and Societal Change During the Late Antique Little Ice Age from 536 to Around 660 AD." *Nature Geoscience* 9, no. 3: 231–36.

Burstyn, Yuval, Belen Martrat, Jordi F. Lopez, Eneko Iriarte, Matthew J. Jacobson, Mahjoor Ahmad Lone, and Michael Deininger. 2019. "Speleothems from the Middle East: An Example of Water Limited Environments in the SISAL Database." *Quaternary* 2, no. 2: 16. doi:10.3390/quat2020016.

Butzer, Karl W. 2005. "Environmental History in the Mediterranean World: Cross-Disciplinary Investigation of Cause-and-Effect for Degradation and Soil Erosion." *Journal of Archaeological Science* 32, no. 12: 1773–800.

Carpenter, Rhys. 1966. *Discontinuity in Greek Civilization (The JH Gray lectures)*. Cambridge: University Press.

Cassis, Marica, Owen Doonan, Hugh Elton, and James Newhard. 2018. "Evaluating Archaeological Evidence for Demographics, Abandonment, and Recovery in Late Antique and Byzantine Anatolia." *Human Ecology* 46, no. 3: 381–98.

Chew, Singh C. 2007. *The Recurring Dark Ages. Ecological Stress, Climate Changes and System Transformation*. Lanham, Maryland: Altamira Press.

Cline, Eric H. 2015. *1177 BC: The Year Civilization Collapsed*. Princeton and Oxford: Princeton University Press.

Cook, Benjamin I., Kevin J. Anchukaitis, Ramzi Touchan, David M. Meko, and Edward R. Cook. 2016. "Spatiotemporal Drought Variability in the Mediterranean over the Last 900 Years." *Journal of Geophysical Research: Atmospheres* 121, no. 5: 2060–74.

Cook, Michael Allan. 1972. *Population Pressure in Rural Anatolia, 1450–1600*. Oxford: Oxford University Press.

Coşgel, Metin M. 2004. "Ottoman Tax Registers (Tahrir Defterleri)." *Historical Methods: A Journal of Quantitative and Interdisciplinary History* 37, no. 2: 87–102.

Crema, Enrico R. 2012. "Modelling Temporal Uncertainty in Archaeological Analysis." *Journal of Archaeological Method and Theory* 19, no. 3: 440–61.

Daley, Timothy J., Elizabeth R. Thomas, Jonathan A. Holmes, F. Alayne Street-Perrott, Mark R. Chapman, Julia C. Tindall, Paul J. Valdes, et al. 2011. "The 8200 yr BP Cold Event in Stable Isotope Records from the North Atlantic Region." *Global and Planetary Change* 79, no. 3–4: 288–302.

Dean, Jonathan R., Matthew D. Jones, Melanie J. Leng, Stephen R. Noble, Sarah E. Metcalfe, Hilary J. Sloane, Diana Sahy, Warren J. Eastwood, and C. Neil Roberts. 2015. "Eastern Mediterranean Hydroclimate over the Late Glacial and Holocene, Reconstructed from the Sediments of Nar Lake, Central Turkey, Using Stable Isotopes and Carbonate Mineralogy." *Quaternary Science Reviews* 124: 162–74.

Drake, Brandon L. 2012. "The Influence of Climatic Change on the Late Bronze Age Collapse and the Greek Dark Ages." *Journal of Archaeological Science* 39, no. 6: 1862–70.

Dünkeloh, Armin, and Jucundus Jacobeit. 2003. "Circulation Dynamics of Mediterranean Precipitation Variability 1948–98." *International Journal of Climatology* 23, no. 15: 1843–66.

Duffy, Andrea E. 2020. "Some Like It Hot: Mediterranean Societies at the End of the Little Ice Age." *Agricultural History* 94, no. 2: 176–204.

Eastwood, Warren J., Melanie J. Leng, Neil Roberts, and Basil Davis. 2007. "Holocene Climate Change in the Eastern Mediterranean Region: A Comparison of Stable Isotope and Pollen Data from Lake Gölhisar, Southwest Turkey." *Journal of Quaternary Science* 22, no. 4: 327–41.

England, Ann, Warren J. Eastwood, C. Neil Roberts, Rebecca Turner, and John F. Haldon. 2008. "Historical Landscape Change in Cappadocia (Central Turkey): A Palaeoecological Investigation of Annually Laminated Sediments from Nar Lake." *The Holocene* 18, no. 8: 1229–45.

Ertem, Özge. 2017. "Considering Famine in the Late Nineteenth Century Ottoman Empire: A Comparative Framework and Overview," in ""The Enormous Failure of Nature": Famine and Society in the Nineteenth Century," edited by Andrew G. Newby, special issue, *Collegium* (Helsinki Collegium for Advanced Studies) 22: 151–72.

Finné, Martin, Jessie Woodbridge, Inga Labuhn, and C. Neil Roberts. 2019. "Holocene Hydro-Climatic Variability in the Mediterranean: A Synthetic Multi-Proxy Reconstruction." *The Holocene* 29, no. 5: 847–63.

Fiorentino, Girolamo, Valentina Caracuta, Lucio Calcagnile, Marisa D'Elia, Paolo Matthiae, F. Mavelli, and Gianluca Quarta. 2008. "Third Millennium BC Climate Change in Syria Highlighted by Carbon Stable Isotope Analysis of 14C-AMS Dated Plant Remains from Ebla." *Palaeogeography, Palaeoclimatology, Palaeoecology* 266, no. 1-2: 51–58.

Gaillard, Marie-José, Shinya Sugita, Florence Mazier, Anna-Kari Trondman, Anna Broström, Thomas Hickler, Jed O. Kaplan, et al. 2010. "Holocene Land-Cover Reconstructions for Studies on Land Cover-Climate Feedbacks." *Climate of the Past* 6, no. 4: 483–99.

Göktürk, Ozan M., Dominik Fleitmann, Seraina Badertscher, Hai Cheng, R. Lawrence Edwards, Markus Leuenberger, Andreas Fankhauser, et al. 2011. "Climate on the Southern Black Sea Coast During the Holocene: Implications from the Sofular Cave Record." *Quaternary Science Reviews* 30, no. 19–20: 2433–45.

Haldon, John, Neil Roberts, Adam Izdebski, Dominik Fleitmann, Michael McCormick, Marica Cassis, Owen Doonan, et al. 2014. "The Climate and Environment of Byzantine Anatolia: Integrating Science, History, and Archaeology." *Journal of Interdisciplinary History* 45, no. 2: 113–61. doi:10.1162/JINH_a_00682.

Harper, Kyle, 2017. *The Fate of Rome: Climate, Disease and the End of an Empire*. Princeton and Oxford: Princeton University Press.

Harrison, Sandy P., Patrick J. Bartlein, and I. Colin Prentice. 2016. "What have we Learnt from Palaeoclimate Simulations?" *Journal of Quaternary Science* 31, no. 4: 363–85.

Hassan, Fekri. 2002. *Droughts, Food, and Culture: Ecological Change and Food Security in Africa's Later Prehistory*. New York: Kluwer Academic.

Holling, Crawford Stanley, and Lance H. Gunderson. 2002. "Resilience and Adaptive Cycles." In *Panarchy: Understanding Transformations in Human and Natural Systems*, edited by Lance H. Gunderson and Crawford Stanley Holling, 25–62, Washington, DC: Island Press.

Huntington, Ellsworth. 1907. *The Pulse of Asia: a Journey in Central Asia Illustrating the Geographic Basis of History*. London: Houghton, Mifflin.

Hütteroth, Wolf-Dieter. 1968. *Ländliche Siedlungen im südlichen Inner-Anatolien in den letzten vierhundert Jahren*. Göttingener Geographische Abhandlungen 46. Göttingen: Selbstverl. d. Geograph. Inst. d. Univ. Göttingen.

Inalcık, Halil, and Suraiya Faroqhi. 1997. *An Economic and Social History of the Ottoman Empire (Vol. 1)*. Cambridge: Cambridge University Press.

Izdebski, Adam, Jordan Pickett, Neil Roberts, and Tomasz Waliszewski. 2016. "The Environmental, Archaeological and Historical Evidence for Regional Climatic Changes and their Societal Impacts in the Eastern Mediterranean in Late Antiquity." *Quaternary Science Reviews* 136: 189–208

Jones, Matthew D., C. Neil Roberts, Melanie J. Leng, and Murat Türkeş. 2006. "A High-Resolution Late Holocene Lake Isotope Record from Turkey and Links to North Atlantic and Monsoon Climate." *Geology* 34, no. 5: 361–4.

Jones, Matthew D., Nizar Abu- Jaber, Ahmad al-Shdaifat, Douglas Baird, Benjamin I. Cook, Mark O. Cuthbert, Jonathan R. Dean, et al. 2019. "20,000 Years of Societal Vulnerability and Adaptation to Climate Change in Southwest Asia." *Wiley Interdisciplinary Reviews: Water* 6, no. 2: e1330.

Kabukcu, Ceren. 2017. "Woodland Vegetation History and Human Impacts in South-Central Anatolia 16,000–6500 cal BP: Anthracological Results from Five Prehistoric Sites in the Konya Plain." *Quaternary Science Reviews* 176: 85–100.

Kaniewski, David, Joël Guiot, and Elise Van Campo. 2015. "Drought and Societal Collapse 3200 Years Ago in the Eastern Mediterranean: A Review." *Wiley Interdisciplinary Reviews: Climate Change* 6, no. 4: 369–82.

Katsianis, Markos, Andrew Bevan, Giorgos Styliaras, and Yannis Maniatis. 2020. "An Aegean History and Archaeology Written through Radiocarbon Dates." *Journal of Open Archaeology Data* 8, no. 1: http://doi.org/10.5334/joad.65.

Kaufman, Darrell, Nicholas McKay, Cody Routson, Michael Erb, Basil Davis, Oliver Heiri, Samuel Jaccard, et al. 2020. "A Global Database of Holocene Paleotemperature Records." *Scientific Data* 7, no. 1: 1–34.

Kuniholm, Peter Ian. 1990. "Archaeological Evidence and Non-Evidence for Climatic Change." *Philosophical Transactions of the Royal Society of London. Series A* 330, no. 1615: 645–55.

Kuper, Rudolph, and Stefan Kröpelin. 2006. "Climate-Controlled Holocene Occupation in the Sahara: Motor of Africa's Evolution." *Science* 313, no. 5788: 803–7.

Labuhn, Inga, Martin Finné, Adam Izdebski, Neil Roberts, and Jessie Woodbridge. 2018. "Climatic Changes and their Impacts in the Mediterranean During the First Millennium CE." *Late Antique Archaeology* 12: 65–88.

Langgut, Dafna, Rachid Cheddadi, Josè Sebastián Carrión, Mark Cavanagh, Daniele Colombaroli, Warren John Eastwood, Raphael Greenberg, et al. 2019. "The Origin and Spread of Olive Cultivation in the Mediterranean Basin: The Fossil Pollen Evidence." *The Holocene* 29, no. 5: 902–22.

Lawrence, Dan, Graham Philip, Hannah Hunt, Lisa Snape-Kennedy, and Tony J. Wilkinson. 2016. "Long Term Population, City Size and Climate Trends in the Fertile Crescent: A First Approximation." *PloS One* 11, no. 3: e0152563.

Lawrence, Dan, Alessio Palmisano, and Michelle W. de Gruchy. 2021. "Collapse and Continuity: A Multi-Proxy Reconstruction of Settlement Organization and Population Trajectories in the Northern Fertile Crescent During the 4.2 kya Rapid Climate Change Event." *Plos One* 16, no. 1: e0244871.

Marcott, Shaun A., Jeremy D. Shakun, Peter U. Clark, and Alan C. Mix. 2013. "A Reconstruction of Regional and Global Temperature for the Past 11,300 Years." *Science* 339, no. 6124: 1198–201.

McCormick, Michael, Ulf Büntgen, Mark A. Cane, Edward R. Cook, Kyle Harper, Peter Huybers, Thomas Litt, et al. 2012. "Climate Change During and After the Roman Empire: Reconstructing the Past from Scientific and Historical Evidence." *Journal of Interdisciplinary History* 43, no. 2: 169–220.

McCorriston, Joy. 1997. "Textile Extensification, Alienation, and Social Stratification in Ancient Mesopotamia." *Current Anthropology* 38, no. 4: 517–35.

McCoy, Mark D. 2017. "Geospatial Big Data and Archaeology: Prospects and Problems Too Great to Ignore." *Journal of Archaeological Science* 84: 74–94.

Naumann, Edmund. 1893. *Vom Goldnen Horn zu den Quellen des Euphrat: Reisebriefe, Tagebuchblätter und Studien über die asiatische Türkei und die anatolische Bahn.* Berlin: Oldenbourg Wissenschaftsverlag.

Özel, Oktay. 2004. "Population Changes in Ottoman Anatolia During the 16th and 17th Centuries: The 'Demographic Crisis' Reconsidered." *International Journal of Middle East Studies* 36, no. 2: 183–205.

Palmisano, Alessio, Andrew Bevan, and Stephen Shennan. 2017. "Comparing Archaeological Proxies for Long-Term Population Patterns: An Example from Central Italy." *Journal of Archaeological Science* 87: 59–72.

———. 2018. "Regional Demographic Trends and Settlement Patterns in Central Italy: Archaeological Sites and Radiocarbon Dates." *Journal of Open Archaeology Data* 6. 10.5334/joad.43

Palmisano, Alessio, Dan Lawrence, Michelle W. de Gruchy, Andrew Bevan, and Stephen Shennan. 2021. "Holocene Regional Population Dynamics and Climatic Trends in the Near East: A First Comparison Using Archaeo-Demographic Proxies." *Quaternary Science Reviews* 252: 106739. Dataset and R scripts available at http://doi.org/10.5281/zenodo.4322979.

Parker, Geoffrey. 2013. *Global Crisis: War, Climate Change, and Catastrophe in the Seventeenth Century.* New Haven and London: Yale University Press.

Ponton, Camilo, Liviu Giosan, Tim I. Eglinton, Dorian Q. Fuller, Joel E. Johnson, Pushpendra Kumar, and Tim S. Collett. 2012. "Holocene Aridification of India." *Geophysical Research Letters* 39, no. 3. 10.1029/2011GL050722

Roberts, Neil. 1990. "Human-Induced Landscape Change in South and South-West Turkey During the Later Holocene." In *Man's Role in the Shaping of the Eastern Mediterranean Landscape*, edited by Sytze Bottema, Gertie Entjes-Nieborg, and Willem van Zeist, 53–67. Rotterdam: A. A. Balkema.

———. 2018. "The Land of the Hittites: Airs, Waters and Places." In *The Geography and Landscape of the Hittites*, edited by Mark Weeden and Lee Z. Ullmann, 7–17. Leiden: Brill.

Roberts, Neil, Samantha L. Allcock, Hannah Barnett, Anne Mather, Warren J. Eastwood, Matthew Jones, Nick Primmer, Hakan Yiğitbaşıoğlu, and Boris Vannière. 2019b. "Cause-and-Effect in Mediterranean Erosion: The Role of Humans and Climate upon Holocene Sediment Flux into a Central Anatolian Lake Catchment." *Geomorphology* 331: 36–48.

Roberts, Neil, Mathew D. Jones, Abdelfattah Benkaddour, Warren J. Eastwood, Maria Letizia Filippi, Michael R. Frogley, Henry F. Lamb, et al. 2008. "Stable Isotope Records of Late Quaternary Climate and Hydrology from Mediterranean Lakes: The ISOMED Synthesis." *Quaternary Science Reviews* 27, no. 25–26: 2426–41.

Roberts, Neil, Jane M. Reed, Melanie J. Leng, Catherine Kuzucuoğlu, Michel Fontugne, Jacques Bertaux, Henk Woldring, et al. 2001. "The Tempo of Holocene Climatic Change in the Eastern Mediterranean Region: New High-Resolution Crater-Lake Sediment Data from Central Turkey." *The Holocene* 11, no. 6: 721–36.

Roberts, C. Neil, Jessie Woodbridge, Alessio Palmisano, Andrew Bevan, Ralph Fyfe, and Stephen Shennan. 2019a. "Mediterranean Landscape Change During the Holocene: Synthesis, Comparison and Regional Trends in Population, Land Cover and Climate." *The Holocene* 29, no. 5: 923–37.

Roffet-Salque, Mélanie, Arkadiusz Marciniak, Paul J. Valdes, Kamilla Pawłowska, Joanna Pyzel, Lech Czerniak, Marta Krüger, C. Neil Roberts, Sharmini Pitter, and Richard P. Evershed. 2018. "Evidence for the Impact of the 8.2-kyBP Climate Event on Near Eastern Early Farmers." *Proceedings of the National Academy of Sciences* 115, no. 35. 8705–9.

Rosen, Arlene M. 2007. *Civilizing Climate. Social Responses to Climate Change in the Ancient Near East.* Lanham, MD: Altamira Press.

Samartin, Stéphanie, Oliver Heiri, Fortunat Joos, Hans Renssen, Jörg Franke, Stefan Brönnimann, and Willy Tinner. 2017. "Warm Mediterranean Mid-Holocene Summers Inferred from Fossil Midge Assemblages." *Nature Geoscience* 10, no. 3: 207–12.

Seager, Richard, Nicholas Graham, Celine Herweijer, Arnold L. Gordon, Yochanan Kushnir, and Ed Cook. 2007. "Blueprints for Medieval Hydroclimate." *Quaternary Science Reviews* 26, no. 19–21: 2322–36.

Shanahan, Timothy M., Nicholas P. McKay, Konrad A. Hughen, Jonathan T. Overpeck, Bette Otto-Bliesner, Clifford W. Heil, John King, Christopher A. Scholz, and John Peck. 2015. "The Time-Transgressive Termination of the African Humid Period." *Nature Geoscience* 8, no. 2: 140–44.

Shennan, Stephen, Sean S. Downey, Adrian Timpson, Kevan Edinborough, Sue Colledge, Tim Kerig, Katie Manning, and Mark G. Thomas. 2013. "Regional Population Collapse Followed Initial Agriculture Booms in Mid-Holocene Europe." *Nature Communications* 4, no. 1: 1–8.

Sinha, Ashish, Gayatri Kathayat, Harvey Weiss, Hanying Li, Hai Cheng, Justin Reuter, Adam W. Schneider, et al. 2019. "Role of Climate in the Rise and Fall of the Neo-Assyrian Empire." *Science Advances* 5, no. 11: eaax6656.

Tainter, Joseph A. 2006. "Archaeology of Overshoot and Collapse." *Annu. Rev. Anthropol.* 35: 59–74.

Timpson, Adrian, Sue Colledge, Enrico Crema, Kevan Edinborough, Tim Kerig, Katie Manning, Mark G. Thomas, and Stephen Shennan. 2014. "Reconstructing Regional Population Fluctuations in the European Neolithic Using Radiocarbon Dates: A New Case-Study Using an Improved Method." *Journal of Archaeological Science* 52: 549–57.

Touchan, Ramzi, Ünal Akkemik, Malcolm K. Hughes, and Nesat Erkan. 2007. "May–June Precipitation Reconstruction of Southwestern Anatolia, Turkey During the last 900 Years from Tree Rings." *Quaternary Research* 68, no. 2: 196–202.

Turchin, Peter. 2003. *Historical Dynamics: Why States Rise and Fall.* Princeton, NJ: Princeton University Press.

Türkeş, Murat. 1996. "Spatial and Temporal Analysis of Annual Rainfall Variations in Turkey." *International Journal of Climatology* 16: 1057–76.

Türkeş, Murat, and Ecmel Erlat. 2003. "Precipitation Changes and Variability in Turkey Linked to the North Atlantic Oscillation During the Period 1930-2000." *International Journal of Climatology* 23, no. 14: 1771–96.

Turner, Rebecca, Neil Roberts, and Matthew D. Jones. 2008. "Climatic Pacing of Mediterranean Fire Histories from Lake Sedimentary Microcharcoal." *Global and Planetary Change* 63, no. 4: 317–24.

Walsh, Kevin, Jean-François Berger, C. Neil Roberts, Boris Vannière, Matthieu Ghilardi, Antony G. Brown, Jessie Woodbridge, et al. 2019. "Holocene Demographic Fluctuations, Climate and Erosion in the Mediterranean: A Meta Data-Analysis." *The Holocene* 29, no. 5: 864–85.

Wanner, Heinz, Jürg Beer, Jonathan Bütikofer, Thomas J. Crowley, Ulrich Cubasch, Jacqueline Flückiger, Hugues Goosse, et al. 2008. "Mid-to Late Holocene Climate Change: An Overview." *Quaternary Science Reviews* 27, no. 19–20: 1791–828.

Weiberg, Erika, Andrew Bevan, Katerina Kouli, Markos Katsianis, Jessie Woodbridge, Anton Bonnier, Max Engel, et al. 2019. "Long-Term Trends of Land Use and Demography in Greece: A Comparative Study." *The Holocene* 29, no. 5: 742–60.

Weiss, Harvey, ed. 2017. *Megadrought and Collapse: From Early Agriculture to Angkor.* Oxford: Oxford University Press.

Weiss, Harvey, M-A. Courty, Wilma Wetterstrom, François Guichard, Louise Senior, Richard Meadow, and Anna Curnow. 1993. "The Genesis and Collapse of Third Millennium North Mesopotamian Civilization." *Science* 261, no. 5124: 995-1004.

White, Sam. 2011. *The Climate of Rebellion in the Early Modern Ottoman Empire.* Cambridge: Cambridge University Press.

Whittow, Mark. 2003. "Decline and Fall? Studying Long-Term Change in the East." *Late Antique Archaeology* 1: 404-23.

Wilkinson, Tony J. 2003. *Archaeological Landscapes of the Near East.* Tucson, Arizona: University of Arizona Press.

Williams, Alan N. 2012. "The Use of Summed Radiocarbon Probability Distributions in Archaeology: A Review of Methods." *Journal of Archaeological Science* 39, no. 3: 578e589.

Woodbridge, Jessie, C. Neil Roberts, Alessio Palmisano, Andrew Bevan, Stephen Shennan, Ralph Fyfe, Warren J. Eastwood, et al. 2019. "Pollen-Inferred Regional Vegetation Patterns and Demographic Change in Southern Anatolia through the Holocene." *The Holocene* 29, no. 5: 728-41.

Xoplaki, Elena, J. F. González-Rouco, Jürg Luterbacher, and Heinz Wanner. 2004. "Wet Season Mediterranean Precipitation Variability: Influence of Large-Scale Dynamics and Trends." *Climate Dynamics* 23, no.1: 63-78.

Xoplaki, Elena, Jürg Luterbacher, Sebastian Wagner, Eduardo Zorita, Dominik Fleitmann, Johannes Preiser-Kapeller, Abigail M. Sargent, et al. 2018. "Modelling Climate and Societal Resilience in the Eastern Mediterranean in the Last Millennium." *Human Ecology* 46, no. 3: 363-79.

Yiğitbaşıoğlu, Hakan, Jonathan R. Dean, Warren J. Eastwood, Neil Roberts, Matthew D. Jones, and Melanie J. Leng. 2015. "A 600 Year-Long Drought Index for Central Anatolia." *Journal of Black Sea/Mediterranean Environment* Special Issue: 84-88.

I

Insights over the Long Term: Continuity and Change

An Assessment of Long-Term Climate Dynamics across Anatolia Based on the Results of Macrophysical Climate Modeling

Bülent Arıkan

Introduction

The Anatolian Peninsula, also known as Asia Minor, is situated at the eastern terminus of the Mediterranean Basin. Geologically, and to a certain extent culturally, Anatolia is considered to be part of the Near East, yet it sits at the junction of several climate and phytogeographic zones, leading to significantly higher rates of biodiversity than in neighboring regions. These wide variations in climate, flora, and fauna are also responsible for differences in archaeological societies' social, economic, technological, and political organizations.

Climate change currently affects humanity at a global scale. Past climate changes, however, their impacts on local ecology, and how human societies responded, have been the subject of scholarly debate for the last fifty years. As we develop a more comprehensive and scientific approach to past climate changes, the frameworks we use to assess human behavior under climate-induced stress evolve from environmental determinism to possibilism. Environmental determinism postulates a strict approach wherein human societies at a given time remain passive in the face of climatic perturbations. Under this model, human societies respond in only a few ways when faced with such pressures (Weiss and Bradley 2001; Cullen et al. 2000), and today such deterministic approaches continue under the guise of habitat-tracking (Weiss et al. 1993). According to these approaches, humans are mere puppets that will migrate—or risk collapse—only when faced with climate changes. A reflection of such deterministic approaches in paleoclimate is the adoption of supra-regional patterns, which have been used increasingly in the discussion on the fate of archaeological societies. In the last two decades, however, environmental possibilism has emerged as an alternative framework in assessing climate-human interactions. Unlike the deterministic approach, possibilism states that human societies consisted of agents with free will and ingenuity. Under this model, humans may develop tools to minimize risks and mitigate threats from changing climate, leading to an understanding of a broad repertoire of wide-ranging behaviors (Arıkan, Mohr, and Bürgi 2020; Barton et al. 2018; Arıkan 2013; Batiuk 2013). The wide variety of behaviors observed in the archaeological record helps us understand how past societies dealt with past climatic changes. Today, empirical data suggest various technological (crop selection), economic (balancing agriculture with husbandry), and socio-political (hierarchic vs. heterarchic) maneuvers displayed by past human societies (Arıkan 2010). Such behaviors directly relate to minimizing risks associated with climate change.

An important issue in the discussion of past climate changes is the scale and severity of climate oscillations. The first paleoenvironmental reconstructions focused on identifying periods of climate change (e.g., the transition between the Early and Middle Holocene) and phases of climatic instability (COHMAP 1988). In the next phase of research, between the 1990s and 2010s, researchers suggested that phases of climatic instability (e.g., 8.2, 6.2, and 4.2 ka BP events) affected past societies at varying levels (Alder and Hostetler 2015; Martinez-Ruiz et al. 2015). In these discussions, some scholars (Weiss 2017; Staubwasser and Weiss 2006) even suggested that some of these phases (e.g., the 4.2 ka BP event) had a global nature. As archaeological and paleoclimatic research increases yearly, we now need to understand better the scale and severity of each phase of climatic instability (Cleator et al. 2020; Morin et al. 2018). Archaeologists and others state that these phases do not reveal uniform patterns; each phase of climatic instability impacted different parts of the Near East differently. It is impossible, therefore, to argue for a uniform human response across the Near East under climatic stress.

This paper aims to present an assessment of the paleoclimate of Anatolia from 12,000-2000 cal BP / 10,000 BCE-0. The paper does not focus on all aspects of paleoclimate across the whole Anatolian Peninsula nor delve into the archaeological evidence for past climate change. Rather, the paper focuses on average annual precipitation and temperature changes throughout the Early (12,000-6000 cal BP), Middle (6000-2600 cal BP), and Late Holocene (2600-2000 cal BP) in the Amuq Plain, Upper Euphrates, and Upper Tigris basins, north-central Anatolia, and the north Aegean (Fig. 1). These regions reflect almost all of the climatic and other environmental characteristics of Anatolia, while they yield rich archaeological evidence concerning past human-climate interactions. Macrophysical Climate Modeling (MCM) forms the basis of paleoclimatic reconstructions in this paper. The discussion of proxy-based paleoenvironmental reconstruction has not been included because of disparities in data resolution among these regions and the lack of such data from the Amuq Plain. The results of MCM are presented in separate sections for each region, followed by an overall assessment.

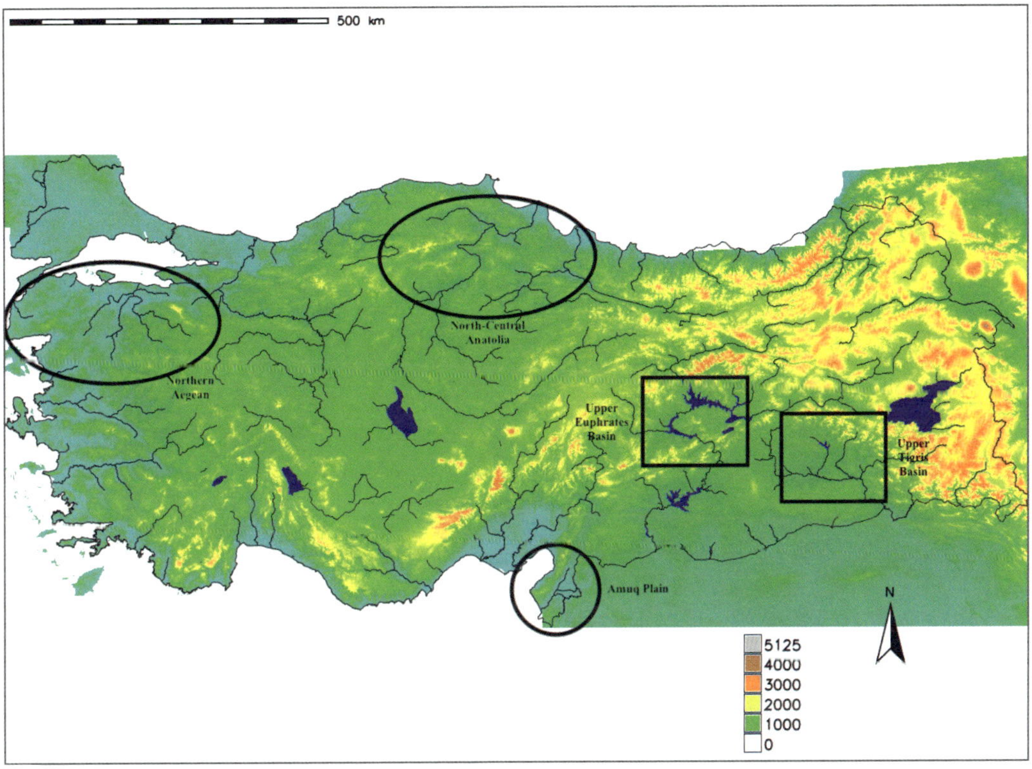

Fig. 1. Digital elevation model of Anatolia with major rivers, lakes, and reservoirs, with the borders of the Republic of Turkey in black lines (30 m resolution showing meters above sea level). Black circle, ovals, and rectangles show the regions mentioned in the text.

Paleoclimate Modeling

MCM produces a synoptic paleoclimate model that calculates changes in the Earth's heat budget (Bryson and DeWall 2007, 8). Such modeling assumes that the heat budget of the planet is related to albedo, which changes only as a result of chemical and physical characteristics of the atmosphere (i.e., volcanism) and Milankovitch Cycles (i.e., axial tilt, eccentricity, and precession), and that there has been no change in albedo or Milankovitch Cycles over the last 40,000 years (Bryson and DeWall 2007, 7–9).

Bryson (2007a, 21–24) suggested that climatic changes over the last 40,000 years resulted from shifts in the locations of upper atmospheric pressure systems (i.e., the Jet Stream and the Intertropical Convergence Zone). The particulars of these changes are discussed elsewhere (Arıkan 2015; Bryson 2007d); however, the changes can be summarized as follows: roughly every decade, the shifting positions of atmospheric pressure systems between the Azores and Iceland affect the climate of the Mediterranean Basin and North America. These decadal shifts cause milder and wetter or cooler and drier winters. The shifts are also known as North Atlantic Oscillation (NAO) positive/negative indices.

Bryson (2007a, 2007b, 2007e) developed an algorithm that calculated decadal changes in the upper atmospheric pressure systems based on observations from the "Climate Normal" (1960–1990). This algorithm observes the relationship between pressure systems' locations in the upper atmosphere and average values of annual precipitation, temperature, and other variables. Bryson (2007c) applied the multilinear regression method that retrodicts average precipitation, temperature, and other climatic variables depending on the past locations of atmospheric pressure systems.

The main advantage of MCM is that it runs at a much higher spatial resolution (usually at 10 km grids) than Global Circulation Models (GCMs) (usually at 150–300 km grids). The temporal resolution of MCM is 100 years, compared to the six-hour temporal resolution of GCMs, a trade-off of lower temporal for higher spatial resolution. As archaeological data sets are similarly coarse grained, the results of MCM are valuable for interpreting archaeological and historical evidence.

In this paper, MCM results have been presented for five regions, focusing on average annual precipitation (mm) and temperature (°C). For each region, data gathered from local meteorological stations have been used in MCM. The results are averaged for each millennium because the centennial resolution is far too fine grained for the purposes of this paper. Variance for precipitation and temperature has also been calculated for each millennium to explore any meaningful changes through time. To discuss changes in precipitation and temperature around the 8.2, 6.2, and 4.2 ka BP events, a 600-year average of centennial data was calculated. This means, for example, averaging precipitation and temperature values between 8500 and 7900 cal BP to assess the 8.2 ka BP event.

Additionally, the patterns at these target events are compared with the millennial patterns to test whether climatic instability phases show significant deviations from long-term climatic trends. It is important to mention that each region has unique climatic conditions. Consequently, "normal" means different precipitation or temperature values for each region. Historically, as in the modern era, north-central Anatolia is the region that has the driest climate, whereas the north Aegean has the wettest conditions. Other regions rank between these two extremes.

Regional Climate Patterns

The long-term regional climate patterns are presented through changes in precipitation (Fig. 2) and temperature (Fig. 3). The graphs show comparisons of all five regions covered here.

Upper Euphrates Basin

Long-term (12,000–2000 cal BP) average annual precipitation in the Upper Euphrates Basin drops sharply from 950 mm to 700 mm by 8000 cal BP and then to 600 mm by 3500 cal BP (Fig. 2). The median for this time range is 575 mm, while much higher values

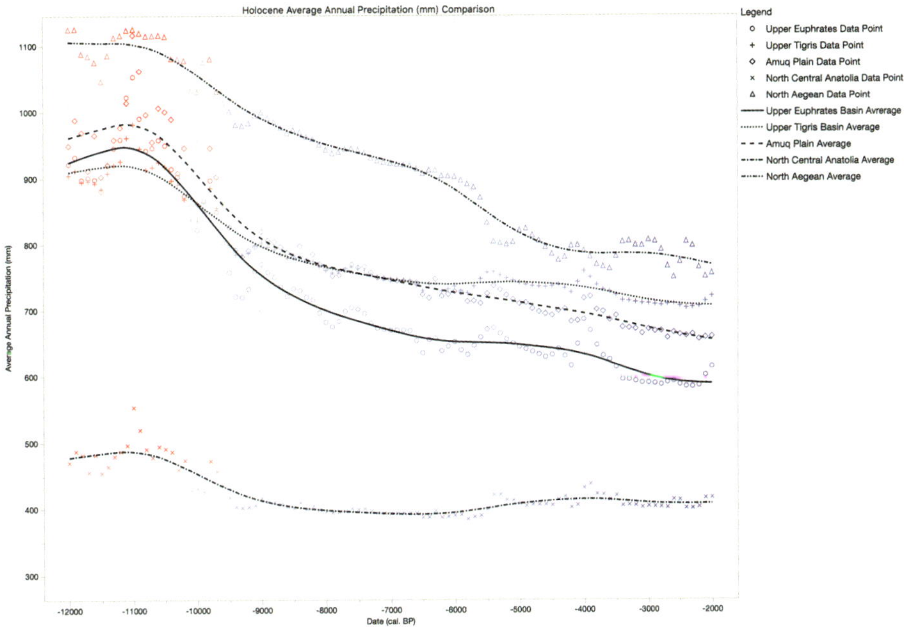

Fig. 2. Average annual precipitation (mm) per century between 12,000-2000 cal BP for the regions discussed in the text. Black lines represent smoothing of data points. Colors of data points show whether they are above (red) or below (blue) median values.

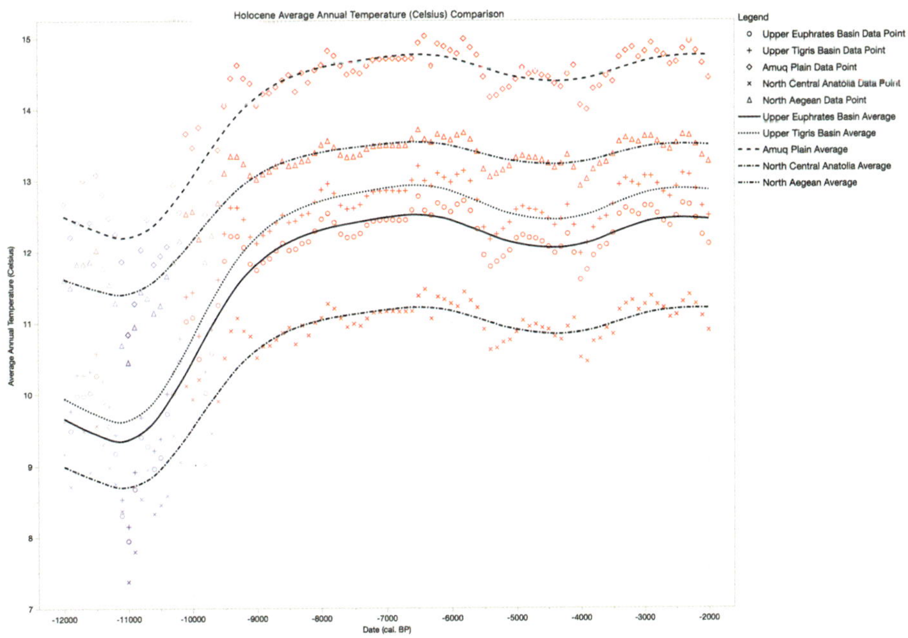

Fig. 3. Average annual temperature (Celsius) per century between 12,000-2000 cal BP for the regions discussed in the text. Black lines represent smoothing of data points. Colors of data points show whether they are above (red) or below (blue) median values.

occur in the Early Holocene. During the Middle and Late Holocene, precipitation does not deviate from the median significantly. Average annual temperature rises sharply from 9.7°C to 12.5°C by 8000 cal BP, after a slight cooling (0.5°C) between 6000 and 3000 cal BP, and warming resumes after that date (Fig. 3). The median temperature is 10.3°C; however, average annual temperature remains above the median throughout the Middle and Late Holocene. These changes in precipitation and temperature suggest quick drying and gradual warming in the region. The first half of the Early Holocene seems to be a period of above-median precipitation and below-median temperature values, which only stabilizes after 9000 cal BP. After this date, warmer and drier climate sets in, and it continues to get warmer and more arid until about 6000 cal BP. Between 6000 and 4000 cal BP precipitation remains stable and temperature drops, which leads to slightly cooler conditions. Warmer and arid conditions resume with dropping precipitation and increasing temperature after 4000 cal BP (see Arıkan 2015 for an extensive discussion on the paleoclimatic changes of the region).

Upper Tigris Basin

Long-term paleoclimatic patterns in the Upper Tigris Basin are similar to those in the Upper Euphrates Basin. At the beginning of the Early Holocene, average annual precipitation is 910 mm and drops to 760 mm by 8000 cal BP. Following this sharp decline, precipitation remains stable in the second half of the Early Holocene and Middle Holocene (around 745 mm) (Fig. 2). A gradual but persistent decline (around 700–725 mm) takes place around 4000 cal BP. The median for precipitation from 12,000–2000 cal BP is 800 mm, and the great majority of centennial precipitation values remain close to that value. The only exception is the first half of Early Holocene, when above-median precipitation values emerge. Temperature changes are also gradual and punctuated; Early Holocene average annual temperature is 10°C, which increases to 12.5°C by 8000 cal BP and remains close to that value until the beginning of the Middle Holocene (ca. 6000 cal BP) (Fig. 3). Between 6000 and 3000 cal BP, MCM results show slight cooling (around 0.5°C), and temperatures return to a 12.5°C average thereafter. The median annual temperature is 10.5°C, and centennial values are below the median until 9000 cal BP. After that however, centennial temperature values always remain above the median. The overall pattern of the Upper Tigris Basin shows fast-changing precipitation patterns for the first half of Early Holocene. Between 11,100 and 10,400 cal BP centennial precipitation seems to be erratic, as changes in the median suggest. Early Holocene changes in precipitation become less dramatic after 8000 cal BP, and drier conditions become the norm compared to the Early Holocene. Around 9000 cal BP, warmer conditions set in, and the temperature continues to increase. Temperature changes complement the overall paleoclimatic picture of the Early Holocene in the region, with warm and dry conditions emerging after 7000 cal BP (see Arıkan 2015 for an extensive discussion on the region's paleoclimatic changes).

Amuq Plain

Long-term changes in the average annual precipitation of the Amuq Plain show a continuous decline without reaching a plateau. At the beginning of the Holocene, average annual precipitation is 960 mm, which decreases to 800 mm at 9000 cal BP and to 730 mm at 6000 cal BP (Fig. 2). By the beginning of the Late Holocene (2600 cal BP), the average annual precipitation is 660 mm. The median for precipitation is 800 mm, and above-median values emerge in the first half of Early Holocene. Until 6500 cal BP, centennial precipitation values move towards the median, and after that date values remain below the median. Changes in average annual temperature of the Amuq Plain show a more complicated pattern, which is also similar to the previously discussed regions. At the beginning of the Early Holocene, the average annual temperature is 12.5°C; by 9000 cal BP, it increases to 14.2°C, reaching 14.5°C by 6000 cal. BP (Fig. 3). Between 6000 and 3000 cal BP, a slight drop in average annual temperature takes place, but temperature increases again after 3000 cal BP. The median temperature is 14.5°C. During the first half of the Early Holocene, centennial temperature values remain below average, but by 9000 cal BP, centennial temperature values exceed the median and stay above it for the rest of the period. Long-term precipitation changes reflect a gradual but persistent drop in the Amuq Plain. Coupled with the dramatic increase in temperature over a short period, the Amuq Plain becomes warm and dry by the end of Early Holocene.

North-Central Anatolia

Long-term precipitation patterns in north-central Anatolia differ significantly from previously described regions. Average annual precipitation at the beginning of the Early Holocene is 455 mm, and by 8000 cal BP, it drops to 400 mm (Fig. 2). This persistent decline reaches a plateau and remains stable until 5500 cal BP (Middle Holocene). Although parts of eastern Anatolia show a continuing drop in precipitation during the Middle Holocene, there is a slight increase (about 25 mm) between 5500 and 3000 cal BP. Additionally, the Late Holocene in north-central Anatolia does not show a significant drop in precipitation. Median precipitation is 430 mm; Early Holocene centennial precipitation values are slightly higher than the median, but after 9500 cal BP, centennial precipitation remains below the median. The average annual temperature in the region is 9°C at the beginning of the Holocene (Fig. 3) and increases to 11°C by 8000 cal BP. During the slightly wetter portion of the Middle Holocene (5500–3000 cal BP), a slight drop in temperature (about 0.25°C) is noteworthy. The median temperature is 9.25°C. Early Holocene centennial temperatures remain below the median, but after 9000 cal BP, these values rise above the median. In north-central Anatolia, drier and warmer conditions emerge at the end of Early Holocene; however, soon after the onset of the Middle Holocene, conditions become slightly wetter and slightly cooler than in eastern Anatolia (see Arıkan and Yıldırım 2018 for a detailed discussion on the Middle

Holocene paleoclimate). Consequently, north-central Anatolia offers a different set of environmental conditions for Middle Holocene cultures and civilizations. Additionally, the Late Holocene climate was not as dry and warm as in eastern Anatolia.

North Aegean

Average annual precipitation at the beginning of the Holocene is 1100 mm and drops sharply through time: it reaches 950 mm at 8000 cal BP, 875 mm at 6000 cal BP, and 790 mm at 4000 cal BP (Fig. 2) (see Gauthier 2016 for discussion of the impacts of Early Holocene aridity on the region's Neolithic societies). Thereafter, precipitation values reach a plateau around 770–780 mm until 2800 cal BP. By the beginning of the Late Holocene (ca. 2600 cal BP), precipitation starts to drop again. The median precipitation is 875 mm, and centennial precipitation values remain above the median in the first half of the Early Holocene. In the second half of the Early Holocene and the early part of the Middle Holocene, centennial precipitation values are close to the median, but they fall below the median after 5500 cal BP. The average annual temperature at the beginning of Holocene is 11.6°C; by 9000 cal BP, it increases to 13°C, and at 6000 cal BP, it reaches 13.5°C (Fig. 3). A slight drop (about 0.25°C) in temperature occurs between 5500 and 3000 cal BP, but temperatures return to 13.5°C after 3000 cal BP. The median temperature over this period is 12.5°C, and centennial values remain below the median during the first half of Early Holocene. Only after 9000 cal BP do centennial values exceed the median, and they remain thus for the rest of the period. The overall pattern for precipitation and temperature in the north Aegean suggests that it becomes progressively dry and warm starting from the second half of the Early Holocene, and this trend continues until the end of Late Holocene (see Katrantsiotis et al. 2019 for discussion of the climate of the region during the last 5000 years).

General Paleoclimatic Patterns

When long-term precipitation and temperature patterns from 12,000–2000 cal BP are assessed, several points emerge. First, all five regions of the Anatolian Peninsula experience higher average annual precipitation in the first half of Early Holocene; however, precipitation rates drop sharply in all regions in the second half of Early Holocene. Second, the average annual temperature in the Early Holocene increases, and by 8000 cal BP, conditions become drier and warmer. Third, Middle Holocene precipitation patterns show a slight decrease after 5000 cal BP, and this drop marks the beginning of a persistent aridification trend that continues into the Late Holocene. The only exception to this pattern, however, is in north-central Anatolia, where slightly higher precipitation values are modeled. Fourth, all regions show a slight drop in temperature between 5000 and 3000 cal BP. Such a drop means drier and cooler conditions everywhere except north-central Anatolia, which experiences wetter and cooler conditions at this time. Fifth, average precipitation values are higher than the median during the first half of Early Holocene and gradually drop below the median by the end of Late Holocene. As far as average temperature is concerned, it remains below median values during the first half of the Holocene and only gradually exceeds the median by the end of the Late Holocene.

Average precipitation calculated for each millennium from 12,000-2000 cal BP in all regions is shown in Figure 4a. A gradual decline in precipitation is evident for all regions except north-central Anatolia. It is striking that average precipitation shows almost no meaningful change during this 10,000-year period. A slight drop in precipitation occurs in the Upper Euphrates Basin, followed by the Amuq Plain and north Aegean. A crucial statistical measure to assess long-term changes in climatic variables is variance. Calculating the millennial variance for precipitation shows when severe changes in precipitation occur within each phase. Using variance as a tool to assess long-term climate change, this measure allows us to identify periods of sharp changes and climatic stability. Figure 4b shows variance in precipitation, which reveals that sharp drops in precipitation occur in the Upper Euphrates Basin and Amuq Plain from 9000-6000 cal BP. Upper Tigris Basin precipitation values display significantly less variance during the same period. In the north Aegean, precipitation variance first drops and then increases, although these changes are minimal. In north-central Anatolia, however, variance in precipitation is low throughout the 10,000-year period.

Changes in average temperature at the millennial-scale are shown in Figure 5a. The Amuq Plain warms up far more rapidly than the rest of the regions considered here. The Upper Tigris Basin and north Aegean show similar patterns, but the rate of change in these regions is slower than in the Amuq Plain. The Upper Euphrates Basin, on the other hand, shows a continuous but gradual warming trend. The rate of change in average temperature at the millennial-scale is the lowest in north-central Anatolia. Calculating the variance in temperature at the millennial-scale reveals striking patterns (Fig. 5b). Variance remains the lowest in the Upper Euphrates Basin. Variance in temperature is the highest for the Upper Tigris Basin, especially from 11,000-6000 cal BP. Other regions show patterns similar to the Upper Tigris Basin, yet with lower variance values.

Inter-Regional Paleoclimate Patterns

This section discusses how precipitation and temperature changed through time in the regions under discussion. The cross-comparison of regions in terms of paleoclimatic patterns in the Early, Middle, and Late Holocene will provide a comprehensive picture of how each region was similar or unique in terms of past climatic conditions. The comparison of each region is based on statistical measures (e.g., variance), which help summarize the general characteristics of the paleoclimatic patterns for that specific region. Variance values highlight these events, and almost all regions display such examples in the Early and Middle Holocene.

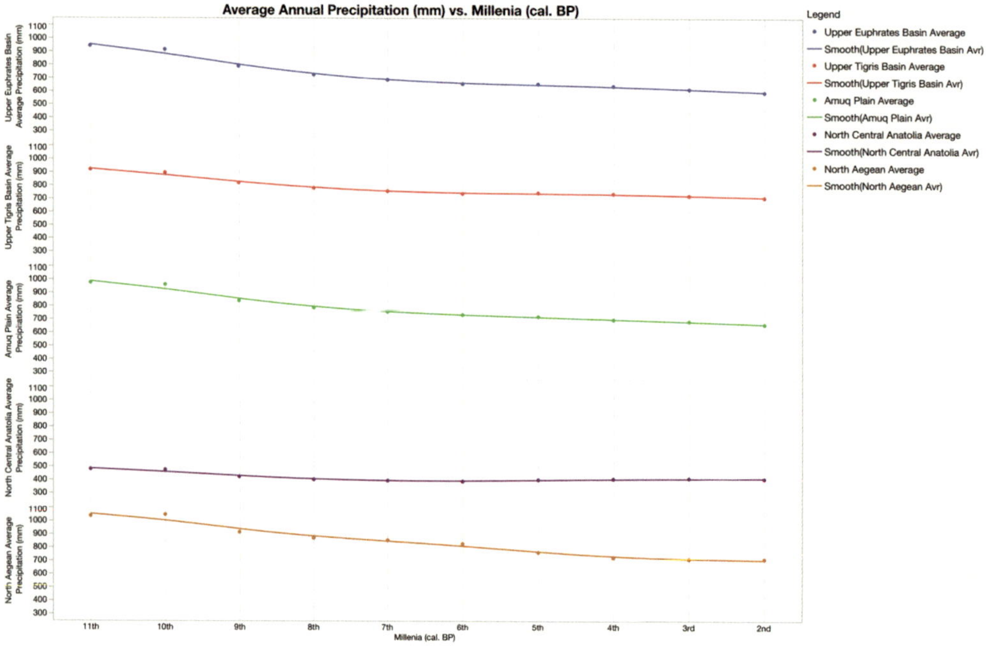

Fig. 4a. Average precipitation (mm) per millennium between 12,000-2000 cal BP in all regions. Lines represent smoothing of data points.

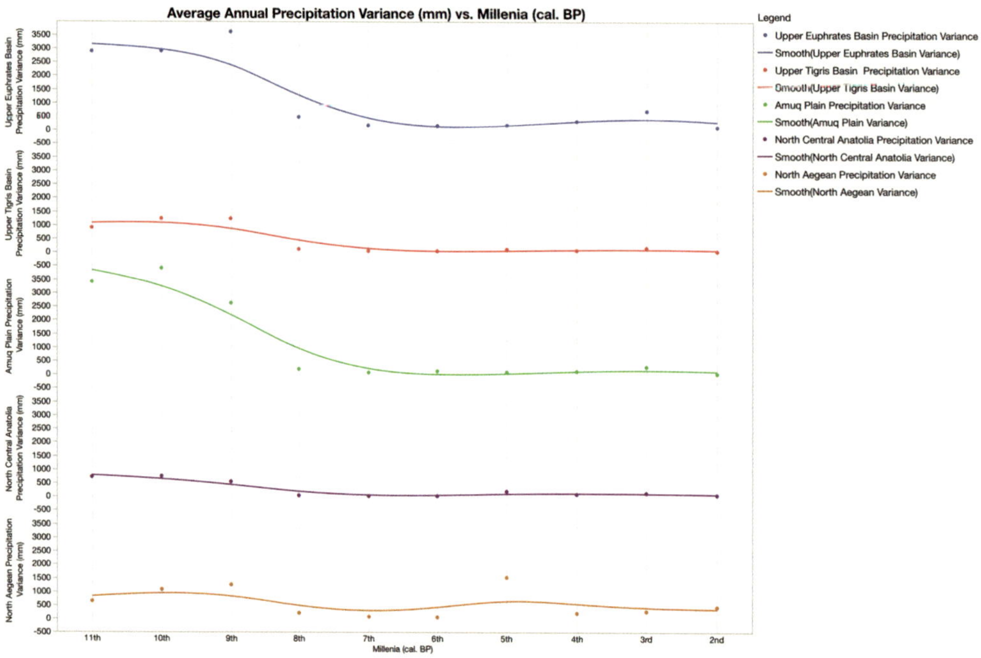

Fig. 4b. Variance in average precipitation (mm) per millennium between 12,000-2000 cal BP in all regions. Lines represent smoothing of data points.

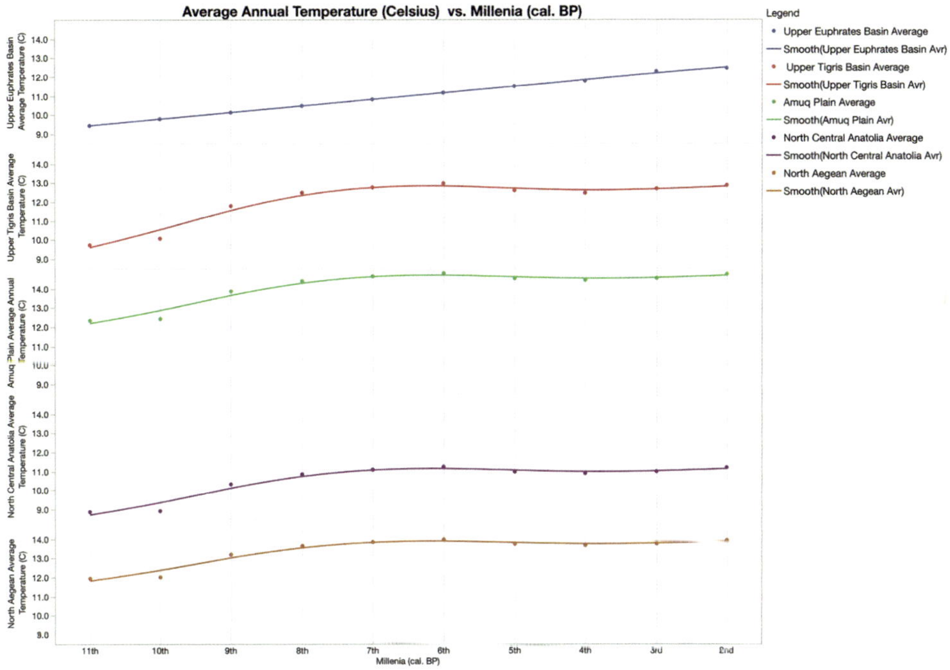

Fig. 5a. Average temperature (°C) per millennium between 12,000–2000 cal BP in all regions. Lines represent smoothing of data points.

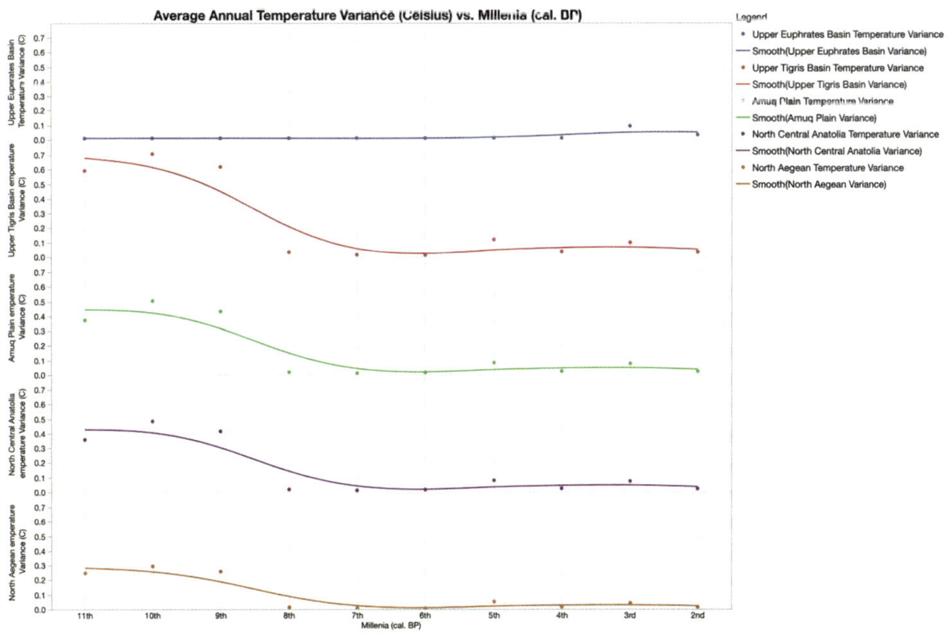

Fig. 5b. Variance in average temperature (°C) per millennium between 12,000–2000 cal BP in all regions. Lines represent smoothing of data points.

Changes in Precipitation

Inter-regional comparisons for precipitation and temperature are based on the simultaneous display of average, median, and variance values at the millennial scale (Fig. 6). Changes in precipitation through time and associated statistics clearly show that the Upper Euphrates Basin and Amuq Plain show greater variance in Early Holocene precipitation, and in this period, average precipitation values are above the median. The Upper Tigris Basin and north Aegean show narrower variance, while Early Holocene average precipitation values also remain above the median. North-central Anatolia shows the lowest variance in precipitation during the Early Holocene.

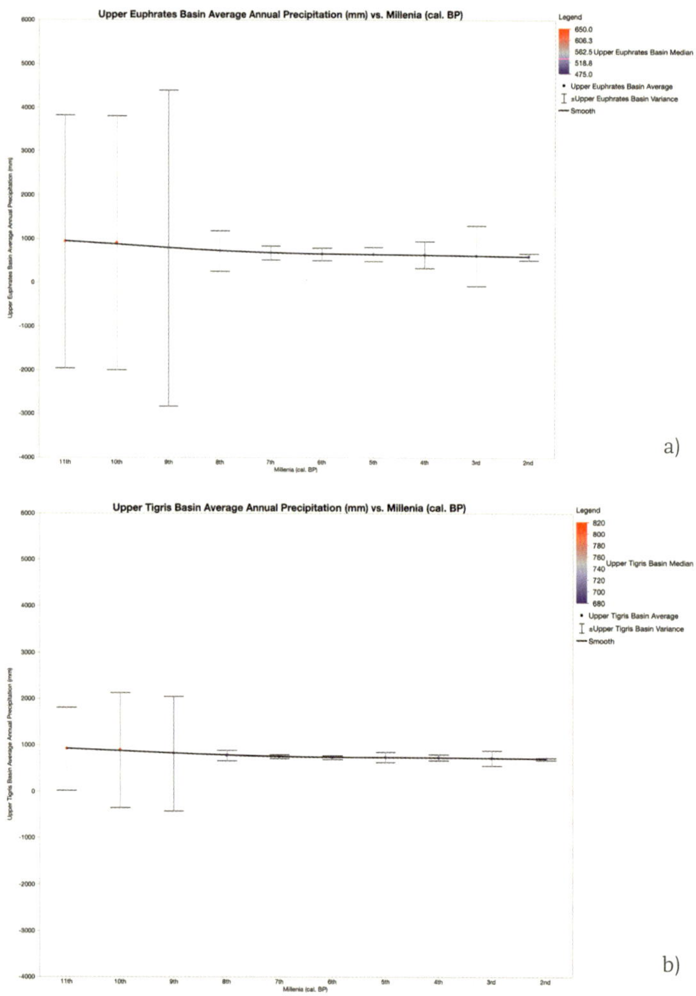

Fig. 6. Average precipitation (mm) per millennium between 12,000–2000 cal BP in a) the Upper Euphrates Basin, b) the Upper Tigris Basin, c) the Amuq Plain, d) north-central Anatolia, and e) the north Aegean. Horizontal lines represent smoothing of data points, vertical lines show variance ranges, and colors of data points show deviations from the median.

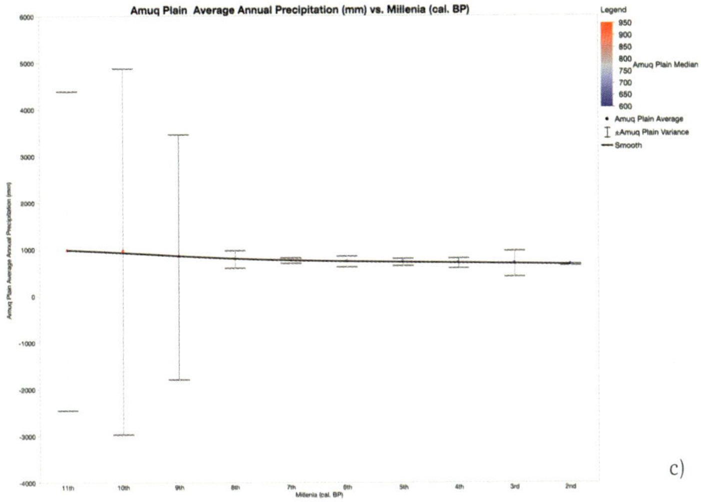
Amuq Plain Average Annual Precipitation (mm) vs. Millenia (cal. BP)
Amuq Plain Average Annual Precipitation (mm)
Millenia (cal. BP)
6000
5000
4000
3000
2000
1000
0
-1000
-2000
-3000
-4000
11th
10th
9th
8th
7th
6th
5th
4th
3rd
2nd
Legend
950
900
850
800
750
700
650
600
Amuq Plain Median
Amuq Plain Average
±Amuq Plain Variance
Smooth

c)

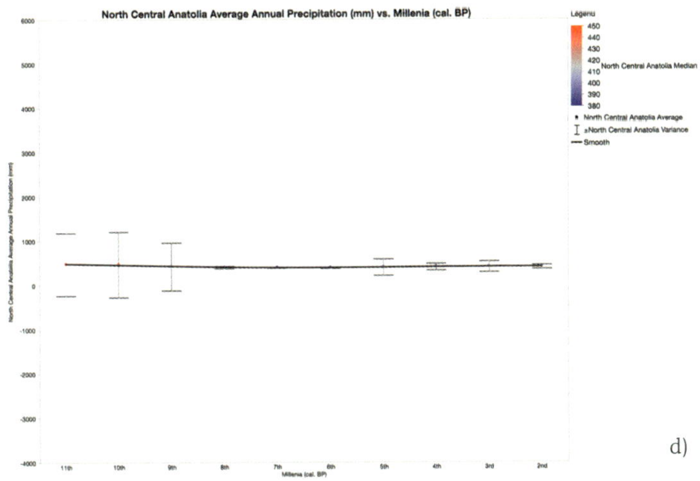
North Central Anatolia Average Annual Precipitation (mm) vs. Millenia (cal. BP)
North Central Anatolia Average Annual Precipitation (mm)
Millenia (cal. BP)
6000
5000
4000
3000
2000
1000
0
-1000
-2000
-3000
-4000
11th
10th
9th
8th
7th
6th
5th
4th
3rd
2nd
Legend
450
440
430
420
410
400
390
380
North Central Anatolia Median
North Central Anatolia Average
±North Central Anatolia Variance
Smooth

d)

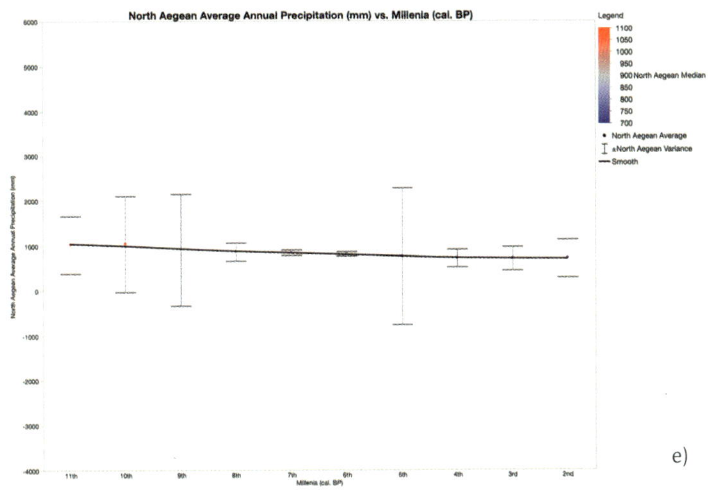
North Aegean Average Annual Precipitation (mm) vs. Millenia (cal. BP)
North Aegean Average Annual Precipitation (mm)
Millenia (cal. BP)
6000
5000
4000
3000
2000
1000
0
-1000
-2000
-3000
-4000
11th
10th
9th
8th
7th
6th
5th
4th
3rd
2nd
Legend
1100
1050
1000
950
900
850
800
750
700
North Aegean Median
North Aegean Average
±North Aegean Variance
Smooth

e)

Middle Holocene patterns that start after 6000 cal BP show slight increases in variance, while average precipitation values remain below the median. Precipitation variance is by far the highest in the north Aegean; in fact, its Middle Holocene precipitation variance is the highest of all the regions considered. In the Upper Euphrates Basin, precipitation variance increases through the Middle Holocene, similar to what is observed in the Amuq Plain. Such increases in variance at different times of the Middle Holocene in different regions may point to the settling of local climate dynamics.

Late Holocene patterns starting around 2600 cal BP show an increase in precipitation variance for the north Aegean. When Middle and Late Holocene precipitation variances are considered together, these oscillations may indicate the onset of Late Holocene aridity. Precipitation variance remains low throughout the periods in north-central Anatolia, mainly because average precipitation values for this region are the lowest of all five regions. The region was on the border of semi-arid conditions when the Holocene began, and the precipitation pattern shows minimal changes over time compared to the other regions.

In terms of precipitation, the Upper Euphrates Basin, Amuq Plain, and north Aegean are the most dynamic regions. Precipitation patterns change through time and, more importantly, significantly large variances in certain periods may indicate phases of instability. Such unstable periods generally indicate transitions from one climate dynamic to another (e.g., the Early-Middle Holocene transition or the Middle-Late Holocene transition).

Long-term precipitation changes have significant consequences for archaeological societies that rely solely on traditional agricultural methods for subsistence. General trends such as drops in precipitation need to be considered; however, short-lived extreme events may be more important than general trends.

Changes in Temperature

Figure 7 shows changes in average temperature per millennium from 12,000-2000 cal BP. The Upper Euphrates Basin shows continuous warming throughout the Holocene with minimal variance. This region shows the highest variance for precipitation but reveals highly stable patterns for temperature. Average temperature values initially remain below the median, and they start to exceed the median only later. The Upper Euphrates Basin warms up throughout the Holocene, while precipitation oscillates in the Early and Middle/Late Holocene. The pattern of temperature change displayed in the Upper Tigris Basin is typical of all the remaining regions. All four regions experience Early Holocene warming around 9000-7000 cal BP, and average temperatures exceed and stay above the median.

Temperature variance is higher in the Early Holocene, especially from 11,000-8000 cal BP, when variance then drops. This drop in variance also correlates with a plateau in average temperature from 7000-4000 cal BP. This plateau suggests the onset of temperatures in these regions that were similar to modern temperatures. There are slight increases in variance from 5000–2000 cal BP; although these changes do not translate into large-scale patterns, we have to consider that such jitters in variance may point to short-lived increases in temperature.

These changes may have had significant consequences for archaeological societies, especially in the Middle Holocene. Middle Holocene (especially, Bronze Age) societies relied heavily on intensive agropastoral subsistence methods. In the Upper Euphrates and Upper Tigris basins and north-central Anatolia, Early to Late Holocene temperatures warm by 3°C. This change is limited to 2°C in the Amuq Plain and north Aegean. A difference of one degree is significant, and such warming also affects long-term evapotranspiration rates. Overall, north-central Anatolia displays the most stable climatic characteristics (i.e., precipitation and temperature) between 12,000 and 2000 cal BP.

The 8.2 ka BP Event

One of the most critical phases of climatic instability in the Holocene, arguably, is the 8.2 ka BP event, identified by many researchers as a significant cooling phase. Like other sudden climate change events, the 8.2 ka BP event is visible in some proxies, as well as GCMs that model paleoclimatic conditions (Flohr et al. 2016; Gauthier 2016). To examine more closely climatic trends before and after this period of most intense climatic instability, 600 years (8500–7900 cal BP) of annual precipitation are averaged to show changes over time (Fig. 8). Average annual precipitation shows variation in each region: north-central Anatolia is the driest, and the north Aegean is the wettest. In most regions, annual precipitation is above the median. Variance in the same annual precipitation data is the highest in the Upper Euphrates Basin, followed by the Amuq Plain and the north Aegean. Again, variance can be used as an essential measure of climatic stability, with higher variance values pointing to climatic instability. The 8.2 ka BP event occurs in the middle of the Early Holocene, so we do not expect to see any significant, long-term climatic change. Climatic instability is usually associated with transitions from one phase to another (e.g., Early to Middle Holocene). Based on these observations, it is possible to argue that the 8.2 ka BP event manifests itself as a climatically less stable phase when significant variance emerges in annual precipitation. Average annual temperature shows that values remain below the median except in the Amuq Plain (Fig. 9). The Amuq Plain and the north Aegean are the warmest of all five regions, while the Upper Euphrates Basin and north-central Anatolia are the coolest. In terms of variance, the calculated values have small ranges, with almost no variance in the Upper Euphrates or north Aegean and low variance in other regions. Low variance means that temperatures did not vary significantly from 8500–7900 cal BP. Although paleoclimate literature proposes a significant cooling event in the ancient Near East associated with the 8.2 ka BP event, no significant cooling in the regions discussed here is identifiable in the MCM results. On the other hand, there is evidence for significant variance in precipitation, which may be a factor of climatic stress.

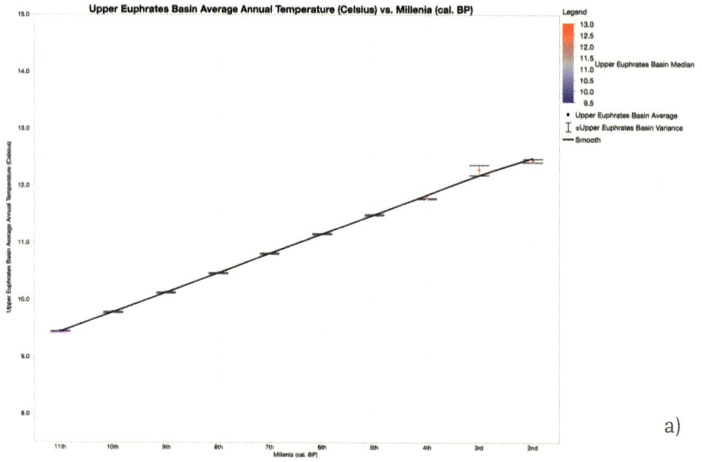

a)

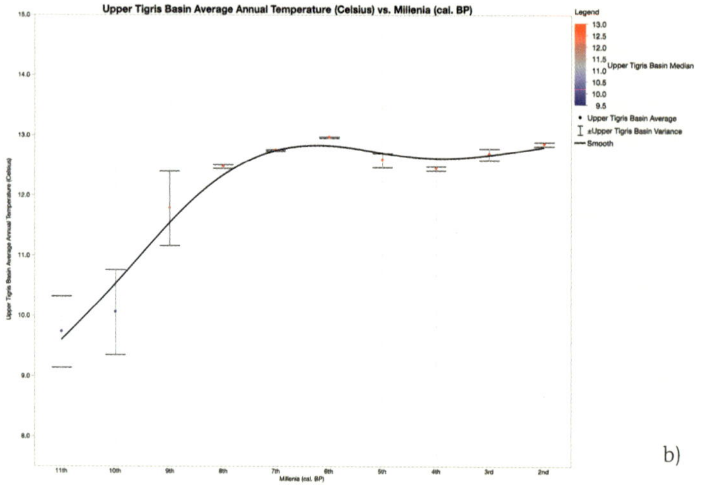

b)

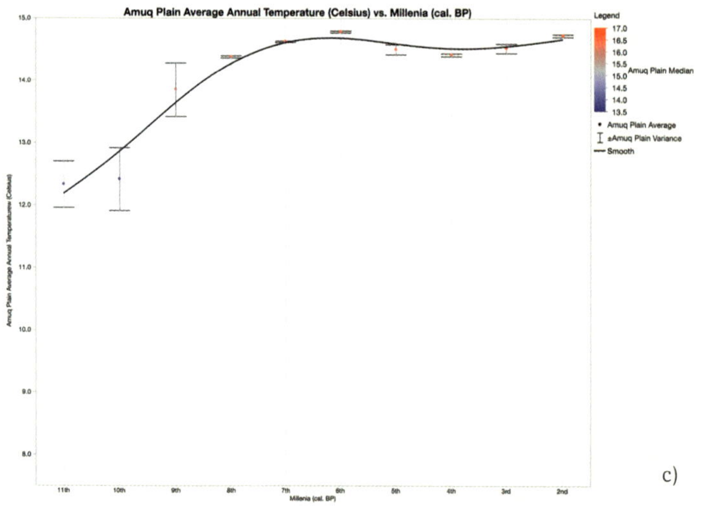

c)

Fig. 7. Average temperature (ºC) per millennium between 12,000–2000 cal BP in a) the Upper Euphrates Basin, b) the Upper Tigris Basin, c) the Amuq Plain, d) north-central Anatolia, and e) the north Aegean. Horizontal lines represent smoothing of data points, vertical lines show variance ranges, and colors of data points show deviations from the median.

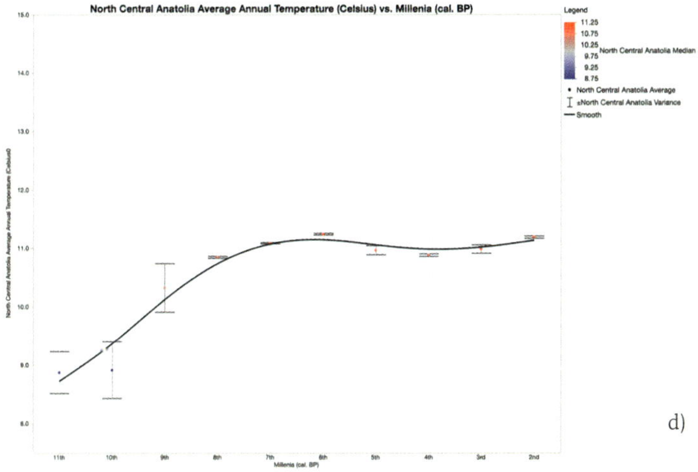

d)

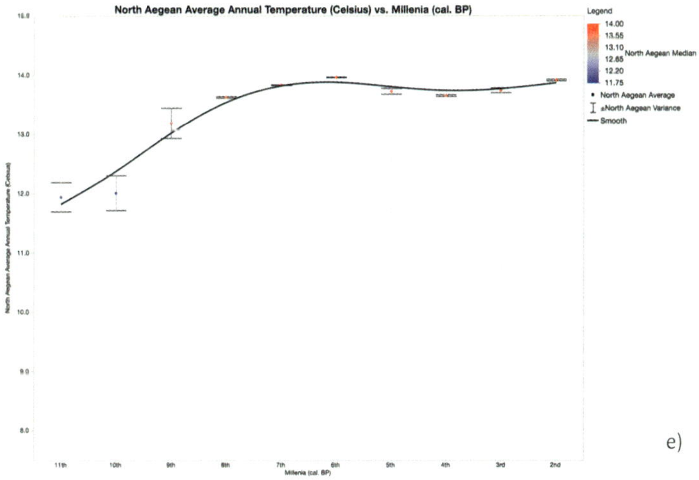

e)

The 6.2 ka BP Event

The 6.2 ka BP event is a less well-known phase of climatic instability, yet it remains important to consider when modeling the paleoclimate of some areas of Anatolia. Average annual precipitation rates from 6500–5900 cal BP (Fig. 8), again calculated by averaging 600 years of annual precipitation rates, show that average precipitation values remained above the median, in general; the north Aegean is the wettest region, whereas north-central Anatolia is the driest. Variance values for the same data are significantly lower than those associated with the 8.2 ka BP event. North-central Anatolia shows almost no variance, the Upper Tigris Basin and north Aegean have low variance, and the Amuq Plain and Upper Euphrates Basin show some variance, but these values do not suggest climatic instability. Average annual temperatures show a pattern similar to that associated with the 8.2 ka BP event (Fig. 9). The Amuq Plain is the warmest region, and it is the only one in which the annual temperature is above the median. Temperature variances are so low that there is almost no change through time. Based on these low variance values for both precipitation and temperature, we can propose that the 6.2 ka BP event remains unidentifiable in MCM results.

The 4.2 ka BP Event

The 4.2 ka BP event is by far the most discussed phase of climatic instability in the ancient Near East. It is suggested that the 4.2 ka BP event is a sudden warming phase occurring when precipitation dropped and temperature increased. Average annual precipitation in the regions of interest show that, except for north-central Anatolia, annual precipitation remains above the median across this period (Fig. 8). Although there is no significant difference among the regions, the north Aegean is the wettest, and north-central Anatolia is the driest. Similar to the variance values associated with the 8.2 ka BP event, high variance values suggest climatic instability resulting from wide-ranging annual precipitation, especially in the Upper Euphrates Basin, Amuq Plain, and north-central Anatolia. The latter is especially important, because north-central Anatolia is also the driest region. Average annual temperatures remain above the median in the Amuq Plain and north Aegean, while they are below the median in other regions (Fig. 9): the Amuq Plain is the warmest and north-central Anatolia the coolest. In terms of variance, the values are low, and no meaningful pattern suggests climatic instability. Based on low temperature variance and high precipitation variance, it is possible to suggest that the 4.2 ka BP event emerges as a phase of climatic instability because of unstable precipitation events. Temperature change does not seem to materialize significantly.

Millennial Patterns and Rapid Climate Change Events

Rapid climate change events such as the 8.2, 6.2, and 4.2 ka BP events are assessed based on 600-year averaged precipitation and temperature trends. Comparing these trends with millennial patterns is essential to understand whether there is any significant deviation between millennial patterns and 600-year trends. For example, to assess the 8.2 ka BP event, the 600-year average is based on data from 8500–7900 cal BP, and the millennial pattern is based on data from 8900–7900 cal BP.

Figure 10 compares millennial precipitation patterns with patterns during phases of rapid climate change for each region. In the 8.2 ka BP event (Fig. 10a), 600-year average and median precipitation show similarities with millennial average and median precipitation. The smoothing lines for average and median values for the 8.2 ka BP event differ from the millennial patterns only in the north Aegean. The patterns for variance in 600-year average and millennial data are the same. This graph clearly shows no meaningful difference in precipitation between millennial (i.e., 8900–7900 cal BP) and 600-year (8500–7900 cal BP) patterns.

Like the 8.2 ka BP event patterns, average and median precipitation associated with the 6.2 ka BP event are also similar, except for the north Aegean, which shows only

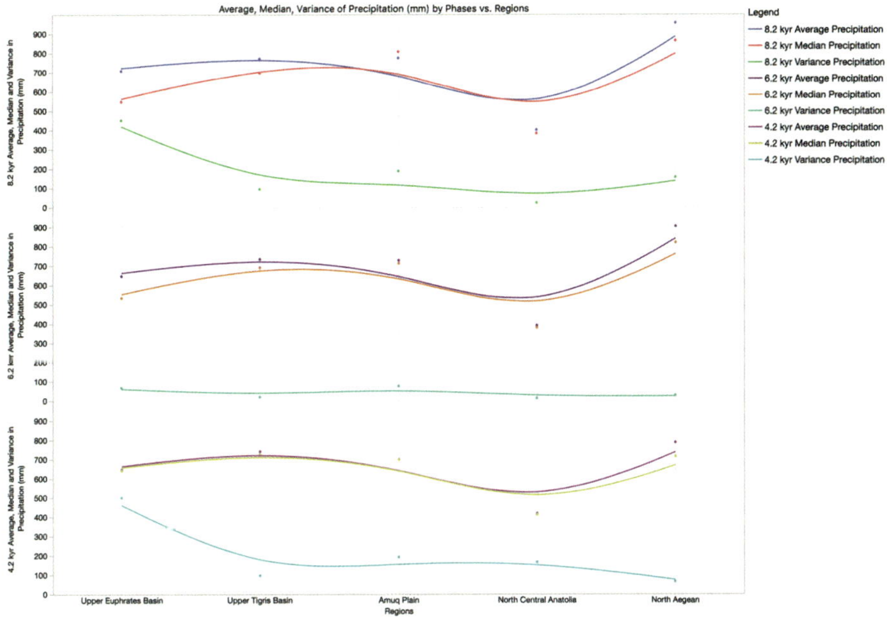

Fig. 8. Average, median, and variance of annual precipitation (mm) around the 8.2, 6.2, and 4.2 ka BP events in all regions of interest. Solid lines represent smoothing of data points for average and median values. Green lines show variance.

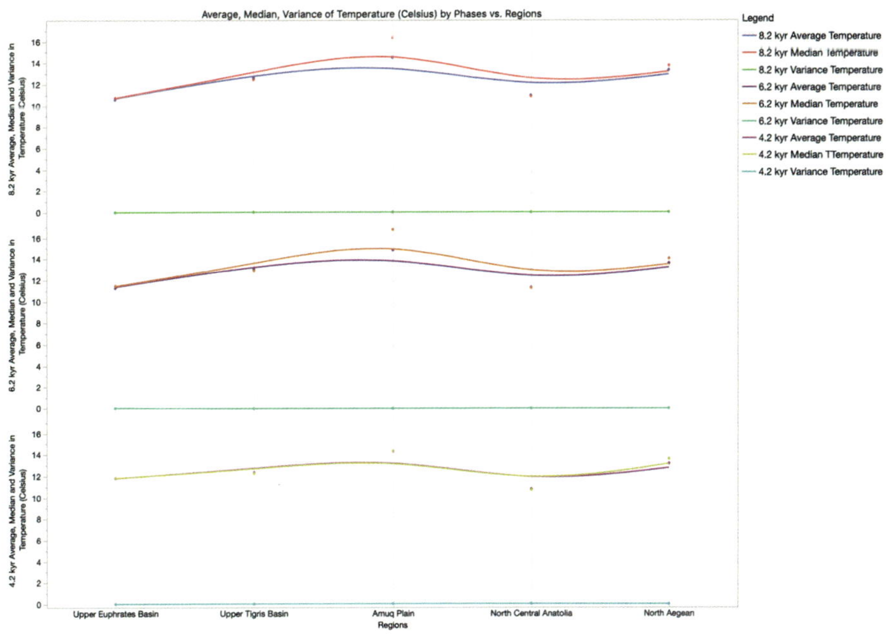

Fig. 9. Average, median, and variance of annual temperature (Celsius) around the 8.2, 6.2, and 4.2 ka BP events in all regions of interest. Solid lines represent smoothing of data points for average and median values. Green lines show variance.

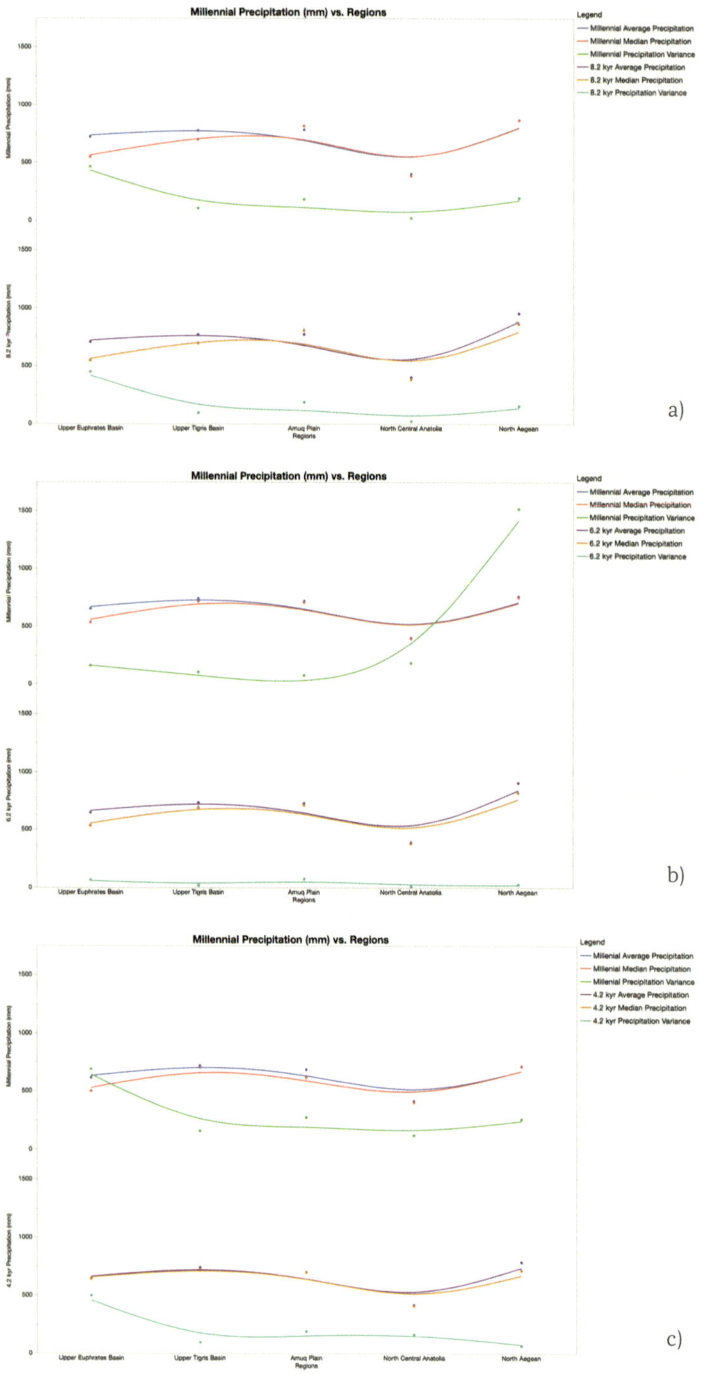

Fig. 10. Comparison of changes in precipitation (mm) per millennium vs. per rapid climate change event for each region: a) comparison with the 8.2 ka BP event, b) comparison with the 6.2 ka BP event, and c) comparison with the 4.2 ka BP event. Graphs show average, median, and variance. Solid lines show smoothing.

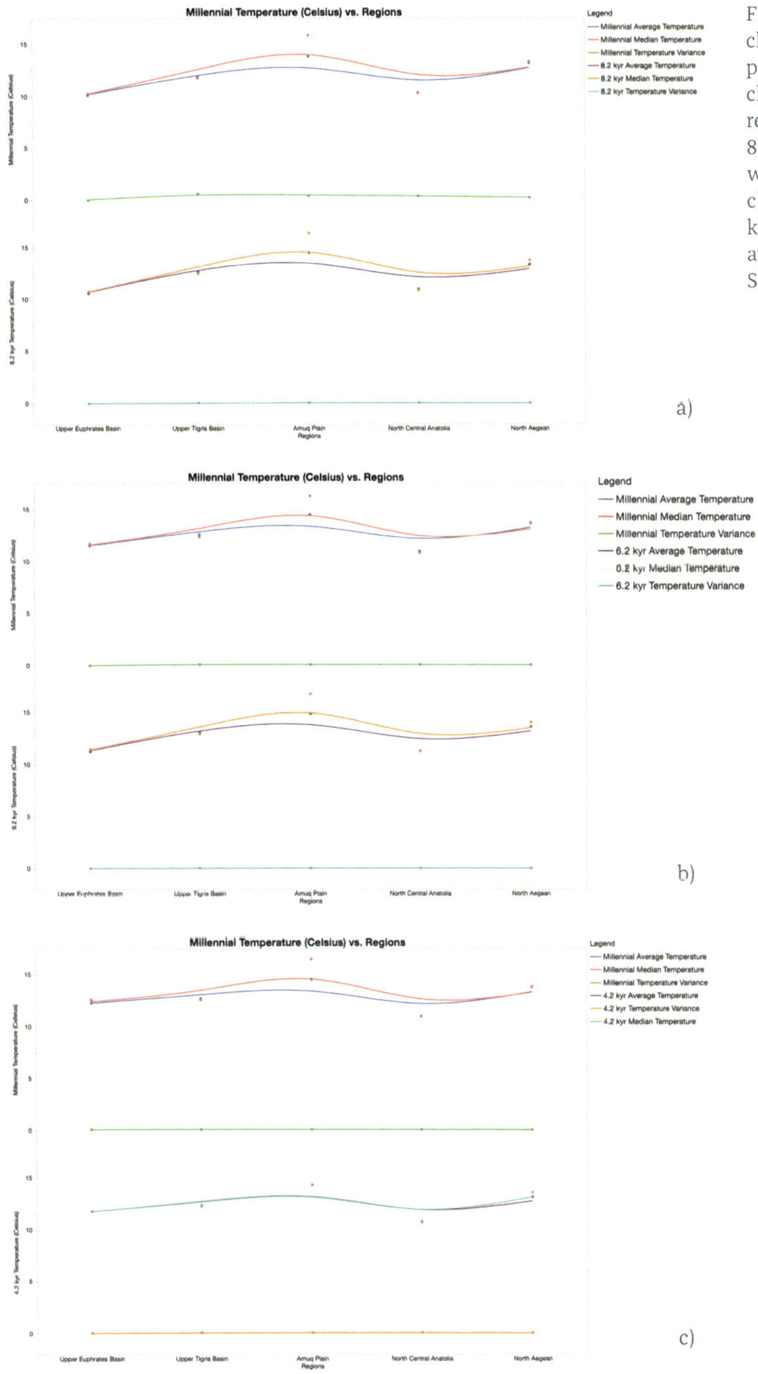

Fig. 11. Comparison of changes in temperature (ºC) per millennium vs. per rapid climate change event for each region: a) comparison with the 8.2 ka BP event, b) comparison with the 6.2 ka BP event, and c) comparison with the 4.2 ka BP event. Graphs show average, median, and variance. Solid lines show smoothing.

a slight deviation (Fig. 10b). The variance seems to remain low (almost zero), while the millennial variance from 6900-5900 cal BP is significantly higher in the north Aegean. This difference between millennial and 600-year average data must result from the higher variance in the 400 years of non-overlap between the datasets.

Contrary to the earlier events, 600-year average and millennial average and median precipitation values associated with the 4.2 ka BP event differ in all regions except for north-central Anatolia, even if such differences are not significant (Fig. 10c). The variance shows the same pattern in all regions, except for a slight dip in the 4.2 ka BP event in the north Aegean. Again, the variance values in the north Aegean during the 400 years of non-overlap between the datasets must be the reason for the slight drop.

Comparing millennial and 600-year patterns for temperature per rapid climate change event in all the regions is also instructive (Fig. 11). Figures 11a and 11b show a comparison of patterns between the 8.2 and 6.2 ka BP events. Neither phase shows significant deviation from millennial patterns, and variance remains low. In both cases, there is a slight difference in average and median temperatures in the Amuq Plain. For the 4.2 ka BP event (Fig. 11c), millennial patterns for average, median, and variance of temperature are similar to those associated with the 8.2 and 6.2 ka BP events. The 4.2 ka BP patterns, however, reveal that in all regions, the difference between average and median disappears. The variance is also low in the 4.2 ka BP data. Low variance suggests that the temperature data does not show any meaningful change during the 4.2 ka BP event.

Conclusion

This survey of MCM paleoclimatic patterns in five separate regions of Anatolia over a 10,000-year period reveals interesting results. First, each region displays unique climatic conditions. Although the Upper Euphrates Basin, Upper Tigris Basin, and Amuq Plain are physically close to one another, these regions' climatic patterns show distinct precipitation and temperature patterns. The same holds for north-central Anatolia and the north Aegean. This pattern suggests that supra-regional characterizations of past climatic patterns need to be re-evaluated. Second, millennial patterns suggest a rapid drop in precipitation and a gradual increase in temperature as soon as the Holocene starts. All regions considered become drier and warmer by 8000 cal BP. Paleoenvironmental reconstructions across the eastern Mediterranean based on multi-proxy evidence corroborate these MCM results. Third, in the grand scheme of Holocene warming and drying, north-central Anatolia seems to be most conservative in preserving the essence of its paleoclimatic characteristics. These MCM results suggest that this region has always been semi-arid, while significant changes occur in other regions from 12,000–2000 cal BP. Fourth, in the five regions covered here, there is some evidence for a 4.2 ka BP rapid warming event; however, MCM results do not support either the 8.2 or the 6.2 ka BP event. When millennial patterns (average, median, and variance) of precipitation and temperature are compared with 600-year-long rapid climate change periods, there is no concrete evidence to pinpoint phases of climatic instability. The high variance in precipitation for only the 4.2 ka BP event may point to changes in precipitation patterns and, hence, climatic instability.

Reconstructing past climatic conditions through modeling (either Global Circulation or Synoptic models) and comparing model results with paleoenvironmental reconstructions based on multi-proxy evidence have become standard research practices in the last two decades. With increases in regional data and advances in computer science (e.g., agent-based simulations of land use), we have a more detailed and high-resolution perspective on paleoclimatic changes, how landscapes changed, and how human-environment interactions diversified through time. It is time to step back from deterministic approaches that stem from discussions of supra-regional paleoclimatic patterns. Instead, focus should be given to human agency in the face of past climatic perturbations. Emphasizing how diverse behaviors emerged in the context of region-specific paleoclimatic changes should define the agenda of environmental archaeology for the decades to come.

Acknowledgments

The paleoclimate of the Upper Euphrates and Tigris basins were modeled during the author's tenure as a post-doctoral researcher at Koç University's Research Center for Anatolian Civilizations (ANAMED) from 2011-2012. Further enhancements to the modeling were made as part of an agent-based modeling project funded by the Scientific and Technological Research Council of Turkey (TÜBİTAK, Grant #113K826). Modeling of the paleoclimate of north-central Anatolia was funded also by TÜBİTAK (Grant #215K130). Modeling of the north Aegean was funded by the Research Fund of Istanbul Technical University (Grant #41991). The author is grateful to all these institutions.

Works Cited

Alder, Jay R., and S. W. Hostetler. 2015. "Global Climate Simulations at 3000-year Intervals for the Last 21,000 years with the GENOM Coupled Atmosphere-Ocean Model." *Climate of the Past* 11: 449–71.

Arıkan, Bülent. 2010. "Reorganization and Risk: Environmental Change and Tribal Land Use in Marginal Landscapes of Southern Jordan." PhD diss., Arizona State University.

———. 2013. "Macrophysical Climate Modeling, Economy, and Social Organization in Early Bronze Age Anatolia." *Journal of Archaeological Science* 43 (C): 38–54.

———. 2015. "Modeling the Paleoclimate (ca. 6,000–3,200 cal. BP) in Eastern Anatolia: The Method of Macrophysical Climate Model and Comparisons with Proxy Data." *Journal of Archaeological Science* 57: 158–67.

Arıkan, Bülent, Franziska Mohr, and Matthias Bürgi. 2020. "Exploring the Common Ground of Landscape Ecology and Landscape Archaeology Through A Case Study from Eastern Anatolia, Turkey." *Landscape Ecology* 136: DOI: 10.1007/s10980-020-01128-z.

Arıkan, Bülent, and Tayfun Yıldırım. 2018. "Paleoclimate, Geology, Geomorphology, and Middle Holocene Settlement Systems in the Delice Valley of North-Central Anatolia." *Journal of Field Archaeology* 43 (8): 570–90.

Barton, C. Michael, J. E. A. Tortosa, O. Garcia-Puchol, J. Riel-Salvatore, N. Gauthier, M. V. Conesa, and G. P. Bouchard. 2018. "Risk and Resilience in Late Glacial: A Case Study from the Western Mediterranean." *Quaternary Science Reviews* 184: 68–84.

Batiuk, Steve D. 2013. "The Fruits of Migration: Understanding the 'Longue Dureé' and the Socio-Economic Relations of the Early Transcaucasian Culture." *Journal of Anthropological Archaeology* 32: 449–77.

Bryson, Reid A., 2007a. "Calculating Potential Evapo-transpiration (PET)." In *A Paleoclimatology Workbook: High Resolution, Site-specific Macrophysical Climate Modeling,* edited by R. A. Bryson and Katherine M. DeWall, 53–56. Hot Springs: Mammoth.

———. 2007b. "Estimation of the Past Latitudes of the Subtropical Anticyclones and the Jet Stream." In *A Paleoclimatology Workbook: High Resolution, Site-specific Macrophysical Climate Modeling,* edited by R. A. Bryson and Katherine M. DeWall, 37–44. Hot Springs: Mammoth.

———. 2007c. "Overview of the Excel Model: Applying the model to a new location." In *A Paleoclimatology Workbook: High Resolution, Site-specific Macrophysical Climate Modeling,* edited by R. A. Bryson and Katherine M. DeWall, 59–68. Hot Springs: Mammoth.

———. 2007d. "The Paradigm of Climatology." In *A Paleoclimatology Workbook: High Resolution, Site-specific Macrophysical Climate Modeling,* edited by R. A. Bryson and Katherine M. DeWall, 19–24. Hot Springs: Mammoth.

———. A. 2007e. "The Present and Past Position of the Intertropical Convergence." In *A Paleoclimatology Workbook: High Resolution, Site-specific Macrophysical Climate Modeling,* edited by R. A. Bryson and Katherine M. DeWall, 45–52. Hot Springs: Mammoth.

Bryson Reid A., and K. M. DeWall. 2007. "An Introduction to Archaeoclimatology. Macrophysical Climate Model." In *A Paleoclimatology Workbook: High Resolution, Site-specific Macrophysical Climate Modeling,* edited by R. A. Bryson and Katherine M. DeWall, 3–11. Hot Springs: Mammoth.

Cleator, Sean F., S. P. Harrison, N. K. Nichols, I. C. Prentice, and I. Roulstone. 2020. "A Method for Generating Coherent Spatially Explicit Maps of Seasonal Paleoclimates from Site-Based Reconstructions." *Journal of Advances in Modeling Earth Systems* 12 (1). DOI: 10.1029/2019MS001630.

COHMAP Project Members. 1988. "Climatic Changes of the Last 18,000 years: Observations and Model Simulations." *Science* 24: 1043–52.

Cullen, Heidi M., P. B. de Menocal, S. Hemming, G. Hemming, F. H. Brown, T. Guilderson, and F. Sirocko. 2000. "Climate Change and the Collapse of the Akkadian Empire: Evidence from the Deep Sea." *Geology* 28: 379–82.

Flohr, Pascal, D. Fleitmann, R. Matthews, W. Matthews, and S. Black. 2016. "Evidence of Resilience to Past Climate Change in Southwest Asia: Early Farming Communities and the 9.2 and 8.2 ka Events." *Quaternary Science Reviews* 136: 23–39.

Gauthier, Nicholas. 2016. "The Spatial Pattern of Climate Change During the Spread of Farming into the Aegean." *Journal of Archaeological Science* 75: 1–9.

Katrantsiotis, Christos, E. Nordström, R. Smittenberg, M. Finne, E. Weiberg, M. Hattestrand, P. Avramidis, and S. Wastegard. 2019. "Climate Changes in the Eastern Mediterranean over the Last 5000 Years and Their Links to the High-Latitude Atmospheric Patterns and Asian Monsoons." *Global and Planetary Change* 175: 36–51.

Martinez-Ruiz, Francisca, M. Kastner, D. Gallego-Torres, M. Rodrigo-Gamiz, V. Nieto-Moreno, and M. Ortega-Huertas. 2015. "Paleoclimate and Paleoceanography Over the Past 20,000 yr in the Mediterranean Sea Basins as Indicated by Sediment Elemental Proxies." *Quaternary Science Reviews* 107: 25–46.

Morin, Efrat, T. Ryb, I. Gavrieli, and Y. Enzel. 2018. "Mean, Variance, and Trends of Levant Precipitation Over the Past 4500 Years from Reconstructed Dead Sea Levels and Stochastic Modeling." *Quaternary Research* 91 (2): 751–67. DOI: 10.1017/qua.2018.98.

Staubwasser, Michael, and Harvey Weiss. 2006. "Holocene Climate and Cultural Evolution in Late Prehistoric-Early Historic West Asia." *Quaternary Research* 66 (3): 372–78.

Weiss, Harvey. 2017. "4.2 ka BP Megadrought and the Akkadian Collapse." In *Megadrought and Collapse: From Early Agriculture to Angkor,* edited by Harvey Weiss, 93-159. Oxford: Oxford University Press.

Weiss, Harvey, and Raymond S. Bradley. 2001. "Archaeology: What Drives Societal Collapse?" *Science* 291, no. 5504: 609–10.

Weiss, Harvey, M. A. Courty, W. Wetterstrom, F. Guichard, L. Senior, R. Meadow, and A. Curnow. 1993. "The Genesis and Collapse of Third Millennium North Mesopotamian Civilization." *Science* 262, no. 5124: 995–1004.

Early to Mid-Holocene Vegetation History and Human Settlement in Anatolia

Ceren Kabukcu
Eleni Asouti

Introduction

In this chapter we discuss the vegetation history and paleoecology of Anatolian woodlands, focusing on insights gained through the analysis of anthracological remains from prehistoric habitation sites. In the semi-arid regions of Anatolia, anthracology provides direct evidence of the presence and spread of pioneer woodland species, which often go undetected by pollen analysis, including the entomophilous Rosaceae (e.g., *Amygdalus*) and poor and/or irregular pollen dispersers (*Juniperus*, *Pistacia*, and riparian taxa such as Salicaceae and *Fraxinus*). We also review anthracological data from Anatolia indicating the development of woodland management practices and anthropogenic woodland habitats by Neolithic agropastoral communities during the early to mid-Holocene period.

Anthracology entails the study of wood charcoals originating from fuelwood and burnt timber found in archaeological sites (Kabukcu and Chabal 2021). In southwest Asia, the systematic study of charred plant remains from archaeological sites began in the 1960s and 1970s, coinciding with an increasing interest in the origins of agriculture and paleoenvironments (Helbaek 1969; Hillman 1972, 1978; Willcox 1974). From its first applications in southwest Asia anthracology has focused on reconstructing past vegetation and climate change (e.g., Kayacık and Aytuğ 1968; Willcox 1974; Western 1969, 1971). In Anatolia, the identification by Kayacık and Aytuğ (1968) of charred juniper and oak structural timber at Gordion prompted them to question then prevailing perceptions of the vegetation of central Anatolia as treeless steppe. Further studies by Aytuğ (1970), Willcox (1974), and Şanlı (1989) increasingly considered archaeological wood charcoals as proxies for prehistoric woodland vegetation, particularly for probing Holocene human impacts on the landscape.

Charcoals originating from domestic fuelwood use yield appropriate sources of data for reconstructing past woodland composition and its changes through time when sampled from archaeological layers containing the discarded residues of routine fuel consumption in domestic fireplaces, accumulated over prolonged periods of time (e.g., waste discard areas, middens, etc.) (Kabukcu 2018a; Kabukcu and Chabal 2021). Such deposits are more likely to represent the proportions of the wood species commonly used as fuel during the lifetime of a settlement. By contrast, charcoals deposited *in situ* within primary contexts (e.g., hearths) represent single, short-lived events (the last burning episode), while charred timber remains may include species selected for their durability and/or structural properties (Kabukcu and Chabal 2021). The study of domestic fuelwood waste from stratified and securely dated archaeological layers holds distinctive potential for reconstructing the vegetation catchments exploited by prehistoric societies for fuelwood. In addition, as wood anatomical features often remain well-preserved after carbonization, the eco-anatomical examination of archaeological charcoals may also provide useful information about climate conditions and human activities affecting tree growth (e.g., drought or frost-damage, herbivore browsing, wood cutting, etc.) (Kabukcu 2018b).

Late Pleistocene and Early Holocene Woodland Vegetation in Anatolia

Climate and associated vegetation changes during the Pleistocene-Holocene transition frame key shifts in human socioeconomics in southwest Asia, manifested in the emergence of regionally diverse plant and animal resource management strategies which culminated in the appearance of the first agropastoral economies during the late ninth and eighth millennia cal BCE (Asouti 2017; Asouti and Fuller 2012). Paleoclimatic records in southwest Asia come from marine cores in the Mediterranean and the Red Sea, in addition to cores from lakes, fluvial sediments, speleothems, pollen records, molluscs, and calcretes (Roberts et al. 2018). Oxygen stable isotope sequences indicate that there was a degree of regional correlation with global trends in temperatures and precipitation (Kallel et al. 1997; Ariztegui et al. 2000; Rohling et al. 2002; Schilman et al. 2002). As with the rest of the northern hemisphere, climatic conditions shifted from cold and arid to warm and wet at the onset of the Lateglacial (GI-1, ca. 14.7–12.65 ka cal BP). This warming and wetting trend was interrupted by the abrupt cooling and arid oscillation of the Younger Dryas (GS-1, ca. 12.65–11.5 ka cal BP), reflected in various terrestrial and marine records (Bottema 1991; Rossignol-Strick 1995; Bar-Matthews et al. 2003; Robinson et al. 2006; Orland et al. 2012). The start of the Holocene (ca. 11.5 ka cal BP) marks a rapid increase in precipitation and temperature across southwest Asia until around 8 ka cal BP (Climatic Optimum) (Bar-Matthews et al. 1999; Roberts et al. 2018; Wick, Lemcke, and Strum 2003; Sbaffi et al. 2004; Ziegler, Tuenter, and Lourens 2010; Stevens et al. 2012; Göktürk et al. 2011). Two short and abrupt cold/dry episodes (at 8.2 ka and 4.2 ka cal BP) are also registered in ice cores (Alley et al. 1997; Johnsen et al. 2001). However, their regional magnitude and duration are debated, especially with regard to their presumed impacts on water resources (cf. Wiersma and Renssen 2006; Asouti 2009; Roffet-Salque et al. 2018; Wainwright and Ayala 2019).

In central and eastern Anatolia Late Pleistocene tree pollen sums are very low compared to herbaceous pollen (Roberts et al. 1999; Pickarski et al. 2015; Litt et al. 2009). The palynological records from Eski Acıgöl, Nar Lake, and Lake Van indicate high values of *Artemisia* and Chenopodiaceae during the Younger Dryas (Roberts et al. 2001, 2016, 2018; Wick, Lemcke, and Strum 2003; Litt et al. 2009). *Artemisia*-dominated steppe was rapidly replaced by grassland vegetation at the start of the Holocene. In the following two millennia the pollen records indicate low levels of tree pollen, comprising mainly deciduous *Quercus* alongside *Pistacia* and *Juniperus*. The first arboreal pollen maximum is not observed until around 8 ka cal BP at Eski Acıgöl (Roberts et al. 2001; Woldring and Bottema 2001/2). At Nar Lake (Roberts et al. 2016), the earliest part of the Holocene registers a more pronounced rate of *Pistacia* spread (phase 3), with a subsequent gradual increase in deciduous oak pollen (phase 4). Deciduous oak continued its expansion in central Anatolia until around 6.5–6 ka cal BP. Significant human impacts on woodland cover can be detected for the first time between 4.5–4 ka cal BP (Roberts et al. 2001; Woldring and Bottema 2001/2). At Lake Van in eastern Anatolia, the onset of large-scale anthropogenic deforestation dates even later, to around 2.5 ka cal BP (Wick, Lemcke, and Strum 2003; Litt et al. 2009). On the other hand, palynological evidence from varved sequences in littoral Anatolia, from Lake Iznik and the Marmara Sea (northwest Anatolia) and the Black Sea (northern Anatolia) indicate a more rapid establishment of woodland vegetation during the early Holocene (Miebach et al. 2016; Valsecchi, Sánchez-Goni, and Londeix 2012; Kwiecien et al. 2009).

The combined pollen and anthracological data suggest that woodland cover was probably sparse on the Anatolian plateau during the Lateglacial and the first two millennia of the Holocene. Herbaceous steppe vegetation prevailed, associated with a low-density tree cover dominated by Rosaceae (mainly wild almond/*Amygdalus*, syn. *Prunus* subgenus *Amygdalus*), alongside pioneer taxa such as *Juniperus* and *Betula* and a very low presence of deciduous oak (*Quercus*) (Roberts et al. 2001; Woldring and Bottema 2001/2002; Litt et al. 2009; Asouti and Kabukcu 2014; Kabukcu 2017). Anthracological data from Epipaleolithic Pınarbaşı (central Anatolia) and the Karain B and Öküzini caves (southwest Anatolia) point to the importance of entomophilous Rosaceae (*Amygdalus* and Maloideae) and *Juniperus* as the pioneer woodland vegetation taxa (Kabukcu 2017; Asouti and Kabukcu 2014; Martinoli 2004). The importance of *Juniperus*, *Pistacia*, and Rosaceae as pioneer woodland taxa is reflected in anthracological assemblages from several sites in southwest Asia and the eastern Mediterranean dated to the Late Pleistocene (Baruch and Goring-Morris 1997; Neef 2004; Aura et al. 2005; Austin 2007; Asouti, Ntinou, and Kabukcu 2018; Asouti et al. 2020; Deckers et al. 2009; Ntinou and Kyparissi-Apostolika 2016; Ntinou 2021). At the start of the Holocene, oaks might have been more abundant in inner Anatolia at sites located at higher altitudes (e.g., Çayönü) (see also Table 1). Charcoal assemblages from early Holocene sites dated to the tenth and ninth millennia cal BCE in east and southeast Anatolia (Çayönü,

	Çayönü*[1]	Cafer Höyük*[2]	Göbekli Tepe[3]	Körtik Tepe[4]	Hallan Çemi[5]	Gusir Höyük[6]	Pınarbaşı Epipaleo-lithic[7]	Pınarbaşı Aceramic Neolithic[7]	Boncuklu[7]	Aşıklı Höyük[8]	Can Hasan III[7]	Çatalhöyük East***[9]
Quercus dec. tp.	++++	++	1	18.39	X	2.57		0.36	4.76	X	2.59	39.47
Betula						9.61						
Juniperus							20.96	0.27	1.08		3.20	12.48
Pistacia	+++	+	104	0.18	X	29.73	3.83	0.82	3.89	X	16.64	2.34
*Amygdalus/Prunus***	+-+		59	5.31	X	39.24	51.91	95.36	9.84	X	28.55	2.24
Maloideae						0.19	11.48		0.12	X	1.07	1.59
Acer				3.40		1.68						0.14
Rhamnus				1.20		0.39						0.06
Celtis		+		0.56		0.19	1.09			X	0.30	5.23
Ulmus											1.83	4.36
Ulmaceae							0.55		0.65		34.66	14.42
Alnus						0.09						0.03
Salicaceae		+++		43.97	X		0.55		65.84	X	5.04	12.37
Fraxinus	+++	+		11.57	X	16.15	1.64		0.97			2.38
Tamarix	+++			14.75					4.00			0.20
Chenopodiaceae				0.21		1		0.27	0.43	X	0.30	0.58
Total ID charcoal	**(N=60)**	**(N=84)**	**164**	**2827**	**125**	**1009**	**183**	**366**	**925**	**1854**	**655**	**10651**

Table 1. Key woodland taxa represented in published anthracological assemblages from Late Pleistocene and Early Holocene sites in inland Anatolia. Taxon frequencies are expressed as % fragment counts (decimal numbers), or raw counts (integer numbers). X denotes presence in charcoal assemblages with uncertainty regarding the precise quantitative values as currently reported. Total ID charcoal values represent identified charcoal fragment counts for the taxa listed (integer numbers) or the number of samples studied (N = number of samples) when taxon counts are not reported. Dominant taxa are highlighted in grey. Data sources: [1]van Zeist and de Roller 1991/2; [2]Willcox 1992 [3]Neef 2003; [4]Rössner et al. 2018; [5]Rosenberg et al. 1998; [6]Kabukcu et al. 2021; [7]Kabukcu 2017; [8]Bourguet and Tengberg 2017; [9]Kabukcu, Asouti, and Losh 2021.

(*relative taxon scoring scale for Çayönü and Cafer Höyük: + = presence in 1–25% of samples; ++ = 25–50%; +++ = 50–75%, ++++ = 75–100%; **Amygdalus/Prunus is reported in some assemblages as an amalgamated count. For consistency, this row reflects sums of Amygdalus and Prunus spp. if originally reported separately; ***Sums for Çatalhöyük East include sampled phases from the South and North areas of excavation but exclude the TP area samples).

Hallan Çemi, Körtik Tepe, Göbekli Tepe, Cafer Höyük) include *Quercus*, *Pistacia*, Salicaceae, *Amygdalus*, Maloideae, and *Celtis* (van Zeist and de Roller 1991/2; Willcox 1992, 1999, 2002; Rosenberg et al. 1998; Neef 2003; Riehl et al. 2012; Rössner et al. 2018). Further east along the Tigris Basin, the anthracological data from Gusir Höyük indicate the presence of *Amygdalus*, *Quercus*, *Pistacia*, and *Betula* found alongside riparian taxa such as *Fraxinus* and *Alnus* (Kabukcu et al. 2021). In the Konya plain of central Anatolia, the earliest Holocene sites from which published anthracological data are available include Pınarbaşı, Boncuklu and Can Hasan III, dated to the late tenth and eighth millennia cal BCE, including *Amygdalus*, *Pistacia*, *Juniperus*, Salicaceae, *Ulmus*, *Celtis*, and a small component of deciduous *Quercus* (Kabukcu 2017). Charcoal finds from Aşıklı Höyük in Cappadocia reported in summary form (Bourguet and Tengberg 2017) indicate the presence of a similar range of taxa, with *Pistacia* as the dominant fuelwood taxon followed by *Amygdalus* and, during later phases of the settlement, deciduous *Quercus*.

Çatalhöyük East provides the longest and best-documented anthracological sequence for the timespan corresponding to the seventh millennium cal BCE in central Anatolia (Fig. 1). The earliest phases of occupation reflect a stronger focus on riparian woodland taxa for fuelwood, similar to the prominent use of Salicaceae at neighboring Boncuklu (Kabukcu 2017). In step with the onset of the increasing deciduous oak pollen, dated to around 9000 cal BP in the Eski Acıgöl core, the upper layers of the South G and H phases reflect a sharp increase in the use of deciduous oak wood (Fig. 1). Charcoal remains from the mid-late Neolithic phases (South Q-T; Fig. 1) indicate a shift in the dominant fuel species, with juniper becoming dominant. The Eski Acıgöl pollen data suggest that oak continued to increase during this period until around 6000 cal BP, that is, well after the end of the late Neolithic occupation of Çatalhöyük East. It thus appears likely that the observed switch from oak to juniper as the dominant source of fuelwood may reflect a shift in local preferences (e.g., a greater reliance on juniper used as timber and fuel) rather than the reduced availability of oak in the regional vegetation.

The main Neolithic sequence at Çatalhöyük East demonstrates that a wide range of trees and shrubs were exploited as fuel throughout the lifetime of the site (Fig. 1). Recent vegetation surveys and species distribution modeling (SDM) for key indicator woodland species in the Konya plain indicate significant variation in the distribution of woodland habitats in the plain and its environs during the early Holocene (Asouti and Kabukcu 2014; Collins et al. 2018). While riparian and wetland taxa were available in the immediate vicinity of sites such as Çatalhöyük and Boncuklu, SDM suggests that the Neolithic population regularly exploited more distant dryland ecotones, too, in the wider landscape, especially for the collection of oak and juniper wood (compare Figs. 1 and 2). Riparian and wetland ecotones are represented in the anthracological remains mainly by Salicaceae, Ulmaceae, and *Fraxinus*, while dryland open woodlands are represented mainly by *Amygdalus*, *Pistacia*, and Maloideae (the latter possibly including wild pear) and, further afield, on the north-facing foothills of the Taurus mountains, by *Quercus* and *Juniperus*.

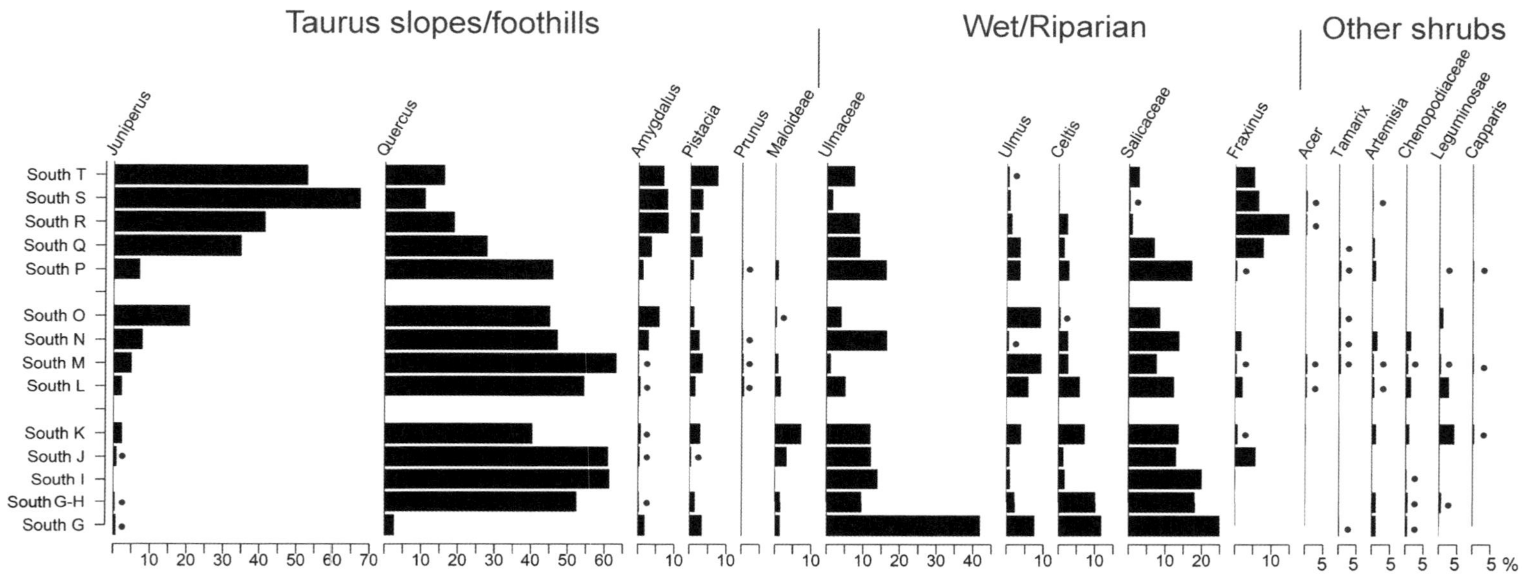

Fig. 1. Anthracological diagram, representing percentage of fragment counts, from all sampled South Area phases at Çatalhöyük East, redrawn after Kabukcu, Asouti, and Losh (2021, figure 4.1, table 4.1).

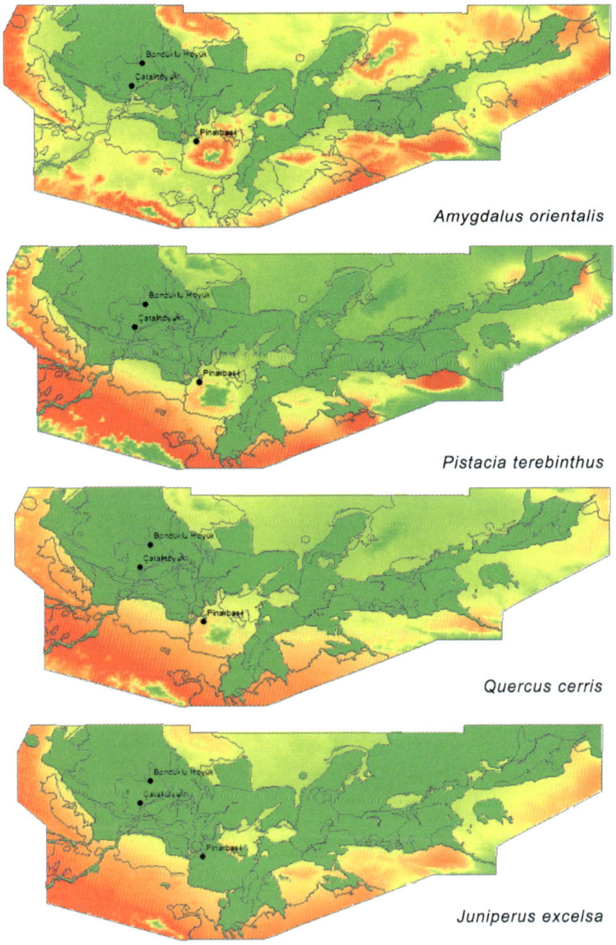

Fig. 2. Species Distribution Modeling (SDM) for selected key indicator tree species in the Konya Plain of south-central Anatolia during the early Holocene (modified after Collins et al. 2018).

Anthracological Evidence for Woodland Management Practices in Neolithic Anatolia

Woodland management practices resulting in the formation of anthropogenic habitats range from the protection of tree stands from browsing, the clearance of herbaceous ground flora via burning or grazing and the selective removal of shrubs, to silvicultural practices such as pruning, coppicing, pollarding, and shredding (Asouti and Kabukcu 2014; for present-day examples of these practices in central Anatolia, see Figs. 3-5). The viability of silvicultural practices relies on the capacity of arboreal species for vegetative propagation from dormant buds and/or root suckers and usually involves cutting cycles of variable duration (from one to several years). As individual trees are already established, the productivity and resilience of managed stands are often higher compared to spontaneous sprouting from seed, particularly in semi-arid environments where temperature tolerances at seedling stage, grazing pressures, and the availability of ground moisture and/or nutrients can all be limiting factors for seed propagation.

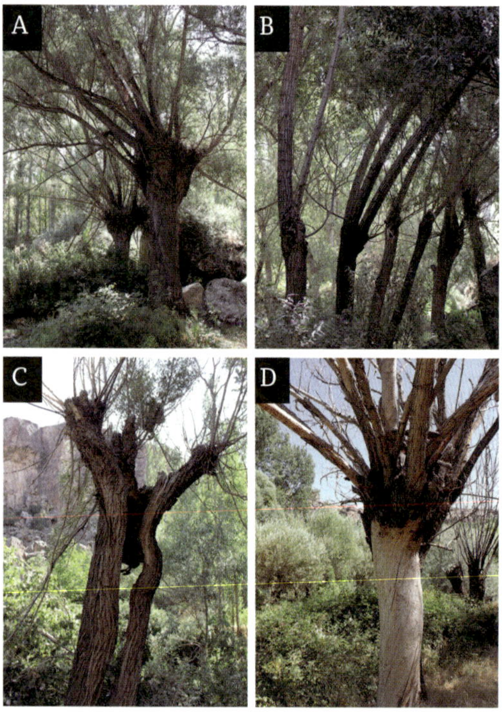

Fig. 3. A–B: Willow pollards of various ages; C–D: Standing deadwood willow pollards (Location: Ihlara Vadisi, Cappadocia. Photos by Eleni Asouti).

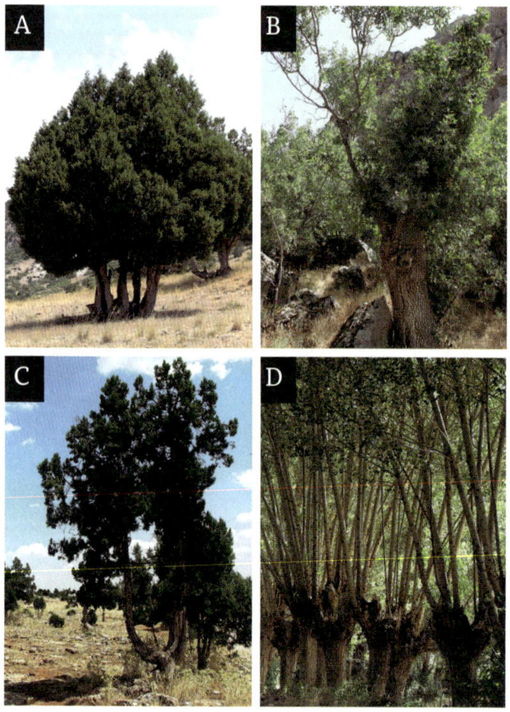

Fig. 4. A: Old juniper coppices in woodland pasture (Cappadocia, near Göllü Dağ); B: *Pistacia* pollard (Ihlara Vadisi); C: Juniper pollard surrounded by coppices (Taurus north-facing slopes); D: Old-growth poplar pollards (Lystra, Taurus) (Photos by Eleni Asouti).

The identification and analysis of woodland management practices in archaeological wood charcoals rests on reconstructing two attributes of prehistoric fuelwood: log size (diameter estimation) and growth rate (annual ring width). For example, the predominant presence of small log sizes may indicate fuelwood collection mainly from coppice stands, while periods of reduced growth (signified by sequences of successive narrow annual growth rings) followed by periods of growth release (wide rings, indicating more rigorous growth) may indicate cycles of cutting/thinning and canopy regrowth associated with coppicing or pollarding (for a detailed description of woodland management indicators in archaeological wood charcoals, see Kabukcu 2018a, 2018b).

The large and well-preserved anthracological assemblage from Çatalhöyük East provides the earliest available empirical evidence for woodland management practices in southwest Asia (Kabukcu 2018b; Kabukcu et al. 2021). This evidence includes quantitative estimations of minimum log diameter and sequential measurements of annual growth ring width for individual wood charcoal specimens. The analysis of average growth rates (Fig. 6A) demonstrates a range of growth forms for both dryland

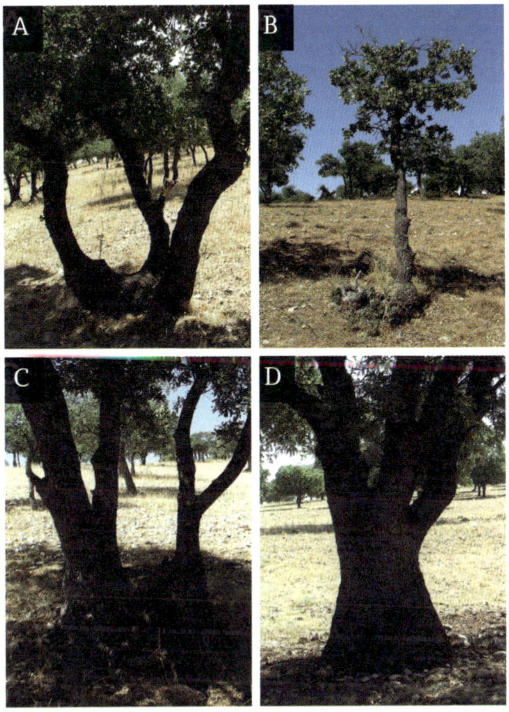

Fig. 5. Managed oaks in Apa Barajı (Taurus): A: old coppice; B: Old coppice stool recently shredded; C: old coppice stump with stools; D: Old pollard (Photos by Eleni Asouti).

and riparian/wetland taxa, particularly in specimens indicating small log sizes (diameter < 10 cm). While some specimens indicate very slow growth rates (< 0.5 mm per annum), others indicate rates > 2 mm per annum. Small-diameter, slow-growing specimens also indicate minor shifts in annual growth rates (Fig. 6B), which may represent saplings or shoots growing under conditions of stress (e.g., closed canopy, defoliation caused by herbivore browsing, etc.) Faster growing, small-diameter specimens, particularly those displaying highly variable annual growth rates, likely represent the development of new shoots following the thinning and/or complete removal of the canopy. The Çatalhöyük East anthracological assemblage contains both suppressed (slow growing) saplings/shoots and fast-growing shoots associated with coppicing cycles. This interpretation is further supported by the fact that the observed variability in growth rates appears to be independent of differences in habitat micro-ecologies: a similar range of growth forms were observed in taxa from both riparian (*Fraxinus*, *Ulmus*) and dryland habitats (*Quercus*, *Celtis*) (Kabukcu 2018b).

Minimum log diameter estimations also indicate the overarching presence of small to medium-sized logs (min. diameter < 15 cm) in the charred fuelwood remains at Çatalhöyük East. This evidence lends additional weight to the possible prevalence of cutting practices involving the harvesting of small diameter wood (coppicing or pollarding). While the fuelwood economy appears to have been consistently focused on small caliber wood (Kabukcu 2018b; Kabukcu et al. 2021), structural timber use involved larger diameter trunks (from ca. 40-50 cm to 80cm; Asouti 2013). Overall, the

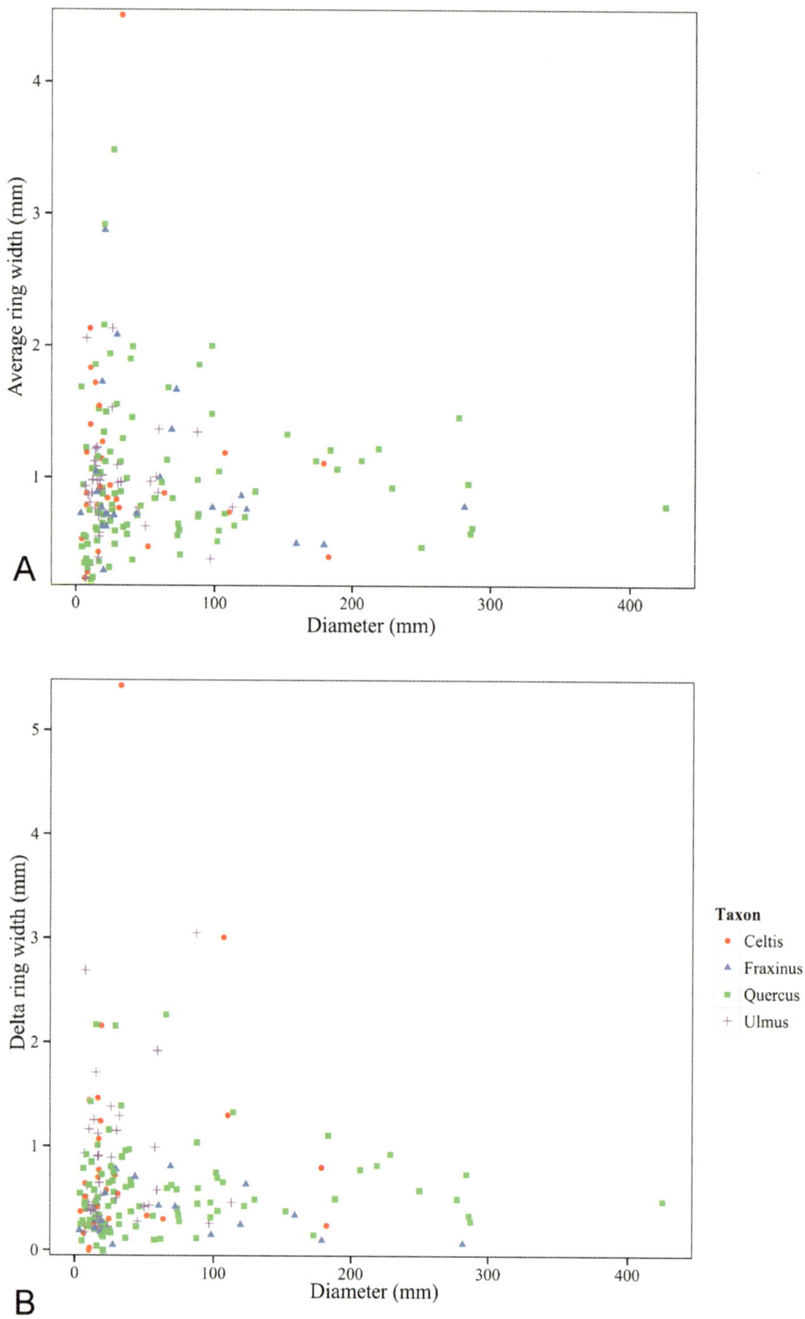

Fig. 6. A: Scatter plot of average ring width and diameter measurements for each specimen from Çatalhöyük; B: scatter plot of delta ring width (difference between the maximum-minimum ring width measurements in each specimen) against diameter calculations from Çatalhöyük.

available evidence suggests the development of enduring and sustainable woodland management practices that routinely encompassed both proximate (riparian/wetland) and more distant habitats on the Taurus foothills some 10–15 km from Çatalhöyük (Fig. 2; Asouti and Kabukcu 2014; Collins et al. 2018).

Conclusions

The combined anthracological and pollen data indicate that human settlement expanded in inner Anatolia during periods characterized by sparse tree cover at the end of the Pleistocene and through the Early Holocene. Early Neolithic communities were adept at managing diverse woodland and grassland ecotones, including both riparian and dryland habitats, that provided critical resources for human sustenance such as arable fields, pastures, and sources of fuelwood and timber. The emergence of large and long-lived agropastoral settlements, such as Aşıklı and Çatalhöyük, during the eighth and seventh millennia cal BCE, following their ninth millennium antecedents (Baird et al. 2018), was not accompanied by vegetation degradation due to deforestation and overgrazing. Human impacts on vegetation from early farming, which remained intensive and localized during this period (Bogaard et al. 2021), clearance for cultivation, and fuelwood collection were probably mitigated by emergent silvicultural practices aimed at enhancing woodland productivity and regeneration, as well as securing the sustainability of plant habitats critical for meeting key subsistence goals, such as fuelwood, fodder, and timber provisioning. Anthracological data and ecological modeling (SDM) also indicate that woodland exploitation encompassed diverse habitats spread across different landscape units. All this likely contributed to the resilience and longevity of settlements such as Aşıklı and Çatalhöyük (Willett et al. 2016). The causes of their transformations at different points in time and their eventual end should be sought in socioeconomic factors, rather than climate change and/or landscape degradation.

The investigation of woodland management practices in prehistoric Anatolia (and across southwest Asia) is still in its infancy. In part, this is the result of limitations in charcoal preservation, especially vis-à-vis the lack of sufficiently large fragments to enable the secure identification and quantification of relevant wood anatomical features (Kabukcu 2018b). However, preservation issues do not diminish the unique potential provided by anthracological assemblages that represent the material residues of prehistoric woodland exploitation. Identifying which species were collected and used as fuel and timber by the prehistoric inhabitants of Anatolia provides only a limited set of information on ancient environments and human-landscape interactions. More than these, archaeological wood charcoal remains hold the ecological signatures (growth conditions and life histories) of the different wood species that were used in the past and of the habitats from which they were collected. Archaeobotanists can record not only species presence and abundance but also the form (stem, round wood) and state (dry or green) of the collected wood, whether it was gathered from closed or open canopy woodlands, and how woodlands were impacted by various environmental stresses, including frost, drought, and human management practices, such as pollarding or coppicing. Contemporary anthracology provides a unique set of sampling and analysis tools with which to disentangle the complex feedback cycles between vegetation, climate, and prehistoric woodland exploitation (Kabukcu 2018b; Dufraisse et al. 2018; Kabukcu and Chabal 2021). The coming years will undoubtedly witness the expansion of advanced anthracological studies in Anatolia and across southwest Asia and the eastern Mediterranean as their paleoecological potential gains further recognition.

Acknowledgments

We would like to thank Chris Roosevelt and John Haldon for their invitation to contribute to this volume, and their invitation to participate in the 15th International ANAMED Annual Symposium, "Winds of Change: Environment and Society in Anatolia."

Works Cited

Alley, Richard B., Paul A. Mayewski, Todd Sowers, Minze Stuiver, Kendrick C. Taylor, and Peter U. Clark. 1997. "Holocene Climatic Instability: A Prominent, Widespread Event 8200 Years Ago." *Geology* 25: 483-86.

Ariztegui, Daniel, Alesssandra Asioli, John J. Lowe, Fabio Trincardi, Luigi Vigliotti, Federica Tamburini, Christina Chondrogianni, Carla Alberta Accorsi, Marta Bandini Mazzanti, Anna Maria Mercuri, Sander van der Kaars, Judith Ann McKenzie, and Frank A. Oldfield. 2000. "Palaeoclimate and the Formation of Sapropel S1: Inferences from Late Quaternary Lacustrine and Marine Sequences in the Central Mediterranean Region." *Palaeogeography, Palaeoclimatology, Palaeoecology* 158: 215-40.

Asouti, Eleni. 2009. "The Relationship between Early Holocene Climate Change and Neolithic Settlement in Central Anatolia, Turkey: Current Issues and Prospects for Future Research." *Documenta Praehistorica* 36: 1-5.

———. 2013. "Woodland Vegetation, Firewood Management and Woodcrafts at Neolithic Çatalhöyük." In *Humans and Landscapes of* Çatalhöyük: *Reports from the 2000-2008 Seasons*, edited by Ian Hodder, 129-62. Los Angeles: Cotsen Institute of Archaeology Press.

———. 2017. "Human Palaeoecology in Southwest Asia during the Early Pre-Pottery Neolithic (c. 9700-8500 cal BC): The Plant Story." In *Neolithic Corporate Identities*, edited by Marion Benz, Hans Georg K. Gebel, and Trevor Watkins. Studies in Early Near Eastern Production, Subsistence, and Environment (SENEPSE), 20: 21-53. Berlin: Ex oriente.

Asouti, Eleni, and Dorian Q. Fuller. 2012. "From Foraging to Farming in the Southern Levant: The Development of Epipalaeolithic and Pre-Pottery Neolithic Plant Management Strategies." *Vegetation History and Archaeobotany* 21, no. 2: 149-62.

Asouti, Eleni, and Ceren Kabukcu. 2014. "Holocene Semi-Arid Oak Woodlands in the Irano-Anatolian Region of Southwest Asia: Natural or Anthropogenic?" *Quaternary Science Reviews* 90: 158-82.

Asouti, Eleni, Maria Ntinou, and Ceren Kabukcu. 2018."The Impact of Environmental Change on Palaeolithic and Mesolithic Plant Use and the Transition to Agriculture at Franchthi Cave, Greece." *PLOS ONE* 13: 11, e0207805.

Asouti, Eleni, Douglas Baird, Ceren Kabukcu, Kate Swinson, Louise Martin, Aroa García-Suárez, Emma Jenkins, and Kamal Rasheed. 2020. "The Zagros Epipalaeolithic Revisited: New Excavations and 14C Dates from Palegawra Cave in Iraqi Kurdistan." *PloS one* 15, no. 9: e0239564.

Aura, Josep Emili, Yolanda Carrión, Elena Estrelles, and Guillem Pérez Jordà. 2005. "Plant Economy of Hunter-Gatherer Groups at the End of the Last Ice Age: Plant Macroremains from the Cave of Santa Maira (Alacant, Spain) ca. 12000–9000 B.P." *Vegetation History and Archaeobotany* 14, no. 4: 542–50.

Austin, Phil. 2007. "The Wood Charcoal Macroremains." In *The Early Prehistory of Wadi Faynan,Ssouthern Jordan,* edited by Bill Finlayson and Steven Mithen, 408–19. Oxford: Council for British Research in the Levant Monographs.

Aytuğ, Burhan. 1970. "Arkeolojik Araştırmaların Işığı Altında İç Anadolu Stebi." İstanbul Üniversitesi *Orman Fakültesi Yayınları* 20, no. 1: 127–43.

Bar-Matthews, Miryam, Avner Ayalon, Aaron Kaufman, and Gerald J. Wasserburg. 1999. "The Eastern Mediterranean Paleoclimate as a Reflection of Regional Events: Soreq Cave, Israel." *Earth and Planetary Science Letters* 166: 85–95.

Bar-Matthews, Miryam, Avner Ayalon, Mabs Gilmour, Alan Matthews, and Chris J. Hawkesworth. 2003. "Sea-land Isotopic Relationships from Planktonic Foraminifera and Speleothems in the Eastern Mediterranean Region and their Implication for Palaeorainfall during Interglacial Intervals." *Geochimica et Cosmochimica Acta* 67: 3181–99.

Baird, Douglas, Andrew Fairbairn, Emma Jenkins, Louise Martin, Caroline Middleton, Jessica Pearson, Eleni Asouti, Yvonne Edwards, Ceren Kabukcu, Gökhan Mustafaoğlu, and Nerissa Russell. 2018. "Agricultural Origins on the Anatolian Plateau." *Proceedings of the National Academy of Sciences* 115, no. 14: E3077–E3086.

Baruch, Uri, and Nigel Goring-Morris. 1997. "The Arboreal Vegetation of the Central Negev Highlands, Israel, at the End of the Pleistocene: Evidence from Archaeological Charred Wood Remains." *Vegetation History and Archaeobotany* 6, no. 4: 249–59.

Bogaard, Amy, Michael Charles, Dragana Filipović, Dorian Q. Fuller, Lara Gonzalez Carretero, Laura Green, Ceren Kabukcu, and Elizabeth Stroud. 2021. "The Archaeobotany of Çatalhöyük: Results from 2009-2017 Excavations and Final Synthesis." In *Peopling the Landscape of Çatalhöyük: Reports from the, 2009–2017 Seasons,* edited by Ian Hodder, 91-124. Oxford: Oxbow Books.

Bottema, Sytze. 1991. "Développement de la végétation et du climat dans le bassin méditerranéen oriental à la fin du Pleistocene et pendant l'Holocène." *L'Anthropologie* 95: 695–728.

Bourguet, Cécile, and Margareta Tengberg. 2017. "Dynamics of Vegetation Cover and Exploitation of Wood Resources in Central Anatolia during the Early Neolithic Period: Anthracological Study of Aşıklı Höyük (Cappadocia, Turkey)." Poster presentation at the 4th *Young Natural History Scientists Meeting,* 7-11 February 2017. Accessed online: http://www.asiklihoyuk.org/wp-content/uploads/2017/07/Bourguet-and-Tengberg-2017.pdf

Collins, Christina, Eleni Asouti, Matt Grove, Ceren Kabukcu, Lee Bradley, and Richard Chiverrell. 2018. "Understanding Resource Choice at the Transition from Foraging to Farming: An Application of Palaeodistribution Modelling to the Neolithic of the Konya Plain, South-Central Anatolia, Turkey." *Journal of Archaeological Science* 96: 57–72.

Deckers, Katleen, Simone Riehl, Emma Jenkins, Arlene Rosen, Andrey Dodonov, Aleksandra N. Simakova, and Nicholas J. Conard. 2009. "Vegetation Development and Human Occupation in the Damascus Region of Southwestern Syria from the Late Pleistocene to Holocene." *Vegetation History Archaeobotany* 18: 329–40.

Dufraisse Alexa, Sylvie Coubray, Olivier Girardclos, Noémie Nocus, Michel Lemoine, Jean-Luc Dupouey, Dominique Marguerie. 2018. "Anthracotypology as a Key Approach to Past Firewood Exploitation and Woodland Management Reconstructions. Dendrological Reference Dataset Modelling with Dendro-Anthracological Tools. *Quaternary International* 463: 232–49.

Göktürk, Ozan M., Dominik Fleitmann, Seraina Badertscher, H. Cheng, R. L. Edwards, M. Leuenberger, A. Fankhauser, O. Tüysüz, and J. Kramers. 2011. "Climate on the Southern Black Sea Coast during the Holocene." *Quaternary Science Reviews* 30: 2433–45.

Helbaek, Hans. 1969. "Plant Collecting, Dry-Farming, and Irrigation Agriculture in Prehistoric Deh Luran." In *Prehistory and Human Ecology of the Deh Luran Plain*, Memoirs of the Museum of Anthropology 1, edited by Frank Hole, Kent Flannery, and James A. Neely, 383–426. Ann Arbor: University of Michigan.

Hillman, Gordon C. 1972. "Archaeo-Botanical Studies." *Anatolian Studies* 22: 17–19.

———. 1978. "On the Origins of Domestic Rye – Secale cereale: The Finds from Aceramic Can Hasan III in Turkey." *Anatolian Studies* 28: 157–74.

Johnsen, Sigfus J., Dorthe Dahl- Jensen, Niels Gundestrup, Jørgen P. Steffensen, Henrik B. Clausen, Heinz Miller, Valerie Masson- Delmotte, Arny E. Sveinbjörnsdottir, and James White. 2001. "Oxygen Isotope and Palaeotemperature Records from Six Greenland Ice Core Stations: Camp Century, Dye-3, GRIP, GISP2, Renland and NorthGRIP." *Journal of Quaternary Science* 16: 299–307.

Kabukcu, Ceren. 2017. "Woodland Vegetation History and Human Impacts in South-Central Anatolia 16,000–6500 cal BP: Anthracological Results from Five Prehistoric Sites in the Konya Plain." *Quaternary Science Reviews* 176: 85–100.

———. 2018a. "Wood Charcoal Analysis in Archaeology." In *Environmental Archaeology. Interdisciplinary Contributions to Archaeology*, edited by E. Pişkin, A. Marciniak, and M. Bartkowiak, 133–54. Springer: Cham. https://doi.org/10.1007/978-3-319-75082-8_7.

———. 2018b. "Identification of Woodland Management Practices and Tree Growth Conditions in Archaeological Fuel Waste Remains: A Case Study from the Site of Çatalhöyük in Central Anatolia, Turkey." *Quaternary International* 463: 282–97.

Kabukcu, Ceren, and Lucie Chabal. 2021. "Sampling and Quantitative Analysis Methods in Anthracology from Archaeological Contexts: Achievements and Prospects." *Quaternary International* 593-594: 6–18.

Kabukcu, Ceren, Eleni Asouti, and Joseph Losh. 2021. "Woodland Vegetation, Fuelwood and Timber Use at Çatalhöyük: The Anthracological Remains from the 1996 to 2017 Excavations." In *Peopling the Landscape of Çatalhöyük: Reports from the 2009–2017 Seasons*, British Institute at Ankara Monograph, Vol. 53, edited by Ian Hodder, 73–90. Oxford: Oxbow Books.

Kabukcu, Ceren, Eleni Asouti, Nadja Pöllath, Joris Peters, and Necmi Karul. 2021. "Pathways to Plant Domestication in Southeast Anatolia based on New Data from Aceramic Neolithic Gusir Höyük." *Scientific Reports* 11, no. 1: 1–15.

Kallel, Nejib, Martine Paterne, Laurent Labeyrie, Jean-Claude Duplessy, and Maurice Arnold. 1997. "Temperature and Salinity Records of the Tyrrhenian Sea during the Last 18,000 Years." *Palaeogeography, Palaeoclimatology, Palaeoecology* 135: 97–108.

Kayacık, Hayrettin, and Burhan Aytuğ. 1968. "Gordion Kral Mezarı'nın ağaç malzemesi üzerinde ormancılık yönünden araştırmalar." İstanbul Üniversitesi *Orman Fakültesi Yayınları* 18, no. 1: 37–54.

Kwiecien, Olga, Helge W. Arz, Frank Lamy, Birgit Plessen, André Bahr, and Gerald H. Haug. 2009. "North Atlantic Control on Precipitation Pattern in the Eastern Mediterranean/Black Sea Region during the Last Glacial." *Quaternary Research* 71: 375–84.

Litt, Thomas, Sebastian Krastel, Michael Sturm, Rolf Kipfer, Sefer Örcen, Georg Heumann, Sven Oliver Franz, Umut B. Ülgen, and Frank Niessen. 2009. "Lake Van Drilling Project 'PALEOVAN', International Continental Scientific Drilling Program (ICDP): Results of a Recent Pre-Site Survey and Perspectives." *Quaternary Science Reviews* 28: 1555–67.

Martinoli, Danielle. 2004. "Reconstruction of Local Woodland Vegetation and Use of Firewood at Two Epipalaeolithic Cave Sites in Southwest Anatolia (Turkey)." In *Foragers to Farmers: Papers in Honour of G. C. Hillman*, edited by A. Fairbairn and E. Weiss, 161–70. Oxford: Oxbow Books.

Miebach, Andrea, Phoebe Niestrath, Patricia Roeser, and Thomas Litt. 2016. "Impacts of Climate and Humans on the Vegetation in Northwestern Turkey: Palynological Insights from Lake Iznik since the Last Glacial." *Climate of the Past* 12: 575–93.

Neef, Reinder. 2003. "Overlooking the Steppe Forest: Preliminary Report on the Botanical Remains from Early Neolithic Göbekli Tepe (Southern Turkey)." *Neo-Lithics* 2, no. 03: 13–16.

———. 2004. “Vegetation and Plant Husbandry.” In *Basta I. The Human Ecology*, edited by H. J. Nissen, M. Muheisen, and H. G. K. Gebel, 187–218. Berlin: Ex oriente.

Ntinou, Maria. 2021. “Vegetation Change in Southeastern Greece during the Late Pleistocene. The Wood Charcoal Record from Klissoura Cave 1 (Peloponnese, Greece).” *Quaternary International* 593–94: 104–17.

Ntinou, Maria, and Nina Kyparissi-Apostolika. 2016. “Local Vegetation Dynamics and Human Habitation from the Last Interglacial to the Early Holocene at Theopetra Cave, Central Greece: The Evidence from Wood Charcoal Analysis.” *Vegetation History and Archaeobotany* 25, no. 2: 191–206.

Orland, Ian J., Miryam Bar-Matthews, Avner Ayalon, Alan Matthews, Reinhard Kozdon, Takayuki Ushikubo, and John W. Valley. 2012. “Seasonal Resolution of Eastern Mediterranean Climate Change since 34 ka from a Soreq Cave Speleothem.” *Geochimica et Cosmochimica Acta* 89: 240–55.

Pickarski, Nadine, Ola Kwiecien, Morteza Djamali, and Thomas Litt. 2015. “Vegetation and Environmental Changes during the Last Interglacial in Eastern Anatolia (Turkey): A New High-Resolution Pollen Record from Lake Van.” *Palaeogeography, Palaeoclimatology, Palaeoecology* 435: 145–58.

Riehl, Simone, Marion Benz, Nicholas J. Conard, Hojjat Darabi, Katleen Deckers, Hassan Fazeli Nashli, and Mohsen Zeidi-Kulehparcheh. 2012. “Plant Use in Three Pre-Pottery Neolithic Sites of the Northern and Eastern Fertile Crescent: A Preliminary Report.” *Vegetation History and Archaeobotany* 21: 95–106.

Roberts, Neil, Samantha L. Allcock, Fabien Arnaud, Jonathan R. Dean, Warren J. Eastwood, Matthew D. Jones, Melanie J. Leng, Sarah E. Metcalfe, Emmanuel Malet, Jessie Woodbridge, Hakan Yiğitbaşıoğlu. 2016. “A Tale of Two Lakes: A Multi- Proxy Comparison of Lateglacial and Holocene Environmental Change in Cappadocia, Turkey.” *Journal of Quaternary Science* 31, no. 4: 348–62.

Roberts, Neil, Stuart Black, Peter Boyer, Warren J. Eastwood, H. I. Griffiths, H. F. Lamb, M. J. Leng, Romola Parish, Jane M. Reed, David R. Twigg, and Hakan Yiğitbaşıoğlu. 1999. “Chronology and Stratigraphy of Late Quaternary Sediments in the Konya Basin, Turkey: Results from the KOPAL Project.” *Quaternary Science Reviews* 18: 611–30.

Roberts, Neil, Jane Reed, Melanie J. Leng, Catherine Kuzucuoğlu, Michel Fontugne, Jacques Bertaux, Henrik Woldring, Sytze Bottema, Stuart Black, Elizabeth Hunt, and Mustafa Karabıyıkoğlu. 2001. “The Tempo of Holocene Climatic Change in the Eastern Mediterranean Region: New High-Resolution Crater-Lake Sediment Data from Central Turkey.” *Holocene* 11: 719–34.

Roberts, Neil, Jessie Woodbridge, Andrew Bevan, Alessio Palmisano, Stephen Shennan, and Eleni Asouti. 2018. "Human Responses and Non-Responses to Climatic Variations during the Last Glacial-Interglacial Transition in the Eastern Mediterranean." *Quaternary Science Reviews* 184: 47–67.Robinson, Stuart A., Stuart Black, Bruce W. Sellwood, and Paul J. Valdes. 2006. "A Review of Palaeoclimates and Palaeoenvironments in the Levant and the Eastern Mediterranean from 25,000 to 5000 Years BP: Setting the Environmental Background for the Human Civilization." *Quaternary Science Reviews* 25: 1517–41.

Roffet-Salque, Mélanie, Arkadiusz Marciniak, Paul J. Valdes, Kamilla Pawłowska, Joanna Pyzel, Lech Czerniak, Marta Krüger, C. Neil Roberts, Sharmini Pitter, and Richard P. Evershed. 2018. "Evidence for the Impact of the 8.2-kyBP Climate Event on Near Eastern Early Farmers." *Proceedings of the National Academy of Sciences* 115, no. 35: 8705–09.

Rohling, Eelco, Paul Mayewski, R. Abu-Zied, J. Casford, and A. Hayes. 2002. "Holocene Atmosphere-Ocean Interactions: Records from Greenland and the Aegean Sea." *Climate Dynamics* 18: 573–92.

Rosenberg, Michael, Mark Nesbitt, Richard W. Redding, and Brian L. Peasnall. 1998. "Hallan Çemi, Pig Husbandry, and Post-Pleistocene Adaptations Along the Taurus-Zagros Arc (Turkey)": *Paléorient* 24, no. 1: 25–41.

Rossignol-Strick, Martine. 1995. "Sea-Land Correlation of Pollen Records in the Eastern Mediterranean for the Glacial-Interglacial Transition: Biostratigraphy Versus Radiometric Time-Scale." *Quaternary Science Reviews* 14: 893–915.

Rössner, Corinna, Katleen Deckers, Marion Benz, Vecihi Özkaya, and Simone Riehl. 2018. "Subsistence Strategies and Vegetation Development at Aceramic Neolithic Körtik Tepe, Southeastern Anatolia, Turkey." *Vegetation History and Archaeobotany* 27: 15–29.

Sbaffi, Laura, Forese Carlo Wezel, Giuseppe Curzi, and Ugo Zoppi. 2004. "Millennial- to Centennial-Scale Palaeoclimatic Variations during Termination I and the Holocene in the Central Mediterranean Sea." *Global and Planetary Change* 40: 201–17.

Schilman, Bettina, Avner Ayalon, Miryam Bar-Matthews, Elisa J. Kagan, and Ahuva Almogi-Labin. 2002. "Sea-Land Paleoclimate Correlation in the Eastern Mediterranean Region during the Late Holocene." *Israel Journal of Earth Sciences* 51: 181–90.

Stevens, Lora R., Morteza Djamali, Valérie Andrieu-Ponel, and Jacques-Louis de Beaulieu. 2012. "Hydroclimatic Variations over the Last Two Glacial/Interglacial Cycles at Lake Urmia, Iran." *Journal of Paleolimnology* 47: 645–60.

Şanlı, İsmet. 1989. "Ahşap ve arkeoloji." İstanbul Üniversitesi *Orman Fakültesi Yayınları* 39, no. 1: 130–41.

Wainwright, John, and Gianna Ayala. 2019. "Teleconnections and Environmental Determinism: Was there Really a Climate-Driven Collapse at Late Neolithic Çatalhöyük?" *Proceedings of the National Academy of Sciences* 116, no. 9: 3343–44.

Western, A. Cecilia. 1969. "An Attempt at the Ecological Interpretation of Charcoals with Special Reference to Material from Jericho." B.Sc. diss, University of Oxford.

———. 1971. "The Ecological Interpretation of Ancient Charcoals from Jericho." *Levant* 3: 31–40.

Wick, Lucia, Genry Lemcke, and Michael Sturm. 2003. "Evidence of Lateglacial and Holocene Climatic Change and Human Impact in Eastern Anatolia: High-Resolution Pollen, Charcoal, Isotopic and Geochemical Records from Laminated Sediments of Lake Van, Turkey." *Holocene* 13: 665–75.

Wiersma, Ane P., and Hans Renssen. 2006 "Model-Data Comparison for the 8.2 ka BP Event: Confirmation of a Forcing Mechanism by Catastrophic Drainage of Laurentide Lakes." *Quaternary Science Reviews* 25: 63–88.

Willcox, George. 1974. "A History of Deforestation as Indicated by Charcoal Analysis of Four Sites in Eastern Anatolia." *Anatolian Studies* 24: 117–33.

———. 1992. "Timber and Trees: Ancient Exploitation in the Middle East: Evidence from Plant Remains." *Bulletin of Sumerian Agriculture* 6: 1–31.

———. 1999. "Charcoal Analysis and Holocene Vegetation History in Southern Syria." *Quaternary Science Reviews* 18: 711–16.

———. 2002. "Evidence for Ancient Forest Cover and Deforestation from Charcoal Analysis of Ten Archaeological Sites on the Euphrates." In *Charcoal Analysis: Methodological Approaches, Palaeoecological Results and Wood Uses*, BAR International Series 1063, edited by S. Thiébault, 141–45. Oxford: Archaeopress.

Willett, Patrick T., Ingmar Franz, Ceren Kabukcu, David Orton, Jana Rogasch, Elizabeth Stroud, Eva Rosenstock, and Peter F. Biehl. 2016. "The Aftermath of the 8.2 Event." In *Climate and Cultural Change in Prehistoric Europe and the Near East*, edited by P. F. Biehl and O. Nieuwenhuyse, 95–116. New York: SUNY Press.

Woldring, Henrik, and Bottema, Sytze. 2001/2. "The Vegetation History of East-Central Anatolia in Relation to Archaeology: The Eski Acıgöl Pollen Evidence Compared with the Near Eastern Environment." *Palaeohistoria* 43/44: 1–34.

Valsecchi, Verushka, Maria Fernanda Sánchez-Goñi, and Laurent Londeix. 2012. "Vegetation Dynamics in the Northeastern Mediterranean Region during the Past 23 000 yr: Insights from a New Pollen Record from the Sea of Marmara." *Climate of the Past* 8: 1941–56.

van Zeist, Willem, and G. J. de Roller. 1991/2. "The Plant Husbandry of Aceramic Çayönü, SE Turkey." *Palaeohistoria* 33: 65–96.

Ziegler, Martin, Erik Tuenter, and Lucas J. Lourens. 2010. "The Precession Phase of the Boreal Summer Monsoon as Viewed from the Eastern Mediterranean (ODP Site 968)." *Quaternary Science Reviews* 29: 1481–90.

Stability through Crisis: Cultural Resilience in the Face of Climatic Fluctuation from 3500 BCE to 1300 CE at Çadır Höyük in North-Central Anatolia

Madelynn von Baeyer

Tevfik Emre Şerifoğlu

Introduction

Çadır Höyük is a medium-sized mound located in the Kanak Su basin in north-central Anatolia (Sorgun, Yozgat), surrounded by fertile agricultural land and overlooked by high hills marking the edges of the basin (Figs. 1–2; Steadman and McMahon 2015, 69–70). The mound was surveyed in 1993 and first excavated in 1994 as part of a salvage project initiated because of the construction of the Gelingüllü Dam, which portended the submersion of both the Kanak Su Basin and the mound itself (Gorny et al. 1995). The artificial lake, however, never affected the mound as expected, and this project that started with two soundings and a step trench was transformed into a long-term one that has seen almost continuous excavation for the past 27 years (Gorny et al. 1999, 2000, 2002; Steadman, McMahon, and Ross 2007; Cassis 2009; Steadman et al. 2013, 2015, 2017). The project was initially sponsored by the University of Chicago, and many other universities joined in over the years, including New Hampshire, SUNY Cortland, Newfoundland, Calgary, Bitlis Eren, Leicester, Hood College, and UCL.

The mound was inhabited at least from the last quarter of the sixth millennium BCE until the eleventh century CE. Over the years, the excavated area expanded to include all slopes and the summit of the mound, together with a short-term excavation of an area north of the mound named the North Terrace. The length of occupation at Çadır Höyük, as well as the duration of excavations, have made it possible to explore long-term cultural resilience using multiple lines of archaeological evidence. Clearly, the population of Çadır Höyük was able to develop multiple strategies for resilience over the 6000 years of occupation. We researchers now have the chronological depth and the exposure to focus on different aspects of cultural resilience. One question we

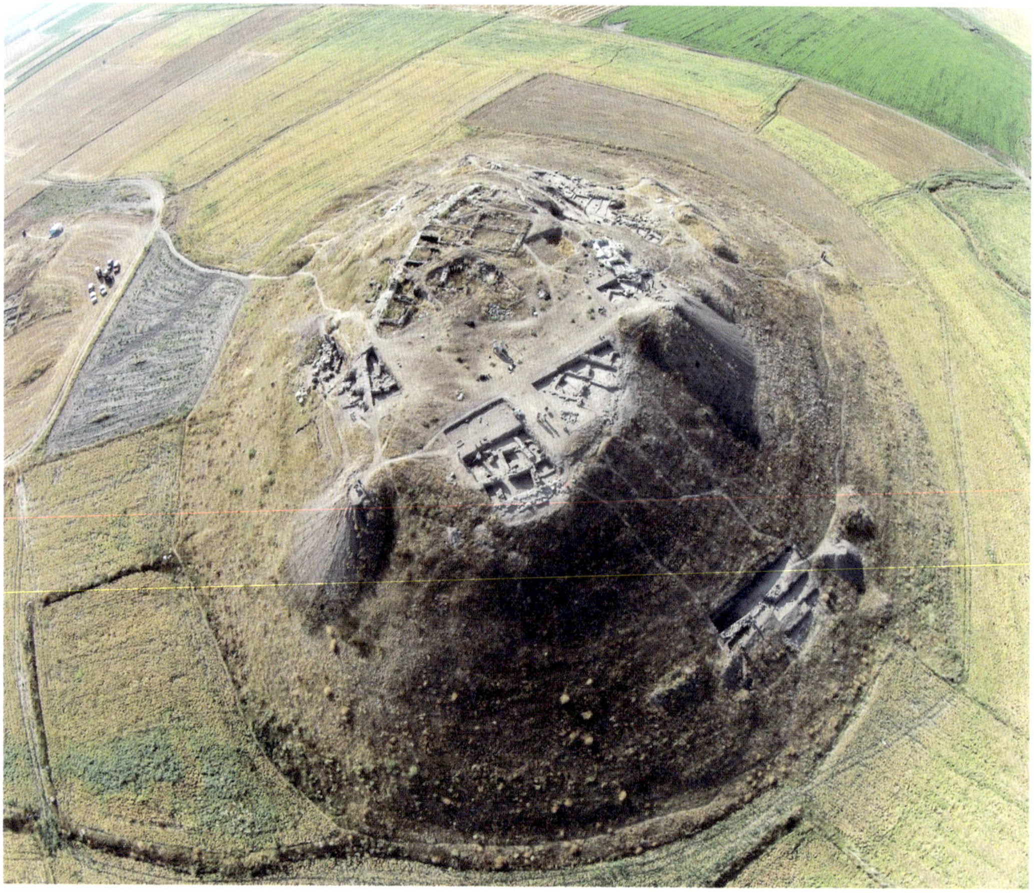

Fig. 1. An aerial view of Çadır Höyük to the southeast.

are beginning to explore is how the population continually adapted to climate change. This chapter outlines our first attempts to bring together multiple lines of evidence from Çadır Höyük—architectural, material, paleoethnobotanical, and zooarchaeological—and examine them against the climatic evidence for three key periods: the transition between Late Chalcolithic (LCh) and Early Bronze Age I (EBA I) around 3000 BCE, the transition between the Late Bronze Age (LBA) and the Early Iron Age (EIA) around 1200 BCE, and the Late Roman and Middle Byzantine periods around 1000 CE. By looking at material, floral, and faunal assemblages together, the goal is to create an integrated model of cultural resilience at Çadır Höyük from the Late Chalcolithic through the Byzantine periods, as well as highlight understudied areas in the broader Çadır Höyük assemblage.

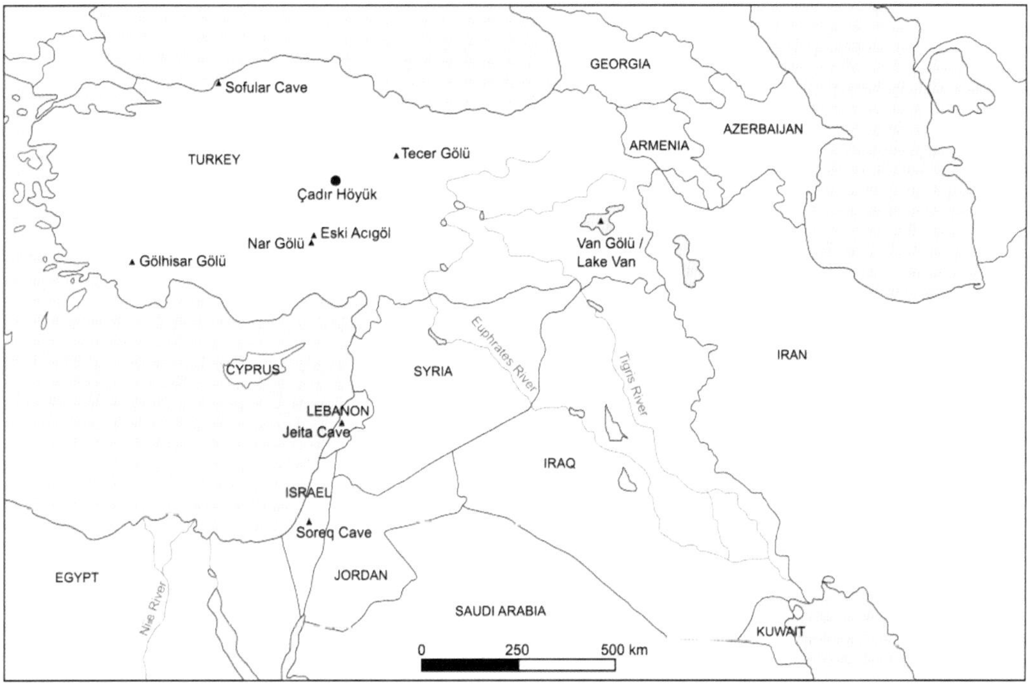

Fig. 2. A regional map of southwest Asia indicating Çadır Höyük and the climatic proxy sites mentioned in the text.

Overview of the Site

The work on the earliest known level at the site, which corresponds to the Chalcolithic period, has focused on the lower southern slopes of the mound, due to the fact that Chalcolithic material was found just below the surface here (Steadman, McMahon, and Ross 2019, 34–37). In this area, we had the opportunity to investigate the whole stratigraphic sequence from the Middle Chalcolithic (last quarter of the sixth millennium BCE) to Early Bronze Age (EB) I (first quarter of the third millennium BCE), revealing extensive architectural remains and a good pottery repertoire that enables comparison with material unearthed or collected by other projects conducted in this part of Anatolia. The eastern slope, where the step trench is located, is where second-millennium BCE investigations have focused (Steadman, McMahon, and Ross 2019, 37–41). The remains of a fortification wall that could be dated to the Late Bronze Age (LBA) were unearthed in this part of the mound. The northern side of the mound had some cyclopean stones, which Ron Gorny identified as architectural features belonging to an LBA gate, and this hypothesis has been confirmed by recent excavations. The LBA casemate-style fortification walls were built on top of a Middle Bronze Age (MBA) wall, which was found in trenches on the other side of the mound on the western slope. Although these fortification walls were investigated in detail, we have not managed to move into the town itself and investigate the actual residential, commercial, or administrative quarters, which has left us with a good set of second-millennium BCE material but with no good architectural contexts. We have also failed to investigate most of the third millennium BCE layers and contexts in any detail, so the EB II and III periods are still more or less a mystery where Çadır Höyük is concerned.

The mound was continuously inhabited at least until the end of the Iron Age (1200–330 BCE), although the Early Iron Age architecture is mostly missing, which points to a socio-economic change and/or a shift in residential area and lifestyle preferences (Steadman, McMahon, and Ross 2019, 41–46). The Early Iron Age levels unearthed on the southern slope of the mound mostly had pits, many of which are shallow, and some shabby walls, and this indicated that the area was mostly used for industrial activities. Within the Middle and Late Iron Age levels, we have found remnants of more extensive architecture, but most of the layers have been disturbed by later activities that destroyed the architectural layouts of these phases. These periods are marked by thick walls, paved streets, and well-made pottery with painted north-central Anatolian types being very common. Parts of a Middle Iron Age wall circling the mound were unearthed during excavations on the northern and western sides of the mound, although the wall seems to have been used by later inhabitants to support other structures and was therefore altered to fit their needs. The summit of the mound did not reveal any extensive Iron Age architecture, but in general, open-space areas with pottery dating to this period were unearthed. It is important to note that in most cases the Iron Age layers were found directly below the Late Roman layers.

The Hellenistic and Roman periods are represented only by archaeological material from a thin horizon of archaeological layers and not from secure contexts (Steadman and McMahon 2017, 106–10). It is quite clear that the extensive engineering work, which reshaped the mound during the Late Roman and Byzantine periods, mostly resulted in the removal of these layers on the mound. However, we suspect that there may have been a Roman and maybe even a Hellenistic lower town stretching from the northern and northeastern slopes of the mound to the fields in that direction. Our work in the North Terrace provided some data that supports this view, as the remains of well-built, extensive structures starting to emerge when the layers just below the Late Roman building were reached.

Our work in the North Terrace revealed a glimpse of rural life in this part of Anatolia during the Late Roman and Byzantine periods. A structure identified as a villa was built north of the mound during the Late Roman period and was probably owned by a land-owning family who employed villagers to manage agricultural activities. This well-built villa was later abandoned, but the building was reused with the addition of shabby interior walls that would enable more people to reside in it. Therefore, we think that the villagers moved in during the Byzantine period and continued cultivating the land around the structure, this time in a more communal way; this probably continued until the Seljuk period. We believe that there are other archaeological layers below the Late Roman structure and, in fact, these may belong to a Roman lower town connected to the mound. Unfortunately, we will be unable to continue our work in this area because of modern land-ownership issues that resulted in the cessation of all archaeological activities in the area.

Remains contemporaneous with the Late Roman and Byzantine North Terrace were also unearthed at the very top of the mound (Steadman, McMahon, and Ross 2019, 46–50). The summit of the mound was encircled by a well-built fortification wall, and our team investigated many rooms built inside it. One structure with an apse is believed to be a chapel, and the remains of the city gate were found at the south side of the summit. It is quite clear that the whole mound was modified as part of a massive engineering project during the Late Roman period, probably removing the Roman, Hellenistic, and Later Iron Age material to achieve a flat summit and steep slopes and, therefore, truncating the stratigraphy so that Late Roman structures were built directly on top of Iron Age layers. The Late Roman structures were modified during the Byzantine period, and the archaeological evidence suggests that the mound was still inhabited during the Seljuk and Danishmend periods, even if it was abandoned shortly after.

Overview of the Zooarchaeological and Paleoethnobotanical Research

Paleoethnobotanical and zooarchaeological studies at Çadır Höyük have been robust throughout the length of the excavation. The biological science research is led by Benjamin Arbuckle, who is also the head of the zooarchaeology team. Most of the zooarchaeological research has been conducted by Arbuckle, Sarah Adcock (second and first millennia BCE), and Katie Tardio (Late Roman and Byzantine). The paleoethnobotanical work was headed by Miriam Chernoff (1994–1999), Alexia Smith (2000–2016), and Madelynn von Baeyer (2016–present). The collection and identification methods for both the zooarchaeological and paleoethnobotanical assemblages have remained the same for the duration of the project, despite the more frequent turnover of paleoethnobotanical specialists. Detailed discussions of project sorting and identification methods can be found in the works of Arbuckle (2009) and von Baeyer (2018), as well as von Baeyer, Smith, and Steadman (2021).

Climate Conditions and Risk Reduction Strategies

It is not surprising that with over 6000 years of occupation, the population of Çadır Höyük would have had to adapt to multiple periods of climate change and fluctuation. This study examines the agricultural and social responses to the three major periods of increased and intense aridity during the fourth through first millennia BCE and the humid periods associated with the Medieval Climate Anomaly.

The Mid-Holocene Climatic Transition

Çadır Höyük's occupation neatly overlaps with the climatic period known as the Mid-Holocene Transition, a period of about 3000 years, starting in the fourth millennium BCE, when the climate in southwest Asia shifted from the humid and wet climate of the Neolithic and Early and Middle Chalcolithic to the much drier climate of the Iron Age (Roberts et al. 2011). This was not a gradual, continual shift, but rather a period punctuated by unusually arid periods interspersed with periods of relatively increased humidity (Roberts et al. 2011). Roberts and colleagues (2011) have identified three major aridification events that left the climate drier than it was before: 3300–3000 BCE (5.3–5.0 ka BP), 2500–1950 BCE (4.5–3.9 ka BP), and 1200–850 BCE (3.2–2.8 ka BP). These periods of aridity, possibly a series of droughts in close succession within each period, correspond at Çadır Höyük to the LCh/EBA transition, the EB II/EB III transition, and the LBA/EIA transition.

Global and regional climate proxies record increased aridity during these periods. Globally, these periods are known as centennial-scale climate events and are associated with increased aridity at low latitudes and cooling at high latitudes (Mayewski et al. 2004). The speleothem record from Soreq Cave in Israel records periods of aridity at 5250–5170 BP and 4200–4050 BP (Bar-Matthews and Ayalon 2011, 166). The speleothem record at Jeita Cave in Lebanon also registers two long arid periods between 5.3–4.2 ka BP and 2.8–1.4 ka BP (Cheng et al. 2015). In Anatolia, the sediment core from Tecer Lake records periods of aridity during all three rapid climate change events (Kuzucuoğlu et al. 2011); Eski Acıgöl records periods of aridity around 5.2 ka and 3.2 ka BP (Turner, Roberts, and Jones 2008); and, Gölhisar records centennial-scale shifts in the evaporation/precipitation balance that fit with the centennial cyclicity of the mid-Holocene rapid climate change events (Eastwood et al. 2007, 334). Interestingly, the climate proxy from Lake Van does not reflect these periods of aridity (Wick, Lemcke, and Sturm 2003). Lake Van remained quite humid until at least the first millennium BCE. Çadır Höyük is not in Lake Van's climatic zone, however, nor is it part of its fluvial system (Türkeş, Koç, and Sariş 2009). It is much more probable that the climate of Çadır Höyük was similar to that of Tecer Lake than Lake Van and therefore experienced these periods of increased aridity punctuated by episodes of relative humidity between the fourth and first millennia BCE.

Unfortunately, it is not yet possible for us to evaluate the effects of the 4.2 ka BP event on the Çadır Höyük community, as the corresponding EB II and III levels have not been fully investigated. This event seems to have triggered a major collapse of socio-economic systems and a movement of populations that manifests itself with destruction horizons at settlements all across the eastern Mediterranean (Şerifoğlu 2017, 161–63; Weiss et al. 1993; Weiss 1997; Staubwasser and Weiss 2006; Cullen et al. 2000). One of our main aims for the future is to reach and reveal larger portions of the relevant levels and the corresponding architecture, as we know that the site was occupied during the entirety of the EBA, based on pottery scatters and mixed material found in various parts of the mound, although it may have been a very small settlement during this period.

The Medieval Climate Anomaly

The final period of increased climate variability occurred at the end of the occupation sequence. The Late Roman to Middle Byzantine period at Çadır Höyük was punctuated by three possible climatic shifts: one in the sixth century CE, which brought increased precipitation followed by a temporary period of increased dryness, and then others in the ninth and tenth centuries CE, a period of once again increased precipitation that corresponds with the larger Medieval Climate Anomaly or Medieval Climate Optimum (Dean et al. 2015). Few regional proxy records exist for central Anatolia during the Late Roman and Byzantine periods (Xoplaki et al. 2016): only the lake cores from Nar (Dean et al. 2013, 2015) and Tecer Lakes (Kuzucuoğlu et al. 2011) come from central Anatolia. In the case of the Late Roman and Middle Byzantine records at Nar and Tecer Lakes, the paleoclimate signatures of each lake core are similar. Both indicate that the end of the Late Roman period was a time of increased rainfall and seasonality and that the Middle Byzantine era of the ninth and tenth centuries CE was another period of increased rainfall and humidity. Tecer Lake's sediment core does not have a clear record between these two humid periods (Kuzucuoğlu et al. 2011), but the Nar Lake sediment core records a brief period of increased dryness (Dean et al. 2015). The increased humidity in the ninth and tenth centuries CE is reflected in the speleothem records of Sofular Cave in the Black Sea region (Göktürk et al. 2011). It seems likely, then, that the Late Roman and Middle Byzantine periods at Çadır Höyük were periods of increased humidity and rainfall.

Risk Reduction Strategies

Risk reduction strategies, defined here as strategies chosen that reduce the chance of loss, can occur as a response to a number of variables that create uncertainty in a society, including political, economic, social, and environmental upheaval. For this case study, we focus on the impact of environmental upheaval or consistency on the populations of Çadır Höyük throughout its occupation. In terms of archaeological records, these adaptation strategies are most often interpreted through archaeobotanical, zooarchaeological, and osteological remains, but changes in settlement size, usage of varying types of food storage facilities (i.e., large storage vessels, pits, or silos), and development and utilization of new, alternative, or a diverse number of food processing and agricultural tools can all be used to interpret risk reduction strategies as well (Table 1). Where Çadır Höyük is concerned, the changing size of the settlement through the ages can be monitored relatively well, implying shifts in population size during times of socio-economic transformations, and food storage facilities also provide good evidence. However, food processing and agricultural tools still need to be studied in further detail to provide sound evidence that would support our interpretations and contribute to our understanding of local adaptive strategies.

Table 1. A list of climatic changes that occurred during Çadır Höyük's occupation, with a selected list of possible risk management strategy responses.

Climatic Change	A Selection of Possible Risk Management Responses	Effected Periods of Çadır Höyük
Increased Aridity	***Architectural*** - Site contraction ***Paleoethnobotanical*** - Shift types of crops away from water dependent crops (e.g., wheat and grapes) - Shift types of cultivars used for food and fodder (e.g., barley is eaten as a food instead of used as fodder) - Expansion of land used for feeding animals ***Zooarchaeological*** - Shift away from highly water-dependent species (e.g., cattle and pigs) - Increased dependence on goats - Shift to a less centralized animal management strategy (e.g., focus on household animal management) - Increased animal provisioning through grazing instead of foddering	• Transition between Late Chalcolithic and Early Bronze Age • Transition between the Early Bronze Age II and III • Transition between the Late Bronze Age and Early Iron Age
Increased Humidity and Rainfall	***Architectural*** - Site expansion ***Paleoethnobotanical*** - Focus on water-dependent crops (e.g., wheat and grapes) - Intensified cultivation (e.g., seasonal crops for multiple growing seasons per year) - Increased foddering - Expansion of land used for cultivation ***Zooarchaeological*** - Intensive strategies that focus on water dependent species (e.g., cattle and pigs)	• Late Roman Period • Middle Byzantine Period

In general, risk management strategies are often discussed in relation to agricultural systems due to the fundamental link between human survival and food. Agricultural systems encompass, contribute to, and shape many more sectors than just foodways. Agricultural risk management strategies usually take one of three forms: 1) diversification that focuses on using and creating multiple different crops, animals, and animal products; 2) intensification strategies that make it possible to create many more agricultural products than previously; and, 3) extensification strategies that use much more land for agricultural output (Marston 2011; Boserup 1965). These strategies can take multiple forms depending on a variety of factors—the political or economic structure of the population, the type of upheaval, and social taboos—and the risk management strategies for cultivation or animal management do not have to be the same or employed at the same time (Marston 2011; Marom et al. 2019). In the following case study, the paleoethnobotanical and zooarchaeological assemblages will be examined and assessed for evidence of the adoption of risk management strategies, or not, based on environmental conditions for the LCh/EBA transition, the LBA/EIA transition, and the Late Roman and Middle Byzantine periods.

The Late Chalcolithic and Early Bronze Age at Çadır Höyük

Our team revealed a continuous stratigraphic sequence with well-defined architectural features during the excavations at the southern and southeastern lower slopes of the mound, which span the end of the Middle Chalcolithic and beginning of the Late Chalcolithic (LCh) to the beginning of the EBA.

The earliest phase of occupation on the mound is what our team refers to as the "Agglutinated Phase" (3800–3600/3500 BCE), which is divided into two sub-phases (Fig. 3; Steadman, McMahon, and Ross 2019, 35–36). This earliest phase is marked by a semi-subterranean building complex consisting of small, single-celled rooms surrounding an internal courtyard, with some rooms containing built-in storage bins and with infant jar burials placed within the walls during construction (Steadman and McMahon 2017, 97–99). The next phase of occupation (3600/3500–3300 BCE) is represented by the "Burnt House and Courtyard," a two-roomed domestic structure (H House), and one non-domestic building which seems to have had a ritual function (Steadman et al. 2008, 2013; Steadman and McMahon 2017, 96–97). This phase was constructed on a mudbrick layer with a deposit rich in pottery, lithics, and animal bone, and also containing the burial of a young infant (Steadman, McMahon, and Ross 2019, 35). The "non-domestic building" contained a semi-circular mudbrick feature with a posthole in its center and small holes with seed deposits around it, two head-sized stones, another posthole, and three circular depressions, two of which had infant/child burials underneath (Steadman and McMahon 2015, 79–81).

A pathway and a platform were unearthed at the western side of the excavated area, which could be dated to 3600–3300 BCE, and a feature composed of mudbrick,

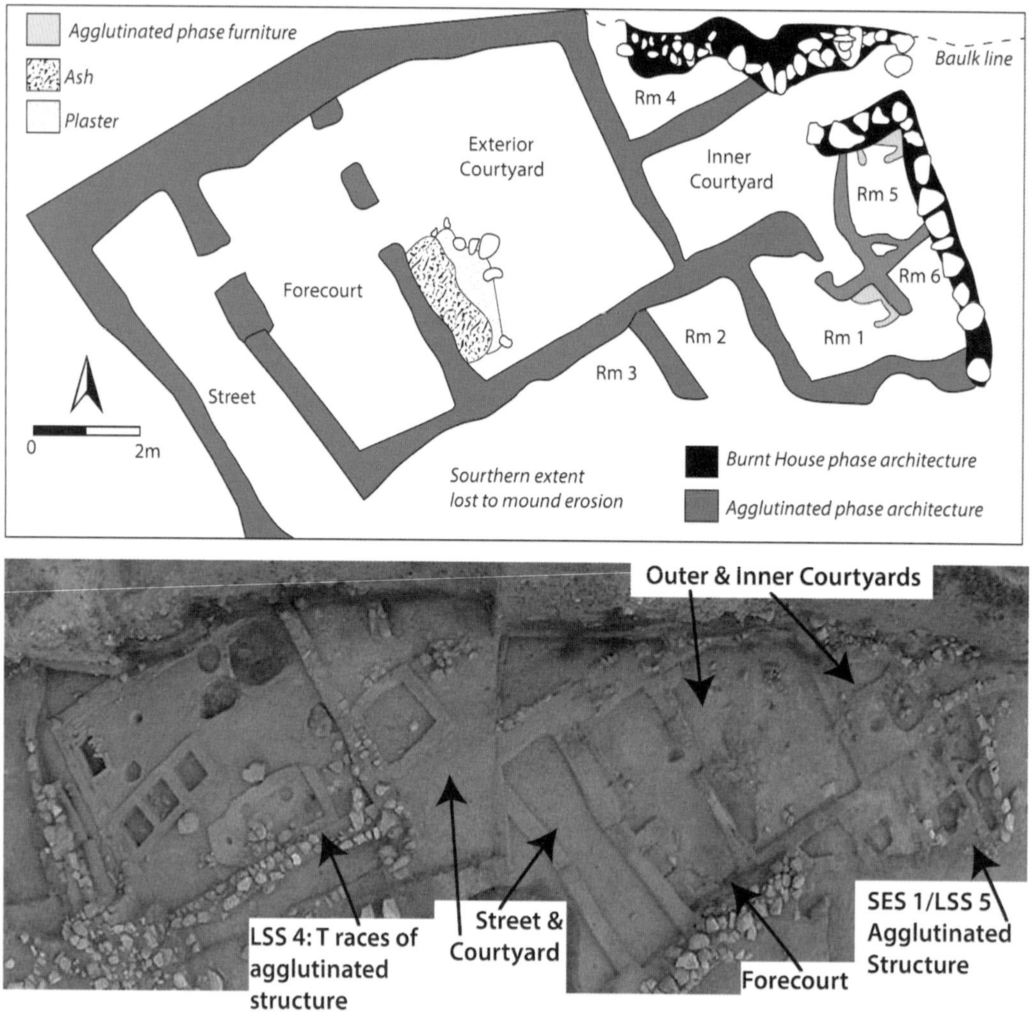

Fig. 3. The architectural plan and an aerial view of the fourth-millennium BCE "Agglutinated Phase."

stone, and plaster was found just to the west of this area (Fig. 3; Steadman and McMahon 2015, 74). A public building with a function associated with pottery manufacture or distribution, based on the high numbers of LCh pottery it contained and the lack of domestic material, was located further to the east and was named the "Omphalos Building" (Steadman et al. 2008). The lowest phase of this building, which is believed to have at least four major floor layers, contained a bull-headed, square ceramic object which may have been a portable hearth—possibly based on Transcaucasian designs—a fruit-stand, a partially surviving storage bowl, and broken, flattened pottery sherds on top of walls, which may have been connected to pottery production in this area (Steadman and McMahon 2017, 100). An enclosure wall and a gate system stretched along the southern and southeastern sides of this LCh complex.

The very end of the LCh (3300–3200/3100 BCE) is marked by a courtyard with three large pits, a kiln, and tools and material found with these that were most probably used for producing pottery. The "Omphalos Building" was still in use during this period, sharing this space with three semi-subterranean structures, of which two had apsidal plans, and had a hearth and a storage bin nearby (Steadman et al. 2013; Steadman and McMahon 2015, 74, 78).

On the slope above the architecture just mentioned lay the LCh to EBA transitional level. This area is represented by small rooms next to an open courtyard, with some containing benches during the earlier phase, and a heavily damaged single-roomed structure, open fire-pits, and an open courtyard during the later phase (Steadman et al. 2008; Steadman and McMahon 2015, 73–74; 2017, 102–3). It seems that the settlement, which we understand to have initially had a terraced lower and upper slope architectural plan with terraces linked by a series of stepped platforms (Steadman, McMahon, and Ross 2019, 36), grew smaller in size and was retracted up the slope, with relocation of ceramic production areas further into the settlement on the south side. Exposed areas belonging to the early phases of the EBA had an industrial function with large ovens and a substantial mudbrick wall, which seems to have served as a perimeter wall encircling the settlement (Figs. 4–5; Steadman and McMahon 2015, 83). The excavated piece of a fruit-stand and animal figurines are similar to examples unearthed at Alişar Höyük (von der Osten 1937, 68: fig. 75: e1617; 79: fig. 85: e2025), reflecting regional contacts.

Paleoethnobotanical Evidence

The most detailed paleoethnobotanical study of an assemblage at Çadır Höyük has been conducted on the LCh assemblage from three architectural units: the Burnt House and Courtyard, the Omphalos Building, and the Southern Courtyard (Steadman et al. 2019b; von Baeyer 2018; von Baeyer, Smith, and Steadman 2021). The study looked at 60 samples from these three areas dating from about 3500–2900 BCE. The samples originate from a variety of contexts, including pits, pyric features, surfaces, and activity areas. Descriptive (e.g., ubiquity, relative abundance, and ratios) and multivariate (e.g., correspondence analysis) statistics were used to find patterns in the paleoethnobotanical assemblage. The raw counts for this project are available for download through OpenContext (von Baeyer 2020).

This analysis highlighted multiple differences between the plant assemblage in each architectural area, as well as a clear shift in agricultural practices through time. The final two phases of the Omphalos Building did not provide any evidence of clear, sustained plant use; the paleoethnobotanical assemblage reflected the background signature of plant use found throughout the site. In contrast, the paleoethnobotanical signatures from the Burnt House and Courtyard and the Southern Courtyard provided clear evidence for plant use (von Baeyer, Smith, and Steadman 2021).

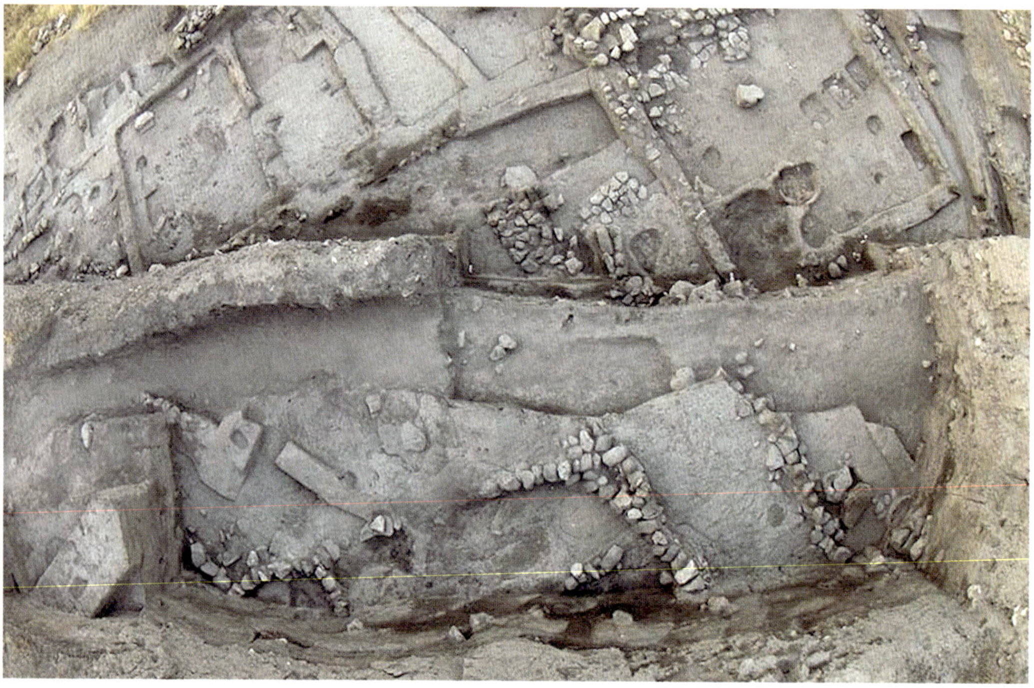

Fig. 4. An aerial view of the Chalcolithic to Early Bronze Age transitional level. North at bottom.

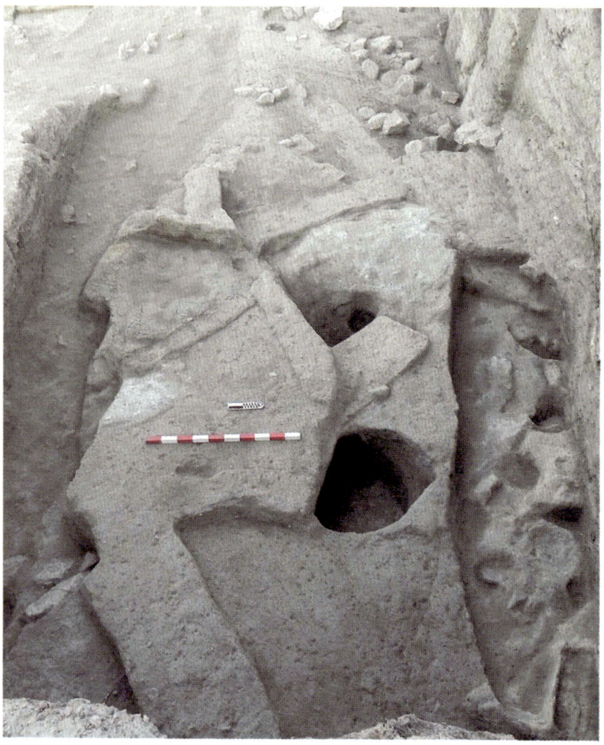

Fig. 5. A view of the area where the earliest Early Bronze Age level was exposed.

The paleoethnobotanical signature of the LCh crops was composed mainly of barley (*Hordeum* sp.), wheat (*Triticum* sp.), lentils (*Lens culinaris* Medik.), bitter vetch (*Vicia ervilia* [L.] Willd.), and a small amount of flax (*Linum usitatissimum* L.). Most of the wheat and the wheat chaff were identified as emmer (*Triticum turgidum* L. spp. *dicoccum* [Schrank] Thell.), with smaller amounts of einkorn (*Triticum monococcum* L.), New Glume Wheat, and free threshing wheat (*Triticum aestivum/durum*) in the assemblage, especially from the Burnt House and Courtyard. Two caches of crops were preserved in the Southern Courtyard, probably as an artifact of a courtyard fire: an *in situ* cache of lentils preserved in a cooking pot, and a cache of barley grains from a secondary refuse deposit (von Baeyer, Smith, and Steadman 2021). The amount of cereal chaff remains (primarily hulled wheat) from the two courtyards suggests that cereal was processed on-site, although no clear crop-processing area was identified (Graham and Smith 2013).

A comparison between the plant signature of the Burnt House and Courtyard and the Southern Courtyard revealed differences in how the population at Çadır Höyük used each space vis-à-vis plant remains. Overall, the function of each space was the same—both were used as outdoor workspaces—although the material and architectural remains associated with the Burnt House and Courtyard mark it as a more elite space than the Southern Courtyard (Steadman et al. 2013, 2015). It was determined that the paleoethnobotanical samples recovered from both courtyards were most probably from dung fuel, except for the caches mentioned above. (It is exceedingly tricky to determine the sample origins for paleoethnobotanical remains, but for a detailed discussion on how this determination was made, see von Baeyer 2018.)

The comparison of paleoethnobotanical samples between the courtyards identified two differing trends in each courtyard: 1) how ashes and refuse from hearths were treated (see Steadman et al. 2019a) and 2) evidence for foddering or grazing animals (see von Baeyer, Smith, and Steadman 2021).

In the Burnt House Courtyard, hearth samples had very few plant specimens and had low species diversity, whereas pit samples had large numbers of specimens and high species diversity. This suggests that the refuse from the hearths was systematically and deposited into the pits in the courtyard. In contrast, in the Southern Courtyard, there was no difference in specimen abundance or species diversity between hearths and pits. Therefore, we cannot assume that the same hearth cleaning took place in the Southern Courtyard (Steadman et al. 2019a).

The second major difference observed between the Burnt House and Courtyard and Southern Courtyard paleoethnobotanical samples was in the weedy plant assemblage (von Baeyer, Smith, and Steadman 2021). Correspondence analysis performed on the weedy species indicated that the grass weeds found in the Burnt House and Courtyard were primarily large grasses, over 1 mm in length, like *Hordeum spontaneum*, *Bromus* sp., and *Lolium perenne*. These grasses are primarily found in agricultural fields (Zohary 1973). In contrast, the samples from the Southern Courtyard were much more likely to contain the very small, less than 1 mm grasses, like *Phleum arenarium*, *Phleum boissieri*, *Phleum exeratum*, *Phleum phleoides*, and *Phalaris* sp. These grasses are not associated with agriculture, but rather grow in fallow fields or away from cultivated fields entirely (Zohary 1973). Since the paleoethnobotanical signature from Çadır Höyük comes from dung fuel, this difference in grass species suggests that there is a difference between what animals were fed between the Burnt House and Courtyard and the Southern Courtyard. The dung from the Burnt House Courtyard has a signature of animal management of on-site feeding, either through foddering or grazing in local agricultural fields. The animal management signature of the dung in the Southern Courtyard reflects off-site grazing, meaning animals were taken away from Çadır Höyük for periods of time to feed (von Baeyer, Smith, and Steadman 2021).

Zooarchaeological Evidence

Benjamin Arbuckle has conducted an extensive analysis on the LCh animal remains from Çadır Höyük (Arbuckle 2009). He found that during the earlier Burnt House and Omphalos Building phases, the economy of animal products was focused on sheep and likely included wool production. Caprines (sheep and goats) represented 65% of the economically important taxa, and sheep (*Ovis aries*) outnumbered goats (*Capra hircus*) almost 2 to 1 (Arbuckle 2009; Steadman et al. 2019a, 2019b). In contrast, pigs (*Sus* sp.) represented only 15% of the assemblage (Steadman et al. 2019b), although the seemingly intentional burial of a two-year-old pig under the floor of the non-domestic building suggests that pigs were culturally important (Steadman et al. 2019a). Cattle (*Bos* sp.), particularly older females, were an integral part of the agricultural system as well and were used for multiple agricultural tasks and secondary products (Arbuckle 2009; Steadman et al. 2019a). At the end of the LCh around 3100 BCE, the animal management strategy at Çadır Höyük is characterized by a more even distribution of livestock species and decrease in wild taxa. Caprines decrease in abundance (52%), while pig (21%) and cattle (23%) increase substantially. Furthermore, instead of sheep dominating the caprine assemblage, goats were more numerous. A shift towards evenness between animal species can be seen as a move towards risk reduction. The increase in pigs is often hypothesized as a strategy to stabilize household production, and the shift to goats is seen as herding risk reduction. The increase in cattle, however, is not a risk reduction strategy due to the high resource cost of raising the species (Dahl and Hjort 1976; Steadman et al. 2019b; Zeder 1991).

The Second and First Millennia BCE at Çadır Höyük

The earliest phase of the second millennium BCE is represented by the MBA defensive wall on the eastern slope, on which first a defensive wall was constructed in the mid-second millennium BCE and then a more substantial one was built during the later phases of the LBA (Figs. 6–7; Steadman et al. 2013; Steadman and McMahon 2015, 75, 84; Steadman, McMahon, and Ross 2019, 38–39). LBA domestic structures and connected

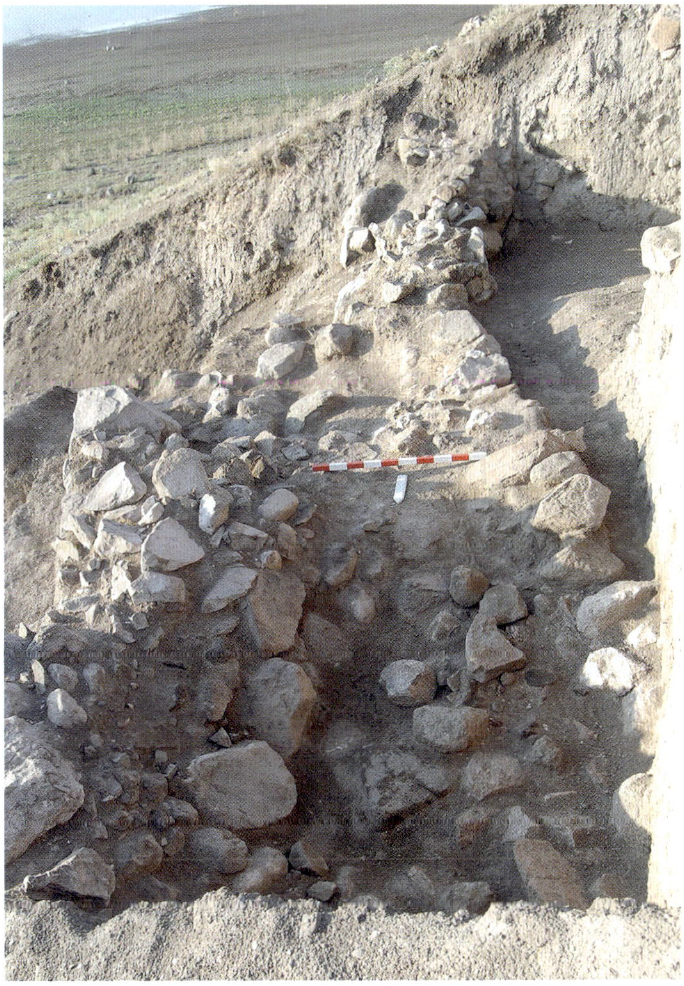

Fig. 6. The earlier Late Bronze Age defensive wall, just below and near the later and more substantial wall.

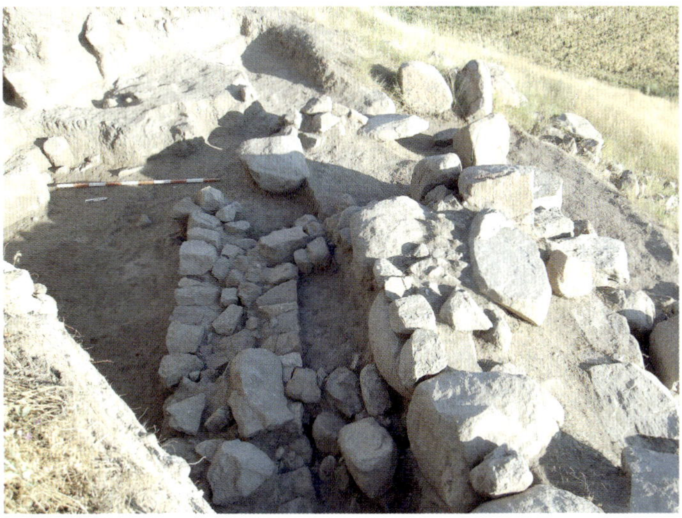

Fig. 7. The later Late Bronze Age casemate fortification wall.

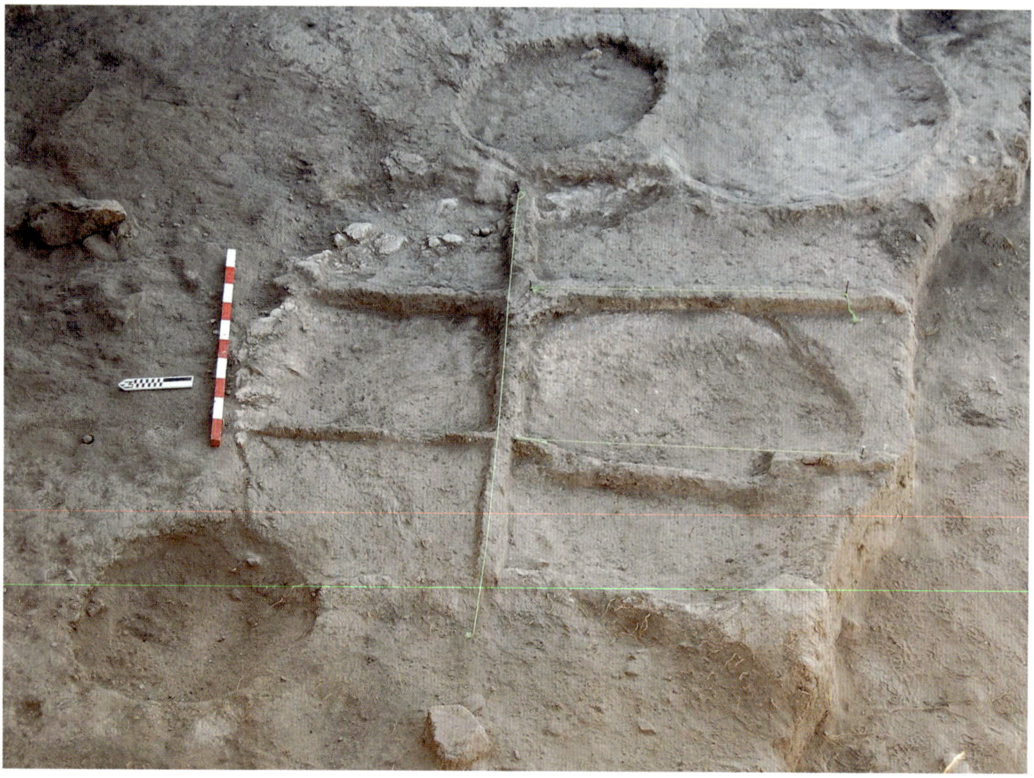

Fig. 8. Late Bronze Age to Early Iron Age transitional period open-air workspaces.

open courtyards were unearthed further up the slope (Gorny 2006, 49–50; Steadman and McMahon 2015, 84–86; 2017, 103).

Excavations on the upper southern slope revealed occupational phases that could be dated to the very end of the LBA. The area had an industrial function, indicated by a metal smelting installation associated with ash deposits, and was divided into smaller sections by mudbrick walls that seem to have been built after the Hittite defensive wall on this part of the mound was demolished (Fig. 8; Steadman and McMahon 2017, 105–6; Steadman, McMahon, and Ross 2019, 40–41).

A full Iron Age sequence was revealed on the upper southern slope of the mound, thanks to a deep trench excavated there (Steadman and McMahon 2015, 76). The Early Iron Age and most of the Middle Iron Age periods are represented by numerous pits, which vary in size, with multiple phases of plastering and usually with a mud lining (Steadman et al. 2013). Some of these were very shallow and may have been used to hold liquids, and some may have actually been a part of circular and semi-subterranean structures (Ross 2010; Steadman and McMahon 2015, 88–89). The excavated areas also had depressions to hold pots or tools, and some mudbrick-lined areas may have been used for heating materials.

The Middle Iron Age level exposed on this side of the mound had an architectural complex with stone foundations, mudbrick features, and thickly-plastered depressions (Steadman and McMahon 2015, 76). This was identified as a possible textile production area, based on the amount of painted and undecorated ceramics, spindle whorls, and unbaked loom weights, and it seems to have continued to be used for this purpose during the early Late Iron Age.

The Late Iron Age at Çadır Höyük is represented by small workspaces and a structure possibly belonging to a Late Iron Age gate (Steadman et al. 2013). A Late Iron Age wall, a room, and an open, paved courtyard were found together with a two-roomed complex located at a slightly lower elevation at the eastern upper edge of the mound, and the extensive wall excavated at the western side of the mound may also belong to this period (Steadman et al. 2017, 228–29). A stone-paved pathway with a rectilinear room just to the south of the complex and two mudbrick-walled rooms used for storage or for daily activities were exposed at the northern upper edge of the mound, both areas probably providing access to the inner town (Fig. 9; Steadman and McMahon 2015, 89–90; 2017, 108; Steadman, McMahon, and Ross 2019, 42). A staircase built during this period seems to have linked the lower slopes to these higher areas to enable access.

Çadır Höyük also appears to have been occupied during the Hellenistic period, based on the existence of pottery from this era, but the later Byzantine construction activities on top of the mound seem to have mostly destroyed this level. A very fine patio covered by mudbrick with a wooden post nearby, a set of stairs linked to a stone pavement that may have provided access to the inner town, and the remains of a wall with an associated packed-mud surface at the upper east and north edge of the mound are more or less the only architectural features that could be clearly dated to this period (Steadman and McMahon 2017, 109–10; Steadman, McMahon, and Ross 2019, 42, 45). A Hellenistic roof tile was found placed on the ground just near the western end of the wall, together with a grinding stone.

Paleoethnobotanical Evidence

Far fewer paleoethnobotanical samples have been analyzed from the LBA and Iron Age than from any other period. Due to research questions, the paleoethnobotanical analysis from these periods focuses on general trends that span the LBA/Early Iron Age (EIA) transition without detailed work on any specific period. The paleoethnobotanical evidence for each of these periods, as well as for the Middle and Late Iron Ages, has simply not been analyzed yet. This will be the focus for future research.

Overall, the LBA/EIA crops are very similar to the LCh ones. The assemblage contains wheat, barley, lentils, pea, and bitter vetch (Smith 2007; Ross et al. 2019a). There is some evidence for crop storage during the LBA based on architectural evidence, but no paleoethnobotanical analysis has confirmed this hypothesis (Gorny et al. 2002; Ross et al. 2019a). Wood-dominated second-millennium BCE hearths and pits indicate the use

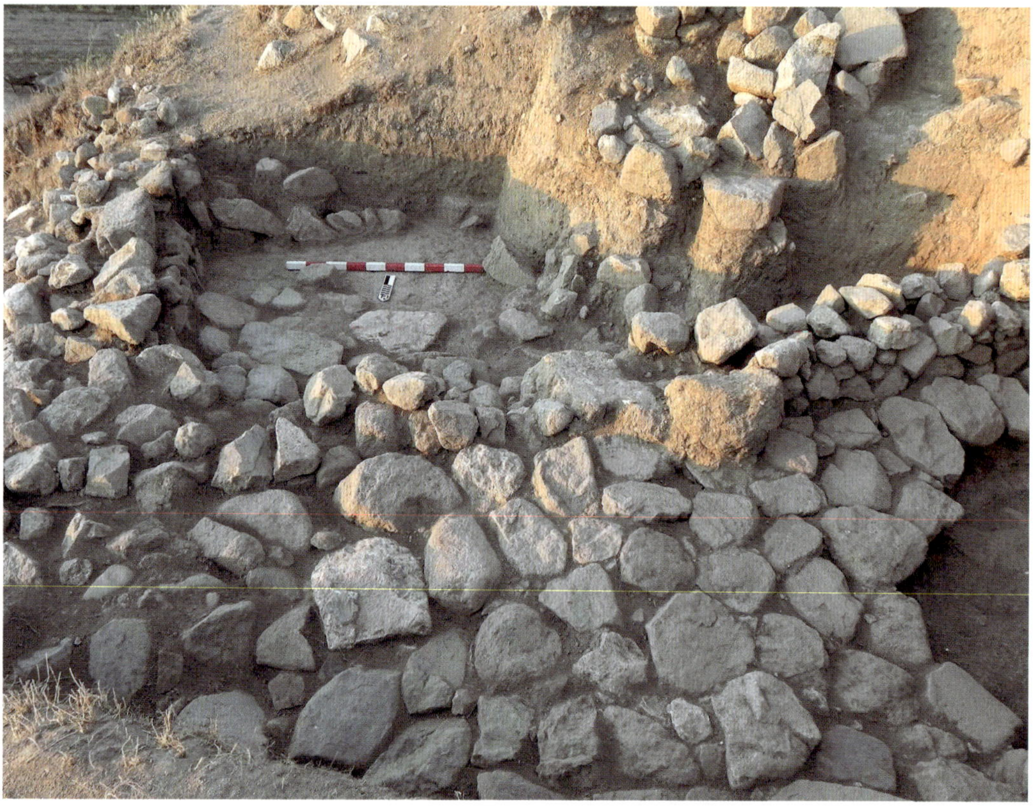

Fig. 9. Late Iron Age stone-paved pathway and the small rectilinear room nearby.

of wood fuel. In contrast, oven deposits contained proportionally more chaff specimens and weeds than hearths. This suggests that ovens used dung fuel to cook, possibly to prepare bread, and hearths used wood fuel, possibly to heat areas (Ross et al. 2019a). Future research into the differing fuel sources and functions is planned. One unique feature of the second and first millennium BCE paleoethnobotanical assemblage is the large wood charcoal assemblage from plaster-lined pits in USS 4 (Ross et al. 2019a). The function of these pits and the ecology signature of the wood charcoal is unknown, but it is a research objective for the future.

Zooarchaeological Evidence

An extensive zooarchaeological analysis by Sarah Adcock has identified multiple areas of change and continuity in animal management practices for caprines, pigs, and cattle in the second and first millennia BCE (Ross et al. 2019a, 2019b). Over 6,500 animal bone specimens from the LBA through the Late Iron Age have been identified (Ross et al. 2019a). The overall animal management trends of the second and first millennia BCE show the following: 1) the proportion of caprines increased from the middle of the LBA

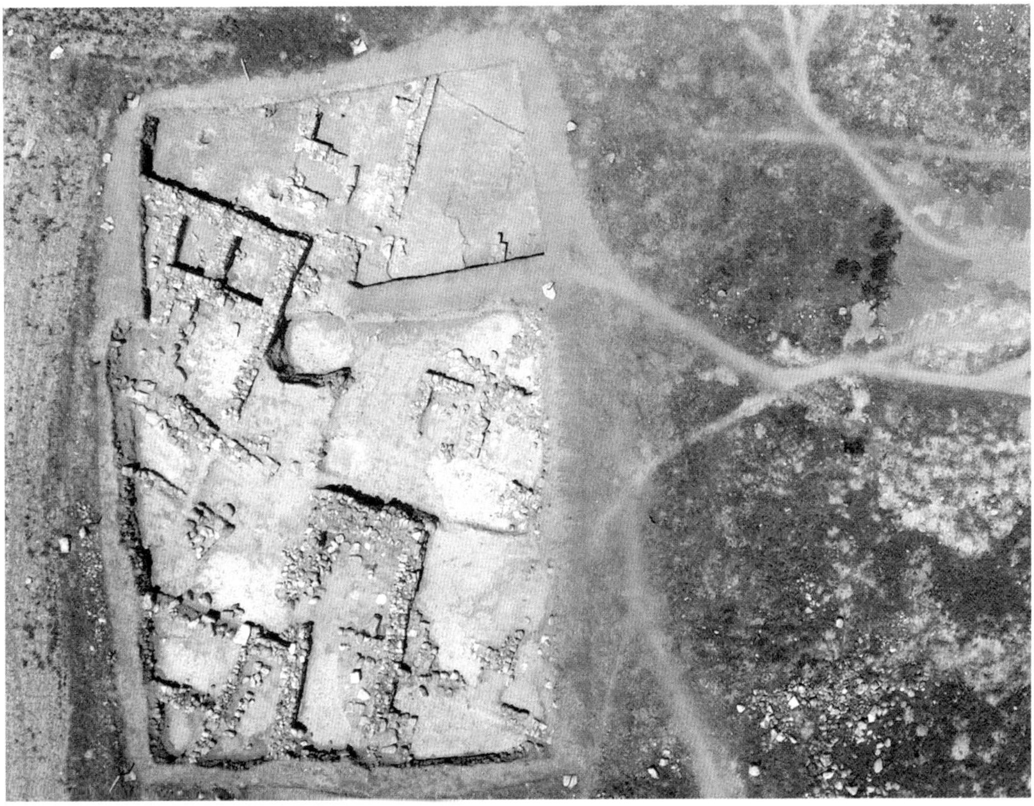

Fig. 10. An aerial view of the North Terrace showing the Late Roman and Byzantine architectural remains.

through the EIA and 2) the proportion of pigs and cattle decreased at the very end of the second millennium BCE. These trends are consistent throughout three quantitative analyses: number of identified specimens, weight in grams, and diagnostic zone (Ross et al. 2019b, 34).

Caprine management in the second and first millennia BCE suggests risk-adverse strategies throughout the period but with markedly different signatures (Adcock 2020). Caprine management during the LBA focused on smaller female sheep, with sheep outnumbering goats by 1.7 to 1 and caprines in general kept well into adulthood, which suggests that wool and fiber were not the primary focus of the local economy. During the EIA, the signature of the caprine assemblage shifted notably. The number of sheep and goats were relatively equal, and more caprines were killed at between two and six months of age. This signature suggests a more generalized animal management strategy was adopted (Adcock 2020). Throughout the LBA and EIA, cattle management remained focused on adult animals and secondary products, with bone weight calculations suggesting that cattle were an important, if not primary focus of the economy, especially during the LBA. Cattle sizes only began to decrease in the Middle Iron Age (Ross et al. 2019a).

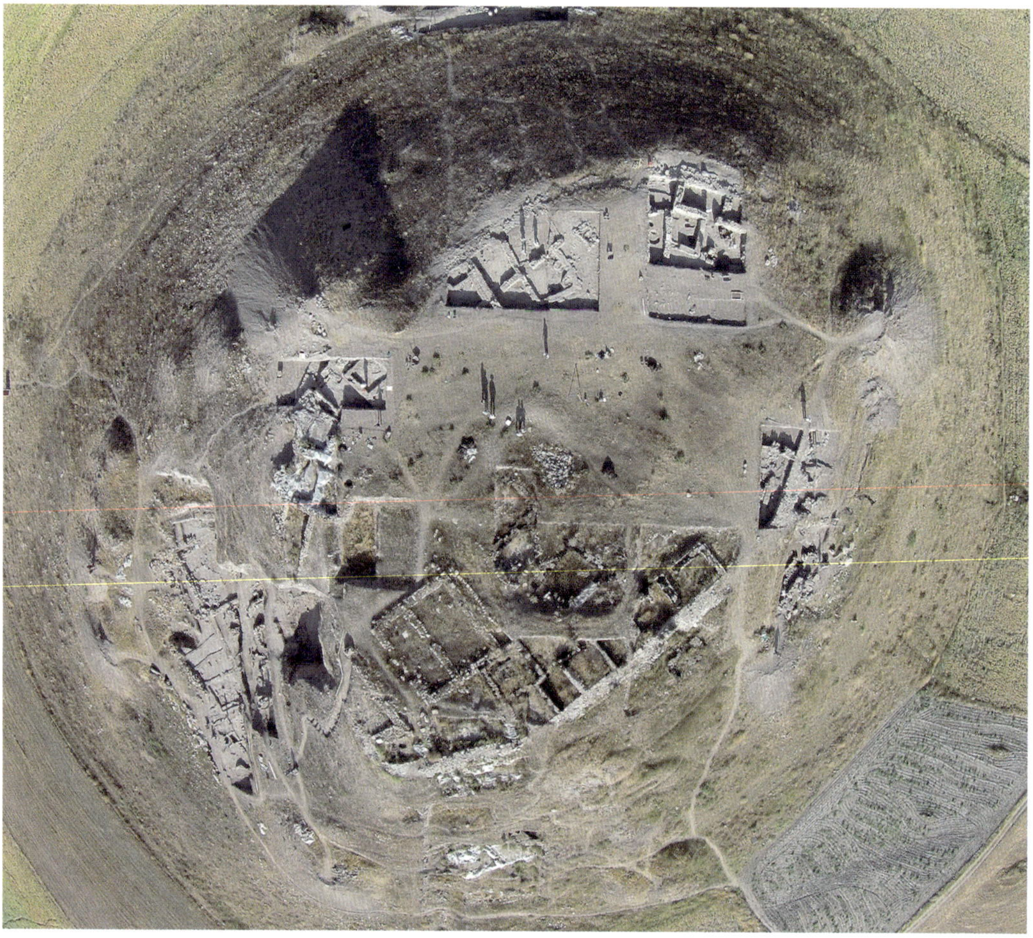

Fig. 11. An aerial view of the mound summit showing the full extent of the Late Roman and Byzantine architectural remains. North at right.

Late Roman and Byzantine Periods at Çadır Höyük

The domestic structure on the north terrace seems to have been first built in the fifth to sixth centuries CE as a spacious, well-built, multi-room residence of a local elite family probably managing the land around this part of the site (Steadman and McMahon 2015, 91, 93). After being abandoned, it was refurbished by dividing large rooms into smaller ones and the introduction of an area we identify as a large kitchen, suggesting that the structure functioned as the residence of several cooperating families from the local farming community until it was abandoned in the eleventh century CE (Fig. 10; Steadman and McMahon 2015, 77, 94).

The summit of the mound reflects a stratigraphic sequence starting with the construction of several structures during the Late Roman period (fourth–seventh centuries CE) and later during the Early Byzantine period (seventh–ninth centuries CE;

Fig. 11; Steadman, McMahon, and Ross 2019, 46). The Late Roman Period is represented by a gate system on the southern edge of the summit, which was modified and used during the Early Byzantine Period, and a stone path or platform excavated nearby that may be a part of the foundations of the earlier defensive wall. The gate was left in place and was in use during the Middle Byzantine Period, although this was not the main gate. The main gate of the tenth-century CE fortification walls that provided access to the settlement was located 30 meters to the east, and it was finally blocked before the brief abandonment period in eleventh century CE.

The massive defensive wall, which had at least two towers and a gate, seems to have been built during the Middle Byzantine Period, probably in the tenth century CE (Steadman and McMahon 2015, 94). The area inside the wall contained a large storage building, a stable, an open courtyard, and a series of rooms, most with packed mudbrick floors built against the fortification wall (Cassis 2009; Steadman and McMahon 2015, 77; 2017, 110–11).

The summit seems to have been occupied by more people, as indicated by the buildings unearthed in these trenches. The summit buildings include the following: a chapel, a second storage building, and multiple buildings with residential purposes constructed hastily before the abandonment some time in the eleventh century CE. Finally, the area may have been used as a Byzantine graveyard during a brief period of abandonment, as indicated by human skeletal remains found at this level (Steadman, McMahon, and Ross 2019, 48–49). The increased level of building activities on the summit just before the abandonment is an implication that local inhabitants were finding it difficult or dangerous to continue to occupy their houses and used the main chapel located outside the fortification walls at a time when Seljuk forces were conducting military raids in this part of Anatolia.

In some of the rooms attached to the defensive wall, metal objects were unearthed that look Seljuk in their general appearance. This is an indication of the final phase on the mound summit that follows the brief period of abandonment is most likely dated to the Seljuk period. During this phase, walls of rooms were rebuilt with mudbrick laid over earlier stone walls. We hypothesize that the settlement may have continued to be inhabited as late as the thirteenth century CE (Steadman and McMahon 2017, 112).

Paleoethnobotanical Evidence

The paleoethnobotanical evidence from the Late Roman and Byzantine occupation at Çadır Höyük was comprised of a preliminary study of 12 paleoethnobotanical samples, ten from the summit and two from the terrace (Cassis et al. 2019). The crop assemblage was consistent with what one expects from a Late Roman and Byzantine settlement and with earlier phases at Çadır Höyük. Cereal cultivation of wheat, barley, and rye (*Secale* sp.) continued to be the primary cultivation practice. The high ubiquity and relative proportions of chaff specimens, primarily bread wheat (*Triticum aestivum* L.), barley rachises, and culm fragments, attested to the importance of cereal cultivation at the site. Non-cereal crops included grapes (*Vitis* sp.), chickpeas (*Cicer arietinum* L.), lentils, and flax. The weedy assemblage from these samples consisted primarily of weeds associated with agricultural fields—*Bromus* sp., *Asperula* sp./*Galium* sp., and small *Lolium* sp.—and *Bolboschoenus glaucus* from the wetlands (Cassis et al. 2019). Although further analysis has to be done to confirm the origin of these samples (i.e., dung fuel or plant processing), the samples are strongly characteristic of dung fuel. If so, the weedy assemblage from the Late Roman and Byzantine phases suggests that animals were foddered or grazed on-site in harvested fields (Cassis et al. 2019).

Zooarchaeological Evidence

Over 6000 Late Roman and Byzantine animal bone specimens have been analyzed from Çadır Höyük. The zooarchaeological assemblage from the North Terrace represented the bulk of the evidence from the Late Roman and Early Byzantine periods. This assemblage was primarily composed of domestic livestock, mainly sheep, goat, cattle, and pig, with fewer numbers of fowl, horse, and donkey (Cassis et al. 2019). Using the number of identified specimen count, caprines made up slightly more than one third of the domestic livestock assemblage, with sheep slightly outnumbering goats seven to five, suggesting that wool production was a part of the agricultural economy. Cattle and pigs were represented in roughly equal numbers, with slightly over 20% each. In contrast to the number of identified specimen counts, the bone weight calculations of the assemblage were heavily dominated by cattle, indicating how important cattle were to agriculture, even if their numbers did not dominate the livestock assemblage (Cassis et al. 2019).

The assemblage from the summit dated to the Middle Byzantine period and showed a different livestock signature than that of the Late Roman assemblage (Cassis et al. 2019). Cattle and caprines each accounted for roughly a third of the assemblage in the early Middle Byzantine period, while pigs accounted for only 12% of the assemblage. The relative abundance of caprines decreased throughout the Middle Byzantine period, while the relative abundance of pigs remained the same. An unusual mass casualty event at the very end of the occupation of the summit in the eleventh century CE provided a rare look at a "living herd" assemblage. This assemblage was dominated by cattle remains at 57% of livestock specimens identified to genus, 22% caprines, 11% pigs, 1.6% fowl, and 0.6% equids. The cattle assemblage from the living herd was composed of adult male and female animals and a few calves. The composition of this cattle herd suggests that at the end of the Middle Byzantine period, cattle was the primary livestock supporting an agricultural system designed for both local needs and external markets (Cassis et al. 2019).

Cultural Resilience in the Face of Climate Fluctuations: Paleoethnobotanical and Zooarchaeological Evidence

Elements of the paleoethnobotanical and zooarchaeological evidence from the LCh/EBA transition, the LBA/EIA transition, and the Late Roman and Byzantine periods suggest that plant and animal management strategies could have been shaped in part by the climate of those periods.

The relationship between the climate and the LCh paleoethnobotanical assemblage has been previously published (see von Baeyer 2018; von Baeyer, Smith and Steadman, 2021). In short, the shift from foddering and on-site grazing to off-site grazing occurs around 3200 BCE, which corresponds with the beginning of the 5.2 ka BP rapid climate event that brought a period of markedly increased aridity to central Anatolia. Increased aridity would have reduced agricultural yields, particularly wheat, and would have made plant cultivation a more precarious activity. Consequently, the amount of food produced through cultivation would have reduced or decreased in quality. Ethnographic work has observed that when plant cultivation is stressed, the categories of plants eaten as food is expanded (Jones 1998). By taking animals off-site to graze, the amount of on-site plants available for human consumption would have increased. Furthermore, expanding the grazing area for herds was, and still is, a common risk-reduction strategy in southwest Asia (Marston 2011). By increasing the number of wild species fed to animals, the overall food procurement strategies at Çadır Höyük became more stable, because the types of food that were eaten were diversified (von Baeyer 2018). The zooarchaeological evidence from the LCh at Çadır Höyük supports this hypothesis of risk reduction as well. At the end of the LCh, the animal profile shifted away from intensive wool production to focus on individual household livestock strategies reliant on pigs, cattle, and goats, a shift that has been hypothesized to be a risk-reduction strategy (Steadman et al. 2019a).

With the expansion of the food base at Çadır Höyük, social mores around plant refuse also were relaxed. Hearths were no longer cleaned as rigorously or frequently. This shift could correlate to a shift in labor organization at Çadır Höyük, because rigorous control over where plants were processed and used has been linked to centralized control (Hastorf 1991). For the case of Çadır Höyük, a relaxing of plant use areas could suggest a decentralization of the site. These two trends indicate that the population at Çadır Höyük was able to maintain cultural resilience during a period of aridity that strained their cultivation by relaxing a number of behaviors around plant and animal management. This flexibility allowed farmers to continue to practice agriculture, albeit in a slightly transformed manner, and with this continuity of agricultural practices, the culture at Çadır Höyük continued.

The evidence from subsequent periods on how climate change impacted agricultural practices at Çadır Höyük has not been studied as extensively, but a few trends in the zooarchaeological record could have been linked to the climate. Around the end of the LBA, another rapid climate event that brought in a period of aridity occurred, known as the 3.2 ka BP event. Before this event at Çadır Höyük, the animal management strategy focused on intensive wool and fiber production from sheep, whereas after the collapse and at the beginning of the climate event, the caprine management strategy shifted towards goats (Ross et al. 2019b, 34). Goats are better adapted to arid environments, because they have less need for water than sheep, cattle, or pigs, the herds grow quickly, and they are more effective milk producers than sheep (Dahl and Hjort 1976; Marom et al. 2019). These qualities also make goats a better species to rebuild animal management practices. Rebuilding animal herds might have been necessary during the EIA after the collapse of the Hittite Empire. At the moment, it is impossible to say whether or not changes in subsistence strategies in the second and first millennia BCE are linked to cultural or climate changes or to both. This will be the focus of future research.

The zooarchaeological assemblage of the Late Roman and Byzantine periods of occupation at Çadır Höyük also reflect what one would expect based on the climate data. These are the only periods, so far, that show a clear reliance on cattle for agricultural practices. Cattle have large resource and water needs (Dahl and Hjort 1976; Marom et al. 2019), and, based on paleoclimate reconstructions, the Late Roman and Middle Byzantine periods were generally periods of increased rainfall and humidity (Dean et al. 2015). In the paleoethnobotanical assemblage, these periods are the first to have evidence for grape cultivation; this is significant because grape has relatively high water needs (Zohary and Spiegel-Roy 1975). These findings, however, are still very preliminary, and future research has to be done to fully explore the relationship between climate and agricultural practices in the Late Roman and Byzantine periods.

It is notable that, overall, that agricultural practices seem to have reacted simultaneously to cultural and environmental factors when necessary. This speaks to the entrenched nature of agricultural practice in cultural resilience. Agriculture requires certain behaviors and rhythms from farmers that over time create an identity and a culture (van der Veen 2014). Being able to maintain the behavior at the core of the identity, even when it is threatened by climate change or cultural change outside one's control, is the key to resilience. This could be why agriculture never faltered over 6,000 years of occupation at Çadır Höyük: it was literally the linchpin of society.

Conclusion

Based on the archaeological evidence, the communities inhabiting Çadır Höyük from the Middle Chalcolithic until the Seljuk period managed to adapt to changing regional climatic and environmental changes relatively well to secure survival with certain changes in agricultural and animal-husbandry practices. Although agricultural practices and choice of crops were not altered extensively, the choice of animals varied greatly depending on the conditions of the specific period.

Even though local inhabitants adopted different ways to adapt to cultural and climatic change, the socio-economic system during the three major aridification events seems to have been stressed. The archaeological evidence clearly demonstrates that the settlement started to shrink in size at the end of the LCh period, which corresponds to the 3300–3000 BCE (5.3–5.0 ka BP) climatic event. Although our data for the end of the EBA is limited, we understand that the settlement was still inhabited but was more or less a small village with a low population, so the second period of aridification (4.5–3.9 ka BP) also seems to have had a direct impact on the local socio-economic system and lifestyle.

Çadır Höyük was located at the heart of the Hittite political system, and its construction projects, increasing size of the settlement, and changes in subsistence strategies may have been a result of direct political intervention up until the collapse of this political organization. The settlement seems to have shrunk in size during the third aridification period (3.2–2.8 ka BP), and what we observe is a site that is limited to the summit of the mound, with industrial areas surrounding the residential quarter. The settlement area may have slightly expanded during the Middle and Late Iron Age, as manifested by the construction of new buildings and a substantial wall that may have had a defensive purpose.

Çadır Höyük seems to have gradually transformed into a larger rural center during the Late Roman period, although the process may have started already at the end of the Hellenistic or during the earlier Roman period. During the Late Roman period and almost up until the end of the occupation, this was a settlement with a small citadel and a lower town mainly expanding into the fields to the north and northeast of the mound, which enjoyed both the benefits of good economic profit from organized agricultural practices and being on the main trade and communication route that linked this part of Anatolia to neighboring regions further to the south and east. This period of prosperity seems to have come to an end rather abruptly, mainly because of political events that resulted in the collapse of the Byzantine administrative and economic system, and it looks like the inhabitants had to abandon the lower town, with the settlement once again becoming a small site limited to the summit. Finally, the settlement on top of the mound was also abandoned, and, after being temporarily used as a cemetery, the mound seems to have served as a Seljuk outpost with a small population for a short period of time.

Although our plan is to end this long-lived archaeological project in the near future, our work at Çadır Höyük will continue in coming years with the hope of gathering more data about the chronological periods that are still not well represented concerning the long history of occupation and socio-cultural adaptation to climatic and environmental conditions. While we continue to analyze the material gathered during the past seasons, our hope is to expose and explore the second-millennium BCE, Hellenistic, and Roman settlements, in particular, together with the earliest occupational levels that have not yet been reached. This final phase of research and investigation on the mound will enable us to gather more data and will allow us to portray a more complete picture of long-term adaptation and resilience by Çadır Höyük communities to changing conditions.

Acknowledgments

The authors would like the thank the Yozgat Museum and the Turkish Ministry of Culture and Tourism for supporting this work. We would also like to thank the entire excavation team at Çadır Höyük. In particular, we thank Ben Arbuckle, Sarah Adcock, and Katie Tardio for their work on the zooarchaeological assemblage and their invaluable comments and suggestions on how to interpret their research. We want to thank the University of Chicago for their initial support of the project, the University of New Hampshire for their subsequent support, and the University of Leicester for their current support. The archaeobotanical work was supported by the National Science Foundation Doctoral Dissertation Improvement Grant (#1463705, 2015) awarded to Madelynn von Baeyer and Alexia Smith, and an American Research in Turkey (ARIT) fellowship (2014) awarded to Madelynn von Baeyer.

Works Cited

Adcock, Sarah Ellen. 2020. "After the End: Animal Economies, Collapse, and Continuity in Hittite and Post-Hittite Anatolia." Ph.D., The University of Chicago (28026998).

Arbuckle, Benjamin S. 2009. "Chalcolithic Caprines, Dark Age Dairy, and Byzantine Beef: A First Look at Animal Exploitation at Middle and Late Holocene Çadır Höyük, North Central Turkey." *Anatolica* 35: 179–224. https://doi.org/10.2143/ANA.35.0.2038076.

Bar-Matthews, Miryam, and Avner Ayalon. 2011. "Mid-Holocene Climate Variations Revealed by High-Resolution Speleothem Records from Soreq Cave, Israel and their Correlation with Cultural Changes." The Holocene 21, no. 1: 163–71. https://doi.org/10.1177/0959683610384165.

Boserup, Ester. 1965. *The Conditions of Agricultural Growth. The Economics of Agrarian Change under Population Pressure.* New Brunswick, NJ: Aldine Transaction.

Cassis, Marica. 2009. "Çadır Höyük: A Rural Settlement in Byzantine Anatolia." In *Archaeology of the Countryside in Medieval Anatolia*, edited by Jacob J. Roodenberg and Tasha Vorderstrasse, 1–24. Leiden: NINO.

Cassis, Marica, Anthony J. Lauricella, Katie Tardio, Madelynn von Baeyer, Scott Coleman, Sarah E. Adcock, Benjamin S. Arbuckle, and Alexia Smith. 2019. "Regional Patterns of Transition at Çadır Höyük in the Byzantine Period." *Journal of Eastern Mediterranean Archaeology and Heritage Studies* 7, no. 3: 321–49. https://doi.org/10.1177/0959683610384165.

Cheng, Hai, Ashish Sinha, Sophie Verheyden, Fadi H. Nader, Xianglei Li, Pingzhong Zhang, J. J. Yin, et al. 2015. "The Climate Variability in Northern Levant over the Past 20,000 Years." *Geophysical Research Letters* 42, no. 20: 8641–50. http://dx.doi.org/10.1002/2015GL065397.

Cullen, Heidi, Peter B. deMenocal, Sidney Hemming, Gary Hemming, Francis H. Brown, Tom Guilderson, and Frank Sirocko. 2000. "Climate Change and the Collapse of the Akkadian Empire: Evidence from the Deep Sea." *Geology* 28, no. 4: 379–82.

Dahl, Gudrun, and Anders Hjort. 1976. *Having Herds: Pastoral Herd Growth and Household Economy*, edited by Anders Hjort, 87–112. Stockholm: Liber Tryck.

Dean, Jonathan R., Matthew D. Jones, Melanie J. Leng, Hilary J. Sloane, C. Neil Roberts, Jessie Woodbridge, George E. A. Swann, et al. 2013. "Palaeo-Seasonality of the Last Two Millennia Reconstructed from the Oxygen Isotope Composition of Carbonates and Diatom Silica from Nar Gölü, Central Turkey." *Quaternary Science Reviews* 66: 35–44. https://doi.org/10.1016/j.quascirev.2012.07.014.

Dean, Jonathan R., Matthew D. Jones, Melanie J. Leng, Stephen R. Noble, Sarah E. Metcalfe, Hilary J. Sloane, Diana Sahy, Warren J. Eastwood, and C. Neil Roberts. 2015. "Eastern Mediterranean Hydroclimate over the Late Glacial and Holocene, Reconstructed from the Sediments of Nar Lake, Central Turkey, Using Stable Isotopes and Carbonate Mineralogy." *Quaternary Science Reviews* 124: 162–74. http://dx.doi.org/10.1016/j.quascirev.2015.07.023.

Eastwood, Warren J., Melanie J. Leng, Neil Roberts, and Basil Davis. 2007. "Holocene Climate Change in the Eastern Mediterranean Region: A Comparison of Stable Isotope and Pollen Data from Lake Gölhisar, Southwest Turkey." *Journal of Quaternary Science* 22, no. 4: 327–41. http://dx.doi.org/10.1002/jqs.1062.

Göktürk, Ozan Mert, Dominik Fleitmann, Seraina Badertscher, Hai Cheng, R. Lawrence Edwards, Markus Leuenberger, Adelheid Fankhauser, Okan Tüysüz, and Jan Kramers. 2011. "Climate on the Southern Black Sea Coast During the Holocene: Implications from the Sofular Cave Record." *Quaternary Science Reviews* 30, no. 19–20: 2433–45. https://doi.org/10.1016/j.quascirev.2011.05.007.

Gorny, Ronald L. 2006. "The 2002–2005 Excavation Seasons at Çadır Höyük: The Second Millennium Settlements." *Anatolica* 32: 29–54.

Gorny, Ronald L., Gregory McMahon, Samuel Paley, and Lisa Kealhofer. 1995. "The Alişar Regional Project 1994." *Anatolica* 21: 65–100.

———. 2000. "The 1999 Alişar Regional Project Season." *Anatolica* 26: 153–71. https://doi.org/10.2143/ANA.26.0.2015497.

Gorny, Ronald L., Gregory McMahon, Samuel Paley, Sharon Steadman, and Bruce Verhaaren. 1999. "The 1998 Alişar Regional Project Season." *Anatolica* 25: 149–83. https://doi.org/10.2143/ANA.25.0.2015488.

———. 2002. "The 2000 and 2001 Seasons at Çadır Höyük in Central Turkey: A Preliminary Report." *Anatolica* 28: 109–36. https://doi.org/10.2143/ANA.28.0.2011763.

Graham, Philip J., and Alexia Smith. 2013. "A Day in the Life of an Ubaid Household: Archaeobotanical Investigations at Kenan Tepe, South-Eastern Turkey." *Antiquity* 87: 405–17. https://doi.org/10.1017/S0003598X00049024.

Hastorf, Christine A. 1991. "Gender, Space, and Food in Prehistory." In *Engendering Archaeology: Women and Prehistory*, edited by Joan M. Gero and Margaret W. Conkey, 132–59. Oxford: Blackwell.

Jones, Glynis. 1998. "Distinguishing Food from Fodder in the Archaeobotanical Record." *Environmental Archaeology* 1: 95–98. https://doi.org/10.1179/env.1996.1.1.95.

Kuzucuoğlu, Catherine, W. Dorfler, S. Kunesch, and F. Goupille. 2011. "Mid- to Late-Holocene Climate Change in Central Turkey: The Tecer Lake Record." *The Holocene* 21, no. 1: 173–88. https://doi.org/10.1177/0959683610384163.

Marom, Nimrod, Meirav Meiri, Yotam Tepper, Tali Erickson-Gini, Hagar Reshef, Lior Weissbrod, and Guy Bar-Oz. 2019. "Zooarchaeology of the Social and Economic Upheavals in the Late Antique-Early Islamic Sequence of the Negev Desert." *Scientific Reports* 9: 6702. https://doi.org/10.1038/s41598-019-43169-8.

Marston, John M. 2011. "Archaeological Markers of Agricultural Risk Management." *Journal of Anthropological Archaeology* 30, no. 2: 190–205. https://doi.org/10.1016/j.jaa.2011.01.002.

Mayewski, Paul A., Eelco E. Rohling, J. Curt Stager, Wibjörn Karlén, Kirk A. Maasch, L. David Meeker, Eric A. Meyerson, et al. 2004. "Holocene Climate Variability." *Quaternary Research* 62, no. 3: 243–55. http://dx.doi.org/10.1016/j.yqres.2004.07.001.

Roberts, Neil, Warren J. Eastwood, Catherine Kuzucuoglu, Girolamo Fiorentino, and Valentina Caracuta. 2011. "Climatic, Vegetation and Cultural Change in the Eastern Mediterranean During the Mid-Holocene Environmental Transition." *The Holocene* 21, no. 1: 147–62. https://doi.org/10.1177/0959683610386819.

Ross, Jennifer C. 2010. "Çadır Höyük: The Upper South Slope 2006–2009." *Anatolica* 36: 67–87. https://doi.org/10.2143/ANA.36.0.2049239.

Ross, Jennifer C., Gregory McMahon, Yağmur Heffron, Sarah E. Adcock, Sharon R. Steadman, Benjamin S. Arbuckle, Alexia Smith, and Madelynn von Baeyer. 2019a. "Anatolian Empires: Local Experiences from Hittites to Phrygians at Çadır Höyük." *Journal of Eastern Mediterranean Archaeology and Heritage Studies* 7, no. 3: 299–320. https://doi.org/10.5325/jeasmedarcherstu.7.3.0299.

Ross, Jennifer C., Sharon R. Steadman, Gregory McMahon, Sarah E. Adcock, and Joshua W. Cannon. 2019b. "When the Giant Falls: Endurance and Adaptation at Çadır Höyük in the Context of the Hittite Empire and Its Collapse." *Journal of Field Archaeology* 44, no. 1: 19–39. https://doi.org/10.1080/00934690.2018.1558906.

Smith, Alexia. 2007. "Plant Use at Çadır Höyük, Central Anatolia." *Anatolica* 33: 169–84. https://doi.org/10.2143/ANA.33.0.2021758.

Staubwasser, Michael, and Harvey Weiss. 2006. "Holocene Climate and Cultural Evolution in Late Prehistoric-Early Historic West Asia." *Quaternary Research* 66, no. 3: 372–87. https://doi.org/10.1016/j.yqres.2006.09.001.

Steadman, Sharon R., Laurel D. Hackley, Stephanie Selover, Burcu Yıldırım, Madelynn von Baeyer, Benjamin S. Arbuckle, Ryan Robinson, and Alexia Smith. 2019a. "Early Lives: The Late Chalcolithic and Early Bronze Age at Çadır Höyük." *Journal of Eastern Mediterranean Archaeology and Heritage Studies* 7, no. 3: 271–98. https://doi.org/10.5325/jeasmedarcherstu.7.3.0271.

Steadman, Sharon R., and Gregory McMahon. 2015. "Recent Work (2013–2014) at Çadır Höyük on the North Central Anatolian Plateau." In *The Archaeology of Anatolia, Volume I: Recent Discoveries (2011–2014)*, edited by Sharon R. Steadman and Gregory McMahon, 69–97. Newcastle upon Tyne: Cambridge Scholars Publishing.

———. 2017. "The 2015-2016 Seasons at Çadır Höyük on the North Central Anatolian Plateau." In *The Archaeology of Anatolia, Volume II: Recent Discoveries (2015-2016)*, edited by Sharon R. Steadman and Gregory McMahon, 93-116. Newcastle upon Tyne: Cambridge Scholars Publishing.

Steadman, Sharon R., Gregory McMahon, Benjamin S. Arbuckle, Madelynn von Baeyer, Alexia Smith, Burcu Yıldırım, Laurel D. Hackley, Stephanie Selover, and Stefano Spagni. 2019b. "Stability and Change at Çadır Höyük in Central Anatolia: A Case of Late Chalcolithic Globalization?" *Anatolian Studies* 69: 21-57. https://doi.org/10.1017/S0066154619000036.

Steadman, Sharon R., Gregory McMahon, and Jennifer C. Ross. 2007. "The Late Chalcolithic at Çadır Höyük in Central Anatolia." *Journal of Field Archaeology* 32: 385-406. https://doi.org/10.1179/009346907791071458.

———. 2019. "Chalcolithic, Iron Age, and Byzantine Investigations at Çadır Höyük: The 2017 and 2018 Seasons." In *The Archaeology of Anatolia, Volume III: Recent Discoveries (2017-2018)*, edited by Sharon R. Steadman and Gregory McMahon, 33-52. Newcastle upon Tyne: Cambridge Scholars Publishing.

Steadman, Sharon R., Gregory McMahon, Jennifer C. Ross, Marica Cassis, Jeffrey D. Geyer, Benjamin S. Arbuckle, and Madelynn von Baeyer. 2013. "The 2009 and 2012 Seasons of Excavation at Çadır Höyük on the Anatolian North Central Plateau." *Anatolica* 39: 113-67. https://doi.org/10.2143/ANA.39.0.2990786.

Steadman, Sharon R., Gregory McMahon, Jennifer C. Ross, Marica Cassis, T. Emre Şerifoğlu, Benjamin S. Arbuckle, Sarah E. Adcock, et al. 2015. "The 2013 and 2014 Excavation Seasons at Çadır Höyük on the Anatolian North Central Plateau." *Anatolica* 41: 87-124. https://doi.org/10.2143/ANA.41.0.3127290.

Steadman, Sharon R., Jennifer C. Ross, Gregory McMahon, and Ronald L. Gorny. 2008. "Excavations on the North-Central Plateau: The Chalcolithic and Early Bronze Age Occupation at Çadır Höyük." *Anatolian Studies* 58: 47-86. https://doi.org/www.jstor.org/stable/20455412.

Steadman, Sharon R., T. Emre Şerifoğlu, Stephanie Selover, Laurel D. Hackley, Burcu Yıldırım, Anthony J. Lauricella, Benjamin S. Arbuckle, et al. 2017. "Recent Discoveries (2015-2016) at Çadır Höyük on the North Central Plateau." *Anatolica* 43: 203-50. https://doi.org/10.2143/ANA.43.0.3262801.

Şerifoğlu, Tevfik Emre. 2017. "The Late Early Bronze Age Regional Catastrophe: A View from Cilicia." In *Questions, Approaches, and Dialogues in Eastern Mediterranean Archaeology: Studies in Honor of Marie-Henriette and Charles Gates*, edited by Ekin Kozal, Murat Akar, Yağmur Heffron, Çiler Çilingiroğlu, Tevfik Emre Şerifoğlu, Canan Çakırlar, Sinan Ünlüsoy, and Eric Jean, 151-78. Münster: Ugarit Verlag.

Türkeş, Murat, Telat Koç, and Faize Sariş. 2009. "Spatiotemporal Variability of Precipitation Total Series over Turkey." *International Journal of Climatology* 29, no. 8: 1056–74. https://doi.org/10.1002/joc.1768.

Turner, Rebecca, Neil Roberts, and Matthew D. Jones. 2008. "Climatic Pacing of Mediterranean Fire Histories from Lake Sedimentary Microcharcoal." *Global and Planetary Change* 63, no. 4: 317–24. https://doi.org/10.1016/j.gloplacha.2008.07.002.

van der Veen, Marijke. 2014. "The Materiality of Plants: Plant–People Entanglements." *World Archaeology* 46, no. 5: 799–812. https://doi.org/10.1080/00438243.2014.953710.

von Baeyer, Madelynn. 2018. "Seeds of Complexity: An Archaeobotanical Study of Incipient Social Complexity at Late Chalcolithic Çadır Höyük, Turkey." PhD diss, University of Connecticut. https://opencommons.uconn.edu/dissertations/1888.

———. 2020. "Late Chalcolithic Archaeobotanical Remains at Çadır Höyük, Turkey." Edited by Madelynn von Baeyer. Open Context. https://doi.org/10.6078/M7NV9GC1.

von Baeyer, Madelynn, Alexia Smith, and Sharon R. Steadman. 2021. "Expanding the Plain: Using Archaeobotany to Examine Adaptation to the 5.2 kya Climate Change Event During the Anatolian Late Chalcolithic at Çadır Höyük." *Journal of Archaeological Science: Reports* 36. https://doi.org/10.1016/j.jasrep.2021.102806.

von der Osten, Hans H. 1937. *The Alişar Höyük: Seasons of 1930-32. Oriental Institute Publications 28, Parts 1-3*. Chicago: University of Chicago Press.

Weiss, Harvey. 1997. "Late Third Millennium Abrupt Climate Change and Social Collapse in West Asia and Egypt." *NATO ASI Series* 49: 711–23.

Weiss, Harvey, Marie-Agnès Courty, Wendy Wetterstrom, François Guichard, Louise Senior, Richard Meadow, and Anne Curnow. 1993. "The Genesis and Collapse of Third Millennium North Mesopotamian Civilization." *Science (New Series)* 261, no. 5124: 995–1004. https://doi.org/10.1126/science.261.5124.995.

Wick, Lucia, Gerry Lemcke, and Michael Sturm. 2003. "Evidence of Lateglacial and Holocene Climatic Change and Human Impact in Eastern Anatolia: High-Resolution Pollen, Charcoal, Isotopic and Geochemical Records from the Laminated Sediments of Lake Van, Turkey." *The Holocene* 13, no. 5: 665–75. https://doi.org/10.1191/0959683603hl653rp.

Xoplaki, Elena, Dominik Fleitmann, Juerg Luterbacher, Sebastian Wagner, John F. Haldon, Eduardo Zorita, Ioannis Telelis, Andrea Toreti, and Adam Izdebski. 2016. "The Medieval Climate Anomaly and Byzantium: A Review of the Evidence on Climatic Fluctuations, Economic Performance and Societal Change." *Quaternary Science Reviews* 136: 229–52. https://doi.org/10.1016/j.quascirev.2015.10.004.

Zeder, Melinda A. 1991. *Feeding Cities: Specialized Animal Economy in the Ancient Near East.* Washington, D.C.: Smithsonian Institution Press.

Zohary, Daniel, and Pinhas Spiegel-Roy. 1975. "Beginnings of Fruit Growing in the Old World." *Science* 187, no. 4174: 319–27. https://doi.org/10.1126/science.187.4174.319.

Zohary, Michael. 1973. *Geobotanical Foundations of the Middle East.* 2 vols. Stuttgart: Gustav Fischer Verlag.

Genesis of the Beyşehir Occupation Phase: Understanding the Socio-Environmental Systems of Anatolia and Interactions from Kültepe-Kanesh and Paleoecological Records

Çetin Şenkul
Mustafa Doğan
Aziz Ören
Ahmet Köse
Fikri Kulakoğlu
Warren J. Eastwood
Uğur Doğan
Hugh Elton

Introduction

Pollen analytical data are an excellent proxy for highlighting vegetation change and human impact across spatial and temporal scales and are invaluable for detecting human agency and how landscapes recover when human impacts have ceased or lessened in intensity. In Anatolia, pollen data show the development and establishment of woodland following the end of the last ice age around 11,700 years ago, as well as episodes of deforestation and human impact of varying intensity (van Zeist, Woldring, and Stapert 1975; Miebach et al. 2016; Shumilovskikh et al. 2016; Ören 2018; Şenkul and Doğan 2018; Şenkul et al. 2018a; Şenkul et al. 2018b). In particular, a marked period of human impact is registered in pollen diagrams from Anatolia, especially in southwest Anatolia, and is known as the Beyşehir Occupation Phase (BOP). Radiocarbon dating shows that the BOP commenced ca. 3200 cal BP (1440 BCE) and ended ca. 1300 cal BP (650 CE). Although research has identified the causes of the end of the BOP linked to increasing instability and insecurity associated with Arab raiding and the Byzantine-Arab wars (England et al. 2008), we are less informed on the cultures responsible for the beginning of the BOP (Roberts 1990). The BOP was characterized by agropastoralism and the incorporation of arboricultural elements including cereals, olive (*Olea*), manna (*Fraxinus ornus*), sweet chestnut (*Castanea sativa*), vine (*Vitis*), and walnut (*Juglans*), which made it a classical example of early Mediterranean polyculture in Anatolia. Although the crop plants at specific BOP sites across Anatolia varied, it always included the Mediterranean agricultural triad of cereals, vines, and olives (Bottema, Woldring, and Aytuğ 1993; Eastwood, Roberts, and Lamb 1998; Vermoere et al. 2002; Roberts et al. 2019; Langgut et al. 2019; Kaniewski et al. 2007; Izdebski 2013; Haldon et al. 2014).

Continuing research on the vegetation history of Anatolia and the BOP in particular has provided much needed information on socio-ecological systems. Because agricultural economic systems, including trade networks, formed as a result of socio-ecological structures, we follow Haldon and colleagues (2014) in reaching consilience through examining archaeological/material and historical/textual data together in a multi- and interdisciplinary collaboration. This approach allows for more nuanced and holistic investigations and a focus on examining how resilient pre- and proto-historic complex societies were to environmental and climatic perturbations (Haldon et al. 2014). Thanks to this approach, ideas were able to be obtained regarding the correlation of archaeological and historical information and how agricultural activities, nutritional habits, and commercial networks of people changed temporally and spatially. In other words, archaeological and palynological information are mutually supportive on the topics of the socio-economic and socio-cultural structure that environmental conditions shaped and the determination of the use of land that these structures, in turn, formed. Furthermore, these types of data provide the opportunity to debate whether environmental change had an impact on the changes in the period that have been determined archaeologically. In addition to this, a previous lack of understanding of Anatolia's non-uniform climate (indicator species representing different ecological settings) on paleoecology in the BOP and the preceding period limited the understanding of the local and regional effects of these periods.

By the end of the Chalcolithic period and the start of the Bronze Age (ca. 5300 cal BP), archaeological data show the establishment of cities, a broad range of agricultural activities including cereal agriculture and animal husbandry, industrial scale mining of tin, copper, and salt, and the development of industrial and artisanal activities such as weaving (Karul 2011). During this timeframe, however, these nascent complex societies were largely autonomous, owing to repeated states of conflict among each other which acted to limit socio-ecological, -economic, and -cultural networks and promoted local, small-scale industrial production (Karul 2011).

New paleoecological research undertaken at Lake Engir, located six kilometers south of the archaeological site of Kültepe-Kanesh, allows for off-site environmental conditions to be reconstructed. Additionally, because both sites are in the same landscape unit, we are able to place the environmental data in an archaeological and socio-cultural context. To this end, this paper re-evaluates the nature and genesis of the BOP insofar as we are able to provide the earliest date for the commencement of the BOP in Anatolia, and we hypothesize that the BOP was intricately associated and tied to the development of socioecological, socio-cultural, and socio-economic trade networks.

Materials, Methods, and Background Information

Data for this paper were drawn from a range of lake and marsh sites across Anatolia where the BOP is evidenced (Fig. 1; Table 1). The distribution of sites with pollen evidence across Anatolia is uneven. In particular, there is a relatively higher density of pollen sites in southwestern and northwestern Anatolia. The Cappadocia region of central Anatolia is relatively well represented, but there is only sparse pollen data for western Anatolia, the central upland plateau region, and eastern Anatolia. On the other hand, archaeological sites of the Chalcolithic to Late Roman periods are relatively evenly distributed across Anatolia (Fig. 2). Because of the relatively sparser distribution of sites yielding paleoenvironmental information, not all archaeological sites lie in close proximity to environmentally-derived pollen data. One of the exceptions to this is Lake Engir, located near the archaeological site of Kültepe-Kanesh, which has yielded and continues to yield an exceptionally rich archaeological and historical archive.

Three important points are considered in the interpretation of the primary and secondary indicator species in the diagrams for the lakes evaluated in the scope of this study. The first is that the lakes from which pollen diagrams were obtained were selected from across the regions of Anatolia. The second is that the indicator species have different ecological characteristics. The third is that the socio-economic and socio-cultural activities, accumulations of agricultural knowledge, and tool-use properties of the societies that lived in the areas in which these lakes were found and in the discussed periods were different from one another.

A type can be found at very high rates in a particular location because the indicator species in the pollen diagram obtained from any lake based on this situation is in an ecological area suitable for this type (*Olea*—Mediterranean-based, *Castanea*—Black-Sea-based, etc.) or because of any other reason for the societies living in that period (*Fraxinus ornus*, like used in the construction of horse-drawn carriages). Contrariwise, the most important point to be considered here is the location in which a species emerged. Also important is the extent to which it was used by the culture and which species are more concentrated in a location as a result of economic activities.

Tables 2 and 3 show detailed data for the range of primary anthropogenic pollen types (*Olea europaea*, *Juglans*, *Fraxinus ornus*, *Castanea*, *Pistacia*, *Vitis*, and Cerealia-T) that reflect agriculture, arboriculture, and secondary anthropogenic pollen types (*Plantago lanceolata*, *Poligonum aviculare*, *Centaura solstitialis*, *Rumex acetosella*, and *Sangiusorba minor*), which reflect the weeds that benefited from human agency during the period of the BOP (3200–1300 cal BP; Table 2) and the proto-BOP period (4500–3700 cal BP; Table 3). *Olea* was an important constituent of the BOP at Elaia, Gölcük, Gölhisar, Bafa, and Engir, while *Juglans* was well represented at Gölhisar and Ladik. *Fraxinus ornus* (manna) was well represented at Beyşehir and Engir, and *Castanea sativa* was well represented at Beyşehir, Gölcük, Abant, and Engir. *Pistacia* was an important constituent at Bafa, while *Vitis* was well represented at Gölcük. Cereals were well represented at all BOP sites across Anatolia. Of the secondary anthropogenic indicators, *Plantago lanceolata*, *Poligonum aviculare*, *Rumex acetosella*, and *Sangiusorba minor* are excellent indicators of grazing and suggest that agropastoral activities were important at most BOP sites, while *Centaura solstitialis*, an indicator of arable fields, was important at all sites and supports the cereal data (Behre 1981; Behre 1990; Bottema and Woldring 1990; Eastwood, Roberts, and Lamb 1998; 1999; Vermoere et al. 2002; Vermoere 2004; Eastwood, Roberts, and Boyer 2007; Bakker et al. 2013; Gaillard 2013).

Lake Engir (1,078 masl) is located in the southwest-northeast trending Sarımsaklı Basin, six kilometers from Kültepe-Kanesh and 20 kilometers northeast of Kayseri (Fig. 1). The volcanic massif of Erciyes Mountain (3,917 masl) is located southwest of the site, Koramaz Mountain (1,920 masl) is to its southeast, and limestone plateaus (100–400 masl) are located to the west and east. A 714 cm sediment core was retrieved from Lake Engir in 2016 and was subsequently subsampled every 16 cm for pollen, which was extracted

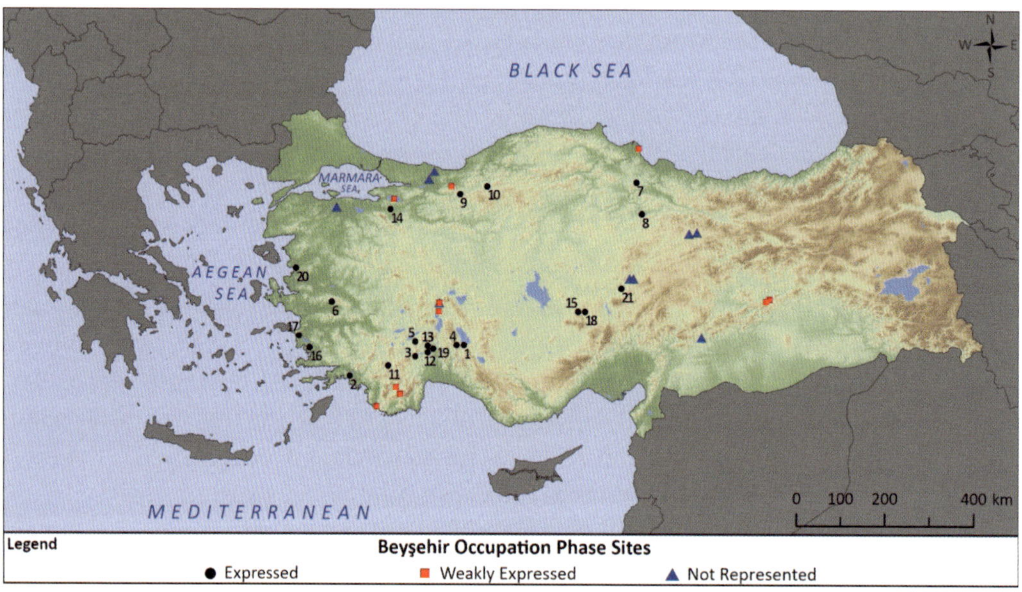

Fig. 1. Map of Anatolia showing selected pollen sites recording a Beyşehir Occupation Phase. Also shown is the location of Lake Engir and Kültepe-Kanesh, located in eastern Cappadocia. Refer to Table 1 for references.

Table 1. Pollen sites in Anatolia reflecting the Beyşehir Occupation Phase. References: [1,2,3]van Zeist, Woldring, and Stapert 1975; [4,5]Bottema and Woldring 1984; [6]Sullivan 1989; [7,8,9,10]Bottema et al. 1993–1994; [11]Eastwood 1997; [12,13]Vermoere et al. 2002; [14]Bottema et al. 2001; [15]Woldring 2001; [16,17]Müllenhoff et al. 2004; [18]England 2006; [19]Kaniewski et al. 2007; [20]Shumilovskikh et al. 2016; [21]Şenkul et al. 2018a.

BEYŞEHİR OCCUPATION PHASE					
Site	**Elevation (m)**	**Begin (cal BP)**	**End (cal BP)**	**Duration (yrs)**	**Basal Age of sequence (cal BP)**
[1]Lake Beyşehir I	1123	3500	1550	1950	6048
[2]Lake Köyceğiz	0	3200	2900	300	4907
[3]Lake Söğüt	1300	3000	1300	1700	20000
[4]Lake Beyşehir II	1123	4446	3249	1197	17100
[5]Lake Pınarbaşı	976	3000	1200	1800	15086
[6]Lake Gölcük	1050	3000	1750	1250	10000
[7]Lake Ladik	800	3200	1300	1900	13822
[8]Lake Kaz	500	2200	1300	900	8157
[9]Lake Abant II	1300	3000	800	2200	11310
[10]Yeniçağa Lake II	976	4430	3500	930	13268
[11]Lake Gölhisar II	930	3200	1300	1900	9500
[12]Gravgaz Marsh 1996	1220	2280	1270	1010	3300
[13]Gravgaz Marsh 1999	1220	2270	1480	790	2478
[14]Yenişehir	228	3266	375	2891	11430
[15]Eski Acıgöl	1271	3500 (BP)	1650 (BP)	1850	14320
[16]Lake Bafa S 1	0	4302	2168	2134	4686
[17]Lake Bafa S 6	0	2681	171	2510	2451
[18]Nar Lake I	1300	1700	1320	380	1700
[19]Bereket Marsh	1410	2250	1550	700	3260
[20]Elaia	1	2800	1200	1600	7500
[21]Lake Engir	1078	1400	690	710	5565

from the sediment matrix using standard laboratory procedures (Faegri and Iversen 1975; Moore, Webb, and Collinson 1991), as described in more detail elsewhere (Şenkul et al. 2020). The sediment core and pollen stratigraphy is supported by eight radiocarbon analyses, providing a relatively robust chronology. The basal age of the sequence is 5565 cal BP. Pollen diagrams were constructed using TILIA (Grimm 1991; 2015).

Archaeological excavations at Kültepe-Kanesh have been carried out since 1948, and so far the earliest settlement has been dated to the Late Chalcolithic Age (5450 cal BP or ca. 3500 BCE; Kulakoğlu 2011). Archaeological data are supported by archaeobotanical, archaeozoological, and textual data in the form of cuneiform tablets. To date, 23,500 cuneiform tablets provide an exceptionally rich archive of a range of historical activities, including commercial and administrative transactions, socio-cultural and economic activities, and legal decisions regarding property sales, land tenure, and land-use changes, as well as the recording of natural incidents such as droughts and famine. These cuneiform tablets offer the earliest broad, direct, and uninterrupted written record of many subjects in Anatolia (Dercksen 2008; Öz 2014). The location of Kültepe-Kanesh in central Anatolia on vital routes connecting Anatolia with the rest of Assyria and Mesopotamia allowed it to act as a gateway, with its sphere of influence reaching into the rest of Anatolia. At the same time, Kültepe-Kanesh played a crucial role not only in Anatolian history but also Mesopotamian history (Larsen 1976; Veenhof 2008). During one of the most important periods at Kültepe, the Assyrian Trade Colonies Period (the first quarter of the second millennium BCE), a systematic, widespread, and organized network of trade was established between Anatolia and northern Mesopotamia (Larsen 1976; Veenhof 2008; Kulakoğlu 2011). During this period, close relationships between Mesopotamia and Anatolia were developed; and, although these were fundamentally commercial in nature, they nonetheless led to important cultural, social, economic, and administrative developments in Anatolia (Larsen 1976; Kulakoğlu et al. 2013). Cuneiform tablets provide explicit detail on the titles and responsibilities of the public officials during this well-organized, socially-complex settlement period (Kulakoğlu 2011). This Assyrian Trade Colonies Period represents the apogee of Kültepe-Kanesh, with an estimated population of around 50,000. The long-range, organized trade colonies that the Assyrians established laid the groundwork for the establishment of the first empire in Anatolia (the Hittites) and for the creation of Anatolian unity by providing for the growth and strengthening of Anatolian city-states (Kulakoğlu 2011).

It has been hypothesized that around 4200 cal BP (ca. 2200 BCE), a global megadrought, the "Meghalayan arid" or "4.2 ka BP event," caused widespread climatic impacts and large-scale societal collapse and abandonment (Weiss et al. 1993; Welc and Marks 2014; Weiss 2016). Archaeological data from Kültepe-Kanesh, however, show that, while Kültepe-Kanesh was less powerful at this time, there was settlement continuity, with the first indications of settlement contraction commencing only around 3920 cal BP (ca. 1970 BCE) (Kulakoğlu 2014). Although there is very good evidence of climate change and aridity in the Cappadocia region via stable isotope data from Nar Lake (Jones et al. 2006; Jones and Roberts 2008; Dean et al. 2013; 2015; 2018; Dean 2014), and while this global aridity event may have had some impact on the region's socio-economic viability (Weiss et al. 1993; Weiss 2015; 2016; Cullen et al. 2000; Drysdale et al. 2006; Welc and Marks 2014), Kültepe-Kanesh did not collapse in any meaningful sense of the term at this time period.

Final abandonment and collapse at Kültepe-Kanesh occurred some 550 years after the 4.2 ka BP event—around 3650 cal BP (1700 BCE). Kültepe-Kanesh is therefore an excellent example of a large, socially complex polity that survived this presumed global event and a good example for adaption and resilience. Following abandonment in 1700 BCE, a 900-year Dark Age or period of discontinuity can be interpreted from the

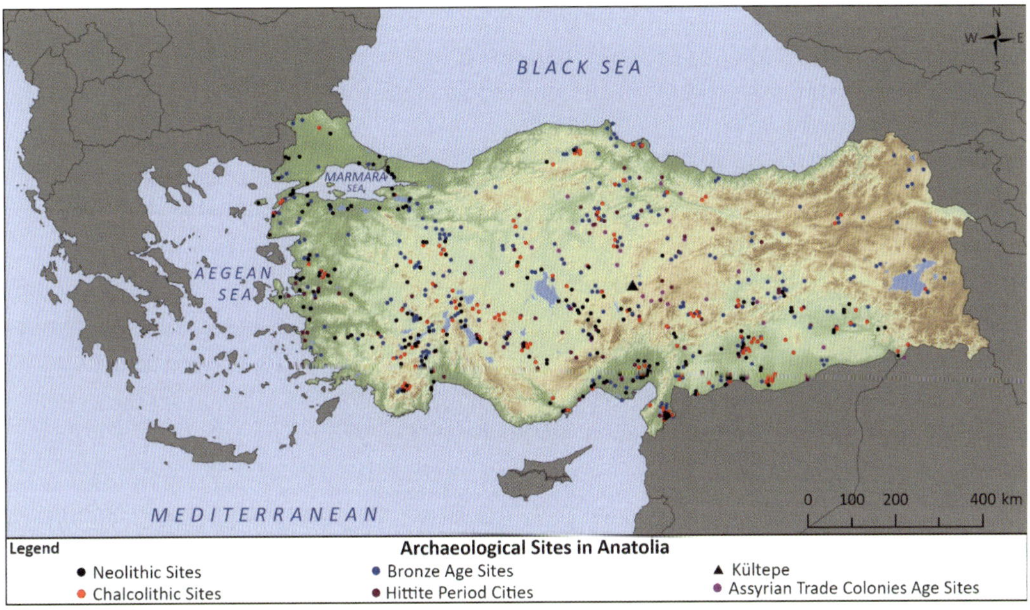

Fig. 2. Map of Anatolia showing selected archaeological sites/city states in Anatolia for the Neolithic to Late Bronze Age periods.

archaeological data, which suggests a further decrease in the importance of Kültepe-Kanesh. Archaeological data for the Iron Age Late Hittite, Hellenistic, and Roman periods once again show some evidence of settlement at Kültepe-Kanesh (Barjamovic 2015), yet population contracted further during the Late Roman, Byzantine, and Seljuk periods (Kulakoğlu 2014).

The Beyşehir Occupation Phase in the Vicinity of Lake Engir

The pollen diagram for Lake Engir (Fig. 3) covers the last 5,565 years and clearly shows the primary and secondary anthropogenic indicators and impacts associated with the BOP. According to the radiocarbon chronology, the BOP commenced at ca. 1400 cal BP (550 CE) and ended at 690 cal BP (1340 CE), lasting 710 cal years. During this timeframe, which represents an intense period of the BOP, the primary indicators *Juglans regia*, *Vitis vinifera*, *Castanea sativa*, *Olea europaea*, and *Fraxinus ornus*, as well as Cereals (*Secale cereale*, *Hordeum*, and *Triticum*), are well represented, together with the secondary indicator species *Sanguisorba minor*, *Plantago lanceolata*, *Polygonum aviculare*, *Rumex acetosella*, and *Centaurea solstitialis*. This correlates well with the cuneiform data from Kültepe-Kanesh which attest to the cultivation of fruit and grain and animal husbandry on the Sarımsaklı plains near Kültepe-Kanesh (Öz 2014). Stable isotope data for Nar Lake indicate enhanced moisture availability for the period ca. 1500–1100 cal BP (Jones et al. 2006; Jones and Roberts 2008; Dean et al. 2013; 2015; 2018; Dean 2014) which covers the BOP recorded at Lake Engir. Increasingly humid conditions would have facilitated and encouraged agricultural activity during the BOP. Arboreal pollen (AP) percentage values before the BOP measured 77.3% and decreased to 16.6% during the BOP, suggesting significant rates of deforestation in the immediate catchment of Lake Engir and on the hills surrounding the site.

The end of the BOP is characterized by a decrease in the primary and secondary anthropogenic indicators and an increase in percentage values of pine. High percentage values of pine in post-BOP assemblages are seen in many pollen records in Anatolia and the eastern Mediterranean (van Zeist, Woldring, and Stapert 1975; Bottema and Woldring 1984; 1990; Baruch 1986; 1990; Roberts 1990; Bottema et al. 1993–1994; Heim et al. 1997; Eastwood, Roberts, and Lamb 1998; Eastwood et al. 1999; Baruch and Bottema 1999; Vermoere et al. 2000; Kaniewski et al. 2007; England et al. 2008). This signals a rewilding of pine woodlands, most probably on the uplands surrounding Kültepe-Kanesh and further afield. While *Pinus brutia* as a species can be discounted, it is possible that *Pinus nigra* or *Pinus sylvestris* may have been the species that recolonized, the latter forming the most southerly distribution of *Pinus sylvestris* woodlands on the Akdağ Mountains, ranging between 40–115 km northeast of Kültepe-Kanesh.

Discussion: Evidence for Proto-Beyşehir Occupation Phase Impacts at Lake Engir

While evidence for human impacts associated with the BOP are clearly expressed in the pollen stratigraphy of Lake Engir for the period of 3200–1300 cal BP, the Engir pollen sequence also provides evidence for previous human impacts, especially for the period 4500–3700 cal BP. We propose here the concept of a proto-BOP in which appear many of the floristic/arboricultural elements that are more commonly associated with the classical or main part of the BOP. These will be discussed alongside archaeological, historical, climatic, and ecological evidence from Kültepe-Kanesh, as this provides a more holistic understanding of human-environment interactions and how the natural environment can be shaped by human agency (Mensing et al. 2018).

The Lake Engir pollen data include primary anthropogenic indicator species for levels prior to the BOP such as *Juglans regia*, *Olea europaea*, *Fraxinus ornus*, *Vitis vinifera*, and Cereals. Of the secondary anthropogenic indicators, *Plantago lanceolata*, *Centaurea solstitialis*, and *Sanguisorba minor* all point to grazing activities (Fig. 3) and are a common feature recorded at other sites of the classical BOP. However, for the period of 4500–3700 cal BP, the primary anthropogenic indicators are poorly represented for some sites in Anatolia (e.g., Yeniçağa, Abant, Kaz, Ladik, Gölcük, Bafa, Gölhisar), yet are well represented at other sites. For example, *Olea* (a primary anthropogenic indicator) is found at sites where a Mediterranean-type climate prevails, such as Beyşehir, Pınarbaşı, Köyceğiz, Bafa, Elaia, and Yenişehir. Most of these coastal sites record the highest percentage values of *Olea* and today are locations for industrial-scale olive cultivation in a true or Eu-Mediterranean climate zone, including Köyceğiz, Bafa, and Elaia, and are consistent with the presence of this species along the Mediterranean coast. Other sites with transitional or continental Mediterranean-type climate regimes include interior and upland sites (e.g., Beyşehir, Pınarbaşı).

The *Olea* percentage value at Engir was 1% between 4500–3700 cal BP. The lake is characterized by a continental Mediterranean-type climate (Kutiel and Türkeş 2005) and, therefore, has climatic conditions that are not suitable for growing olives today.

The fact that there are words meaning "official responsible for oil" in the Kültepe tablets demonstrates that the state played a role in oil production and commerce (Bilgiç 1964). Just as the title can be used with respect to oil products from animals, it could also have been used for oil acquired from olives. According to the tablets, vegetable oils were also widely consumed in Anatolian cuisine. Vegetable oils were generally obtained from olives (Öz 2014). The appearance of *Olea* at Lake Engir (Cappadocia), however, must have occurred with species arriving through trade, as suggested by the densely populated, centralized settlement of Kültepe-Kanesh, where trade routes intersect and enable information exchange.

Juglans has a wider distribution than *Olea*. It is found in more transitional continental Mediterranean climate regimes, as well as montane habitats, and is more cold-tolerant than *Olea*. It is found at Köyceğiz, Pınarbaşı Yenişehir, and Engir in the proto-BOP period (4500–3700 cal BP; Late Chalcolithic to Late Bronze Age) (Table 3). Although its past distribution is mainly restricted to the coastal regions of Anatolia (Bottema and Woldring 1990; Rikli 1943–1948), its presence in the Engir pollen sequence could indicate that it was brought to Kültepe-Kanesh along the dense network of trade routes operating at this time. The cultivation of fruit and nut trees is historically attested in the cuneiform tablets recovered from Kültepe-Kanesh, and the words "garden," "gardener," and "fruit" appear often (Bılgiç 1964), shedding light on some of the socio-economic activities associated with the various fruits, nuts, and vegetables that were cultivated in irrigated fields within the territory of the site.

Fraxinus ornus (ash) typically belongs to the European-Siberian floristic province and has a natural distribution along the temperate Black Sea mountains. It is not natural to the Kültepe-Kanesh region, yet its presence in the Lake Engir pollen sequence indicates that it could be cultivated ecologically in this area. It has been interpreted traditionally as being useful for the production of manna (i.e., "manna from heaven") and used as an astringent exudate with medicinal qualities akin to aspirin (Bottema, Woldring, and Aytuğ 1993) during classical BOP times (Vermoere et al. 2000). The Kültepe-Kanesh tablet data suggest that ash was important for wheel production and battle chariots, in particular. For later periods, for example during Roman times, it was useful in the production of arrows.

Although *Castanea sativa* and *Pistacia* are not recorded at Lake Engir in proto-BOP assemblages, *Castanea* was important at Gölhisar, Elaia, Yenişehir, and Ladik, where it measured 1–3% percentage values, while *Pistacia* was important at Gölhisar, Elaia, and Gölcük, where it measured 3–5% percentage values (Table 3).

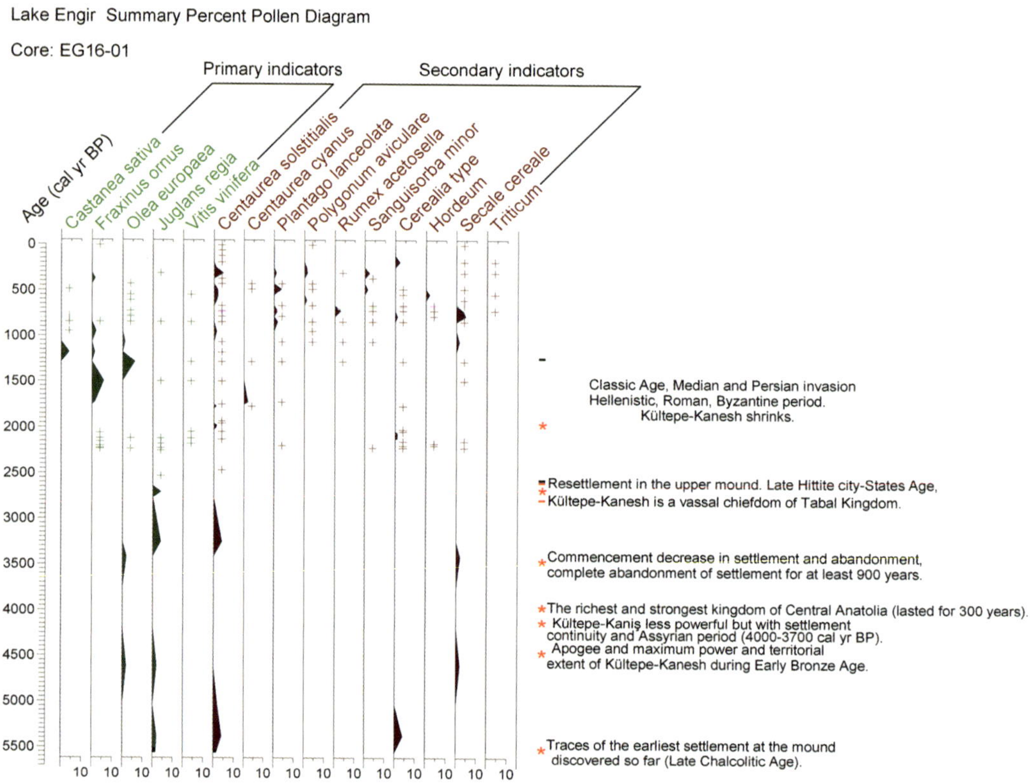

Fig. 3. Lake Engir indicator pollen species and Kültepe-Kanesh chronology.

Vitis is recorded at low percentage values for many of the proto-BOP sites (Beyşehir, Pınarbaşı, Köyceğiz, Yenişehir, Abant, Ladik, and Engir; Table 3). Although low percentage values of ca. 1% suggest that *Vitis* may have been cultivated only in very limited areas, modern pollen-vegetation studies show that *Vitis* produces and disseminates only very small quantities of pollen and is therefore severely underrepresented in the paleoecological record (Faegri and Iversen 1975; Vermoere et al. 2000; 2003). Once again, the tablet data from Kültepe-Kanesh are particularly informative and show that the cultivation of grapes, vineyards, and winemaking (viticulture) were important economic activities in ancient Anatolia (Bilgiç and Günbattı 1995), attesting to the presence of *Vitis* around Lake Engir, in addition to information pertaining to harvesting, wine production, quality control (e.g., wine tasters), and commercial accounts (Larsen 1976).

Cereals (Cerealia-type) have a distinct importance among the primary anthropogenic indicators, as they are the only one recorded at almost all sites. Cerealia-type includes *Secale cereale* (rye), *Hordeum* (barley), *Avena* (oats), and *Triticum* (wheat) and, together with domesticated animals, are a fundamental form of socio-economic activity well suited to the generally semi-arid, treeless rangelands of the interior Anatolian Plateau that have sustained Anatolian cultures since Epipaleolithic and Neolithic times (Öz 2014; Baird et al. 2018; Eastwood et al. 2018). As with *Vitis*, the Cerealia pollen types suffer from toponymical shortcomings. Grains other than the *Secale*-type, which is pollinated by wind, produce low amounts of pollen due to partial self-pollination (Eastwood et al. 2018). Furthermore, the large size and the tendency of *Triticum*, *Hordeum*, and *Avena* pollen grains to remain in their hulls means that cereal pollen grains (apart from *Secale*) are poorly dispersed, usually deposited only locally, and tend to be grossly underrepresented in pollen diagrams (Behre 1990; Eastwood et al. 2018). It is possible, therefore, that a proportion of high percentage values of Cerealia-type pollen may be the result of intensive cereal agriculture within the immediate vicinity of the sampling site (Heim 1963; Behre 1981). Other important crop plants include lentil (*Lens*), pea (*Pisum*), and chickpea (*Vicia*) and are generally not expressed in pollen records due to very low pollen production and dispersal. In the Kültepe texts, the presence of information about grains, barley, grain commerce, and the Assyrian Grain Goddess (Larsen 1976; Bilgiç and Günbattı 1995) fills out our knowledge of these crops.

Secondary anthropogenic indicator pollen types are represented at all proto-BOP sites apart from Gölhisar, Yenicağa, Ladik, and Kaz (Behre 1981; 1990; Bottema and Woldring 1990; Vermoere et al. 2002; Gaillard 2007). The secondary anthropogenic indicator pollen types are important, as they inform about a range of human activities, especially pastoralism and animal husbandry, and are well represented at many sites, including Beyşehir, Pınarbaşı, Köyceğiz, Yenişehir, and Engir. For the landscape around Engir, in particular, the tablet data from Kültepe-Kanesh are particularly informative and show that a large number of artisanal workshops located in the city produced a range of wooden goods, textiles, and metallurgical products for domestic consumption and export (Atıcı 2014), and these were part of an extensive socio-economic system that relied upon primary agricultural practices, such as lumbering and animal husbandry. It is not unreasonable to infer that secondary anthropogenic indicator pollen types for other proto-BOP sites may be interpreted similarly.

Tables 2 and 3 provide the arboreal pollen (AP) values and rate of generalized indicator species in order to be able to rank the effect of humans before and during the BOP (through primary/secondary/all indicator species and AP-rate analysis) and to be able to see the differences between the locations.

In pre-BOP periods (Fig. 4), all indicators reached the highest value in Yenişehir. The primary indicator species reached the highest level at Lake Köyceğiz (in accordance with climatic circumstances), and the secondary indicator species reached the highest value at Yenişehir. The most concentrated human impact based on the indicator species

Table 2. Percentage value of indicator types for the Beyşehir Occupation Phase, 3200–1300 cal BP. Dark gray indicates primary indicator species and light gray indicates secondary indicator species. AP = arboreal pollen. Percentage data are generalized according to the legend classes of the table. References: [1,5]van Zeist, Woldring, and Stapert 1975; [2,4]Bottema and Woldring 1984; [3]Eastwood 1997; [6]Sullivan 1989; [7]Müllenhoff et al. 2004; [8]Shumilovskikh et al. 2016; [9]Bottema et al. 2001; [10,11,12,13]Bottema et al. 1993–1994; [14]Şenkul et al. 2018a.

Site	*Olea*	*Juglans*	*Fraxinus ornus*	*Castanea*	*Pistacia*	*Vitis*	*Cereals*	*Plantago lanceolata*	*Polygonum aviculare*	*Centaurea solstitialis*	*Rumex acetosella*	*Sangiusorba minor*	All Indicators (%)	Min AP (%)
Lake Beyşehir I[1]–II[2]	1–3	1–3	15–20	3–5	-	1–3	3–5	5–10	5–10	15–20	1–3	1–3	85	30
Lake Gölhisar[3]	5–10	5–10	< 1	1–3	1–3	1–3	3–5	1–3	-	-	-	1–3	45	55
Lake Pınarbaşı[4]	3–5	1–3	1–3	1–3	1–3	1–3	3–5	1–3	1–3	5–10	< 1	1–3	41	56
Lake Köyçeğiz[5]	3–5	< 1	< 1	-	1–3	1–3	1–3	1–3	1–3	3–5	< 1	3–5	33	62
Lake Gölcük[6]	15–20	3–5	-	15–20	-	10–15	5–10	-	-	-	-	3–5	75	20
Lake Bafa 1–6[7]	10–15	1–3	-	1–3	3–5	-	1–3	5–10	-	-	-	-	39	25
Elaia[8]	20–25	1–3	1–3	1–3	1–3	1–3	5–10	5–10	1–3	-	1–3	-	66	38
Yenişehir[9]	5–10	3–5	3–5	< 1	-	1–3	1–3	1–3	1–3	10–15	1–3	1–3	54	20
Lake Yenicağa II[10]	1–3	1–3	1–3	-	1–3	< 1	3–5	1–3	1–3	1–3	1–3	1–3	33	55
Abant Lake II[11]	3–5	3–5	< 1	3–5	< 1	< 1	5–10	5–10	1–3	< 1	1–3	< 1	46	51,6
Lake Kaz[12]	3–5	3–5	-	1–3	-	< 1	5–10	3–5	5–10	3–5	< 1	3–5	50	51,6
Lake Ladik[13]	1–3	5–10	-	1–3	< 1	< 1	5–10	5–10	10–15	< 1	1–3	1–3	60	65
Lake Engir[14]	5–10	< 1	5–10	3–5	-	< 1	5–10	3–5	1–3	1–3	1–3	1–3	54	16,6

% < 1 1–3 3–5 5–10 10–15 15–20 20–25

Table 3. Percentage value of indicator types for the Proto-Beyşehir Occupation Phase, 4500–3700 cal BP. Dark gray indicates primary indicator species and light gray indicates secondary indicator species. AP = arboreal pollen. Percentage data are generalized according to the legend classes of the table. References: [1,5]van Zeist, Woldring, and Stapert 1975; [2,4]Bottema and Woldring 1984; [3]Eastwood 1997; [6]Sullivan 1989; [7]Müllenhoff et al. 2004; [8]Shumilovskikh et al. 2016; [9]Bottema et al. 2001; [10,11,12,13]Bottema et al. 1993–1994; [14]Şenkul et al. 2018a.

Site	*Olea*	*Juglans*	*Fraxinus ornus*	*Castanea*	*Pistacia*	*Vitis*	Cereals	*Plantago lanceolata*	*Polygonum aviculare*	*Centaurea solstitialis*	*Rumex acetosella*	*Sangiusorba minor*	All Indicators (%)	Max AP (%)
Lake Beyşehir I[1]–II[2]	<1	-	<1	-	-	<1	1–3	<1	5–10	15–20	<1	-	38	88
Lake Gölhisar[3]	-	-	-	<1	<1	-	1–3	-	-	-	-	-	5	95
Lake Pınarbaşı[4]	<1	<1	<1	-	-	<1	<1	-	<1	15–20	-	<1	27	63
Lake Köyçeğiz[5]	1–3	<1	-	-	-	1–3	1–3	-	<1	5–10	<1	-	22	90
Lake Gölcük[6]	-	-	-	-	3–5	-	-	-	-	-	-	3–5	10	63
Lake Bafa 1–6[7]	1–3	-	-	-	-	-	1–3	1–3	-	-	-	-	9	62
Elaia[8]	1–3	-	-	<1	<1	-	3–5	1–3	-	-	-	-	13	75
Yenişehir[9]	<1	<1	-	<1	-	<1	<1	1–3	1–3	40	-	<1	52	89
Lake Yenicağa II[10]	-	-	<1	-	-	-	-	-	-	-	-	-	1	85
Abant Lake II[11]	-	-	-	-	-	<1	<1	<1	-	-	<1	-	4	78
Lake Kaz[12]	-	-	-	-	-	-	<1	-	-	-	-	-	1	45
Lake Ladik[13]	-	-	-	1–3	-	<1	-	-	-	-	-	-	4	90
Lake Engir[14]	<1	<1	<1	-	-	<1	1–3	<1	-	1–3	-	<1	12	77,3

% <1 1–3 3–5 5–10 10–15 15–20 20–25 40

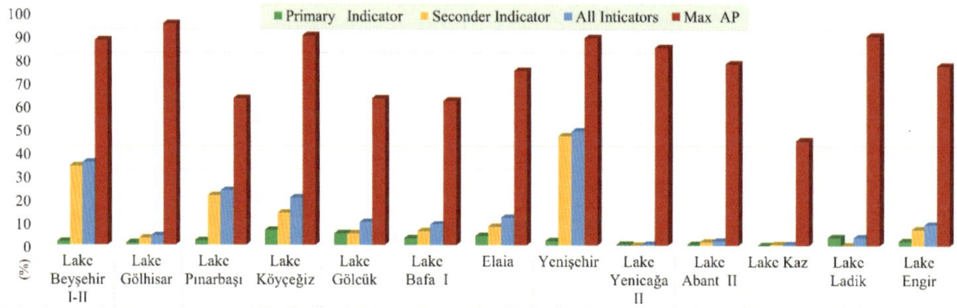

	Lake Beyşehir I-II	Lake Gölhisar	Lake Pınarbaşı	Lake Köyçeğiz	Lake Gölcük	Lake Bafa I	Elaia	Yenişehir	Lake Yenicağa II	Lake Abant II	Lake Kaz	Lake Ladik	Lake Engir
Primary Indicator	1,5	1	2	6,5	5	3	4	2	0,5	0,5	0	3,5	2
Seconder Indicator	34	3	21,5	14	5	6	8	47	0	1,5	0,5	0	7
All Inticators	35,5	4	23,5	20,5	10	9	12	49	0,5	2	0,5	3,5	9
Max AP	88	95	63	90	63	62	75	89	85	78	45	90	77,3

Fig. 4. Indicator types and arboreal pollen (AP) ratio (maximum) value in the proto-Beyşehir Occupation Phase.

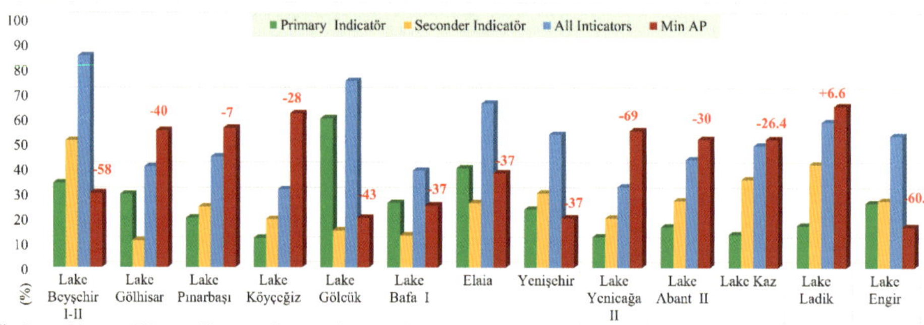

	Lake Beyşehir I-II	Lake Gölhisar	Lake Pınarbaşı	Lake Köyçeğiz	Lake Gölcük	Lake Bafa I	Elaia	Yenişehir	Lake Yenicağa II	Lake Abant II	Lake Kaz	Lake Ladik	Lake Engir
Primary Indicatör	34	29,5	20	12	60	26	40	23,5	12,5	16,5	13,5	17	26
Seconder Indicatör	51	11	24,5	19,5	15	13	26	30	20	27	35,5	41,5	27
All Inticators	85	40,5	44,5	31,5	75	39	66	53,5	32,5	43,5	49	58,5	53
Min AP	30	55	56	62	20	25	38	20	55	51,6	51,6	65	16,6

Fig. 5. Indicator types and arboreal pollen (AP) ratio (minimum) value in the Beyşehir Occupation Phase (minus values indicate the difference between maximum AP and minimum AP between the proto-Beyşehir Occupation Phase and the Beyşehir Occupation Phase).

are viewed in and around Yenişehir, and the human impact does not constitute a pressure on forest cover (AP 89%).

Despite one of the lowest rates compared with the rates of all the indicators at Lake Kaz, the maximum AP rate was lowest at Lake Kaz. According to the Lake Engir data (Fig. 4), the rate of all indicators was low, and the rate of secondary indicator species was twice as great as that of the primary indicator species. The reason that many secondary indicators are seen around Lake Engir originates from the climate conditions of central Anatolia, which are suitable for the cultivation of these plant taxa. Information from the tablets from Kültepe-Kanesh, together with the palynological data showing that cereal cultivation was carried out, confirm the idea that cereals were a fundamental source of nutrition for the people of Kültepe-Kanesh.

In the BOP (Fig. 5), the rate of all indicators reached the highest value at Lake Beyşehir. The primary indicator species reached the highest level at Gölcük (in accordance with climatic circumstances), and the secondary indicator species reached the highest value at Lake Beyşehir. The most concentrated human impact based on indicator species are seen in and around Lake Beyşehir, and the human impact constitutes a pressure on forest cover.

In the pre-BOP period, the AP rate compared with the maximum AP and BOP minimum AP rates (Figs. 4-5) decreased in all locations (between 7-69%), other than at Lake Ladik (+6.6%). The greatest AP decrease in all sample areas was seen in Lake Yeniçağa (69%), Lake Engir (60.7%), and Lake Beyşehir (58%). According to the Lake Engir data (Fig. 5), all indicators increased compared with pre-BOP times, and the rate of primary and secondary indicator species was balanced (27%). When comparing these data, information from the Kültepe-Kanesh tablets, the archaeobotanical record, and increases in indicator species combine to suggest that human impacts on landscapes intensified in the environs of Kültepe-Kanesh during the BOP. Based on the Lake Engir data, the total rate of indicators rose from 12% before the BOP to 54% during the BOP, and the parallel decrease in AP rates shows that the socio-environmental system adapted to socio-cultural and environmental changes. In this period, the conduct of centralized commercial operations at Kültepe-Kanesh provided for the interchange of information and culture, along with trade, and allowed for increasingly dense populations, supported in part by increased grain agriculture. In fact, grains and garden plants are seen in many tablet records that mention, for example, wheat, barley, peas, lentils, vetch, beans, and cucumber (Karul 2011).

Although the presence of indicator species in lakes occurs largely owing to climatic factors, indicator species not suitable to areas like Lake Engir (e.g., *Castanea sativa*, *Olea europaea*, and *Fraxinus ornus*) were also seen. When comparing and correlating archaeological and palynological data, the fact that these species appear or that commerce for their products is known around Kültepe-Kanesh before the BOP period show how socio-cultural and environmental factors of the socio-ecological system were able to harmonize with change in and around Kültepe-Kanesh. The fact that the socio-economic and socio-cultural activities, accumulations of agricultural knowledge, and tool-use properties of the societies that lived in this period were different from one another is shaped largely by environmental factors. It is possible, for instance, that *Olea*, which is cultivated in areas where a Mediterranean climate is dominant, could have been traded within Anatolia with a centralized settlement like Kültepe-Kanesh, the center of networks. The reason that cereals are seen at nearly every location relates to the fundamental nature of cereals in the nourishment of Anatolian peoples from the Neolithic era to the Assyrian Trade Colonies Period, witnessed also in the Kültepe-Kanesh tablets.

The agricultural society that emerged locally and regionally in Anatolia within the Neolithic and Chalcolithic periods encompassed all Anatolia for the first time during the period represented by Kültepe-Kanesh, when the richness and uniqueness of grains, animal husbandry, and gardening cultures specific to Anatolia surfaced. Kültepe itself, owing to its strategic position, had an important role in the formation of this profile. The clear transformation that took place around Anatolia, and specifically in the Lake Engir area, is a precursor to the classic Beyşehir Occupation Phase and is demonstrated not just by palynological records but also by archaeobotanical data and the oldest written documents in Anatolia.

Conclusion

The BOP that has been detected in various areas (southwestern, western, central, and northern Anatolia, the Sea of Marmara, Greece, and Syria) and in differing periods (between 3200–1300 cal BP) was reflected in highly representative fossil pollen evidence from Lake Engir, dating to ca. 3200–1300 cal BP. The Lake Engir pollen data also demonstrate that, in the period preceding the classic BOP, population increases and agricultural knowledge accumulations intensified across Anatolia between ca. 4500–3700 cal BP.

The emergence of an integrated social structure in Anatolia and the establishment of a socio-ecological system that accelerated the development of this structure began to take place at the centralized settlement of Kültepe-Kanesh 4500 years ago. In fact, between 4500–3700 cal BP, the commercial structure established at Kültepe-Kanesh was systematically nourished by a large section of Anatolia for the first time; by paving the way for Anatolia to open outwards, it made it possible to transfer information along with trade and to exchange culture. Thus, along with socio-ecological and socio-economic structures stemming from outside Anatolia penetrating Anatolia through Kültepe-Kanesh, the system led to much information produced throughout Anatolia reaching Kültepe-Kanesh.

The web of relationships established thanks to the flow of mutual information and the temporal and spatial transformation situated upon this web allowed for Kültepe-Kanesh to become a center that harbored a population of ca. 40,000, about which abundant information regarding everything from agriculture to mining, from gardening to transportation, and from woodworking to remote organized commerce, was recorded on tablets. The discovery during this period of new resources and species with agricultural potential for Anatolia, as well as the growing need for these things, laid the foundations for the further intensification of exchanges of information and commercial networks across geographic regions. At the same time, the fact that no contemporary area within or outside Anatolia shows evidence for interregional/intercultural exchanges/relationships at this level constitutes one of the main reasons that Kültepe-Kanesh became Anatolia's most significant initial capital. Just as this accumulation of information at Kültepe-Kanesh can be supported with palynological and archaeobotanical data, so too is it determined with the help of cuneiform tablets.

As a result of this study, we can say that a socio-ecological, socio-economic, and socio-cultural structure that would encompass all of Anatolia began to be established at Kültepe-Kanesh around 4500 years ago, 1700 years before the classic BOP.

Acknowledgments

This study was supported by the Scientific and Technological Research Council of Turkey (TÜBİTAK) under the project titled "Holocene Environmental Change in the Light of the Fossil and Modern Pollen Analysis of Kültepe (Kayseri) Environment" (Grant #114Y578). We would like to thank all our supporters for their continued contributions.

Works Cited

Atıcı, Levent. 2014. "Food and Ethnicity at Kültepe-Kanesh: Preliminary Zooarchaeological Evidence." In *Current Research at Kültepe/Kanesh: An Interdisciplinary and Integrative Approach to Trade Networks, Internationalism, and Identity, Journal of Cuneiform Studies Supplement Series 4,* edited by Levent Atıcı, Fikri Kulakoğlu, Gojko Barjamovic, and Andrew Fairbairn, 195–211. Atlanta, GA: Lockwood Press.

Baird, Douglas, Andrew Fairbairn, Emma Jenkins, Louise Martin, Caroline Middleton, Jessica Pearson, Eleni Asouti, et al. 2018. "Agricultural Origins on the Anatolian Plateau." *Proceedings of the National Academy of Sciences* 115, no. 14: E3077–E3086. doi:10.1073/pnas.1800163115.

Bakker Johan, Etienne Paulissen, David Kaniewski, Jeroen Poblome, Véronique De Laet, Gert Verstraeten, Marc Waelkens. 2013. "Climate, People, Fire and Vegetation: New Insights into Vegetation Dynamics in the Eastern Mediterranean since the 1st Century AD." *Climate of the Past* 9: 57–87.

Barjamovic, Gojko. 2015. "Kültepe after Kaneş." In *Proceedings of the 1st Kültepe International Meeting. 19–23 September 2013. Studies Dedicated to Kutlu Emre. KIM 1 (Kultepe International Meetings 1),* edited by Fikri Kulakoğlu and Cécile Michel, 233–42. Subartu XXXV. Turnhout: Brepols.

Baruch, Uri. 1986. "The Late Holocene Vegetational History of Lake Kinneret (Sea of Galilee), Israel." *Paleorient* 12, no. 2: 37–48.

———. 1990. "Palynological Evidence of Human Impact on the Vegetation as Recorded in the Late Holocene Lake Sediments in Israel." In *Man's Role in the Shaping of the Eastern Mediterranean Landscape,* edited by S. Bottema, G. Entjes, and W. Van Zeist, 283–93. Rotterdam: Balkema.

Baruch, Uri, and Sytze Bottema. 1999. "A New Pollen Diagram from Lake Hula: Vegetational, Climatic and Anthropogenic Implications." In *Ancient Lakes: Their Cultural and Biological Diversity,* edited by H. Kawanabe, G. W. Coulter, and A. C. Roosevelt, 75–86. Gent: Kenobi Productions.

Behre, Karl-Ernst. 1981. "The Interpretation of Anthropogenic Indicators in Pollen Diagrams." *Pollen Spores* 23: 225–45.

———. 1990. "Some Reflections on Anthropogenic Indicators and the Record of Prehistoric Occupation Phases in Pollen Diagrams from the Near East." In *Man's Role in the Shaping of the Eastern Mediterranean Landscape,* edited by S. Bottema, G. Entjes-Nieborg, and W. van Zeist, 219–30. Rotterdam: Balkema.

Bilgiç, Emin. 1964. "Three Tablets from the City Mound of Kültepe." *Anatolia* 8: 145–63.

Bilgiç, Emin, and Cahit Günbattı. 1995. "Ankaraner Kültepe-Texte III [AKT III]: Texte der Grabungskampagne 1970." *Freiburger Altorientalische Studien* 3. Stuttgart: Franz Steiner Verlag.

Bottema, Sytze, and Hendrik Woldring. 1984. "Late Quaternary Vegetation and Climate of Southwestern Turkey II." *Palaeohistoria* 26: 123-49.

———. 1990. "Anthropogenic Indicators in the Pollen Record of the Eastern Mediterranean." In *Man's Role in the Shaping of the Eastern Mediterranean Landscape,* edited by S. Bottema, G. Entjes-Nieborg, G., and W. van Zeist, 231-64. Rotterdam: Balkema.

Bottema, Sytze, Hendrik Woldring, and Burhan Aytuğ. 1993. "Palynological Investigations on the Relations between Prehistoric Man and Vegetation in Turkey: The Beyşehir Occupation Phase." In *OPTIMA. Proceedings of the Fifth Meeting, Istanbul, 8-15 September 1986*, edited by Hüsnü Demiriz and Neriman Özhatay, 315-28. Istanbul: Istanbul University.

———. 1993-1994. "Late Quaternary Vegetation History of Northern Turkey." *Palaeohistoria* 35/36: 13-72.

Bottema, Sytze, Hendrik Woldring, and İlhan Kayan. 2001. "The Late Quaternary Vegetation History of Western Turkey." In *The Ilıpınar Excavations II*. Leiden: Nederlands Instituut Voor Het Nabije Oosten Leiden.

Cullen, Heidi M., Peter B. deMenocal, Sidney Hemming, Francis Brown, Tom Guilderson, and Frank Sirocko. 2000. "Climate Change and the Collapse of the Akkadian Empire: Evidence from the Deep Sea." *Geology* 28: 379-82.

Dean, Jonathan R. 2014. "Stable Isotope Analysis and U-Th Dating of Late Glacial and Holocene Lacustrine Sediments from Central Turkey." PhD diss., University of Nottingham.

Dean, Jonathan R., Matthew D. Jones, Melanie J. Leng, Sarah E. Metcalfe, Hilary J. Sloane, Warren J. Eastwood, and Neil Roberts. 2018. "Seasonality of Holocene Hydroclimate in the Eastern Mediterranean Reconstructed using the Oxygen Isotope Composition of Carbonates and Diatoms from Lake Nar, Central Turkey." *The Holocene* 28, no. 2: 267-76.

Dean, Jonathan R., Matthew D. Jones, Melanie J. Leng, Stephen R. Noble, Sarah E. Metcalfe, Hilary J. Sloane, Diana Sahy, Warren J. Eastwood, and C. Neil Roberts. 2015. "Eastern Mediterranean Hydroclimate over the Late Glacial and Holocene, Reconstructed from the Sediments of Nar Lake, Central Turkey, using Stable Isotopes and Carbonate Mineralogy." *Quaternary Science Reviews* 124: 162-74.

Dean, Jonathan R., Matthew D. Jones, Melanie J. Leng, Hilary J. Sloane, C. Neil Roberts, Jessie Woodbridge, George E. A. Swann, Sarah E. Metcalfe, Warren J. Eastwood, and Hakan Yigitbasioglu. 2013. "Palaeo-Seasonality of the Last Two Millennia Reconstructed from the Oxygen Isotope Composition of Carbonates and Diatom Silica from Nar Lake, Central Turkey." *Quaternary Science Reviews* 66: 35-44.

Dercksen, Jan D. 2008. "Observations On Land Use and Agriculture in Kaneš." In *Old Assyrian Studies in Memory of Paul Garelli (Old Assyrian Archives, Studies, Volume 4)*, edited by C. Michel, 140-57. Leiden: Pıhans.

Drysdale, Russell, Giovanni Zanchetta, John Hellstrom, Roland Maas, Anthony Fallick, Matthew Pickett, Ian Cartwright, and Leonardo Piccini. 2006. "Late Holocene Drought Responsible for the Collapse of Old World Civilizations is Recorded in an Italian Cave Flowstone." *Geology* 34, no. 2: 101.

Eastwood, Warren J. 1997. "The Palaeoecological Record of Holocene Environmental Change in Southwest Turkey." PhD diss., University of Wales.

Eastwood, Warren J., Neil Roberts, and Peter Boyer. 2007. "Pollen Analysis at Çatalhöyük." In *Excavations at* Çatalhöyük: *The 1995-1999 Seasons. Monograph of the McDonald Institute and the British Institute of Archaeology at Ankara (BIAA)*, edited by I. Hodder, 573-80. Cambridge: McDonald Institute Monographs.

Eastwood, Warren J., Neil Roberts, and Henry F. Lamb. 1998. "Palaeoecological and Archaeological Evidence for Human Occupance in Southwest Turkey: The Beyşehir Occupation Phase." *Anatolian Studies* 48: 69-86.

Eastwood, Warren J., Neil Roberts, Henry F. Lamb, and J. C. Tibby. 1999. "Holocene Environmental Change in Southwest Turkey: A Palaeoecological Record of Lake and Catchment Related Changes." *Quaternary Science Reviews* 18: 671-95.

Eastwood, Warren, Andrew Fairbairn, Elizabeth Stroud, Neil Roberts, Henry Lamb, Hakan Yiğitbaşıoğlu, Çetin Şenkul, Andrew Mossa, Rebecca Turner, and Peter Boyer. 2018. "Comparing Pollen and Archaeobotanical Data for Chalcolithic Cereal Agriculture at Çatalhöyük, Turkey." *Quaternary Science Review* 202, no. 15: 4-18.

England, Ann. 2006. "Late Holocene Palaeoecology of Cappadocia (Central Turkey): An Investigation of Annually Laminated Sediments from Nar Gölü Crater Lake." PhD diss., University of Birmingham.

England, Ann, Warren J. Eastwood, C. Neil Roberts, Rebecca Turner, and John F. Haldon. 2008. "Historical Landscape Change in Cappadocia (Central Turkey): A Palaeoecological Investigation of Annually-Laminated Sediments from Nar Lake." *Holocene* 18, no. 8: 1229-45.

Faegri, Knut, and Johs Iversen. 1975. *Textbook of Pollen Analysis*. 3rd ed., New York: Hafner Press.

Gaillard, Marie J. 2007. "Archaeological Applications." In *Encyclopedia of Quaternary Science Elsevier*, 2570-95. Amsterdam: Elsevier.

———. 2013. "Archaeological Applications." In *Encyclopedia of Quaternary Science*, edited by Scott Elias and Cary Mock, 2949-72. Second Edition. Amsterdam: Elsevier.

Grimm, Eric C. 1991. *Tilia and Tilia Graph*. Illinois State Museum: Springfield.

———. 2015. *Tilia Software*. Illinois State Museum: Springfield.

Haldon, John, Neil Roberts, Adam Izdebski, and Dominik Fleitmann. 2014. "The Climate and Environment of Byzantine Anatolia: Integrating Science, History and Archaeology." *Journal of Interdisciplinary History* 45, no. 2: 113–61.

Heim Jean. 1963. "Recherches Sur Les Relations Entre la Végétation Actuelle et le Spectre Pollinique Récent Dans Les Ardennes Belges." *Bulletin De La Société Royale De Botanique De Belgique / Bulletin Van De Koninklijke Belgische Botanische Vereniging* 96, no. 1: 5–92.

Heim Christoph, Norbert R. Nowaczyk, Jörg F. W. Negendank, Suzanne Leroy, Zvi Ben-Avraham. 1997. "Middle-East Desertification: Evidence from the Dead Sea." *Naturwissenschaften* 84: 398–401.

Izdebski, Adam. 2013. *A Rural Economy in Transition: Asia Minor from Late Antiquity into the Early Middle Ages*. Journal of Juristic Papyrology, Supplement 18. Warsaw: Raphael Taubenschlag Foundation.

Jones, Matthew D., and C. Neil Roberts. 2008. "Interpreting Lake Isotope Records of Holocene Environmental Change in the Eastern Mediterranean." *Quaternary International* 181: 32–8.

Jones, Matthew D., C. Neil Roberts, Melanie J. Leng, and Murat Türkeş. 2006. "A High-Resolution Late Holocene Lake Isotope Record from Turkey and Links to North Atlantic and Monsoon Climate." *Geology* 34, no. 5: 361–64.

Kaniewski, David, Veronique De Laet, Etienne Paulissen, and Marc Waelkens. 2007. "Long Term Effects of Human Impact on Mountainous Ecosystems, Western Taurus Mountains. Turkey." *Journal of Biogeography* 34: 1975–97.

Karul, Necmi. 2011. *Tarih Öncesinden Demir Çağı'na Anadolu'nun Arkeoloji Atlası*. Istanbul: Doğan Burda Dergi Yayıncılık.

Kulakoğlu, Fikri. 2011. "Kültepe-Kanes: A Second Millennium B.C.E. Trading Center on the Central Plateau." In *The Oxford Handbook of Ancient Anatolia*, edited by G. McMahon, and S. Steadman, 1012–30. New York: Oxford University Press.

———. 2014. "Kanesh after the Assyrian Colony Period: Current Research at Kültepe and the Question of the End of the Bronze Age Settlement." In *Journal of Cuneiform Studies Supplemental Series "Current Research at Kültepe-Kanesh" An Interdisciplinary and Integrative Approach to Trade Networks, Internationalism, and Identity*, edited by Levent Atici, Fikri Kulakoğlu, Gojko Barjamovic, and Andrew Fairbairn, 85–94. Journal of Cuneiform Studies Supplemental Series 2014, Supplemental Series Number 4. Berkeley: Lockwood Press.

Kulakoğlu, Fikri, Kutlu Emre, Ryoichi Kontani, Sabahattin Ezer, and Güzel Öztürk. 2013. "Kültepe-Kaniş, Turkey: Preliminary Report on the 2012 Excavations." *Bulletin of the Okayama Orient Museum* 27: 43–50.

Kutiel, Haim, and Murat Türkeş. 2005. "New Evidence for the Role of the North Sea — Caspian Pattern on the Temperature and Precipitation Regimes in Continental Central Turkey." *Geografiska Annaler: Series A, Physical Geography* 87, no. 4: 501-13.

Langgut, Dafna, Rachid Cheddadi, Josè S. Carrión, Mark Cavanagh, Daniele Colombaroli, Warren J. Eastwood, Raphael Greenberg, et al. 2019. "The Origin and Spread of Olive Cultivation in the Mediterranean Basin: The Fossil Pollen Evidence." *The Holocene* 29, no. 5: 1-21.

Larsen, Mogens T. 1976. *The Old Assyrian City-State and Its Colonies.* Mesopotamia 4. Copenhagen: Akademisk Forlag.

Mensing, Scott A., Edward M. Schoolman, Irene Tunno, Paula J. Noble, Leonardo Sagnotti, Fabio Florindo, and Gianluca Piovesan. 2018. "Historical Ecology Reveals Landscape Transformation Coincident with Cultural Development in Central Italy since the Roman Period." *Scientific Reports* 8: 2138.

Miebach, Andrea, Phoebe Niestrath, Patricia Roeser, and Thomas Litt. 2016. "Impacts of Climate and Humans on the Vegetation in Northwestern Turkey: Palynological Insights from Lake Iznik since the Last Glacial." *Climate of the Past* 12: 575-93.

Moore, Peter D., J. A. Webb, and Margaret E. Collinson. 1991. *Pollen Analysis.* Oxford: Blackwell.

Müllenhoff, Marc, Handl Mathias, Maria Knipping, and Helmut Brückner. 2004. "The Evolution of Lake Bafa (Western Turkey) - Sedimentological, Microfaunal and Palynological Results." In *Geographie Der Meere Und Küsten / Coastline Reports* 1: 55-66.

Ören, Aziz. 2018. "Kültepe (Kayseri) Çevresinin Fosil Polen Analizleri Işığında Holosen Paleocoğrafyası." PhD diss., Ankara University.

Öz, Esma. 2014. *Kültepe Metinleri Işığında Eski Anadolu'da Tarım ve Hayvancılık.* Ankara: Türk Tarih Kurumu Yayınları.

Rikli, Mittelmeerländer. 1943-1948. *Das Pflanzenkleid der Mittelmeerländer.* 3 Bde. Bern: Huber Verlag.

Roberts, Neil. 1990. "Human Induced Landscape Change in South and Southwest Turkey during the Later Holocene." In *Man's Role in the Shaping of the Eastern Mediterranean Landscape*, edited by S. Bottema, G. Entjes-Nieborg, and W. van Zeist, 53-67. Rotterdam: Balkema.

Roberts, Neil, Samantha Allcock, Hannah Barnett, Anna Mather, Warren J. Eastwood, Matthew Jones, Nick Primmer, Hakan Yiğitbaşıoğlu, and Borris Vannière. 2019. "Cause-and-Effect in Mediterranean Erosion: The Role of Humans and Climate Upon Holocene Sediment Flux into a Central Anatolian Lake Catchment." *Geomorphology* 331: 36-48.

Şenkul, Çetin, and Mustafa Doğan. 2018. "Fosil ve Güncel Polen Analizleri Işığında Mucur Obruk Gölü Çevresinin Paleovejetasyon Değişimleri." *Türk Coğrafya Dergisi* 70: 19-28.

Şenkul, Çetin, Fikri Kulakoğlu, Aziz Ören, and Warren Eastwood. 2020. "Paleo-Ecological Conditions of Kültepe and its Vicinity Based on Palynological Data." In *Integrative Approaches to the Archaeology and History of Kültepe-Kanes,* edited by Fikri Kulakoğlu, C. Michel, and G. Öztürk, 123-39. Belgium: Brepols.

Şenkul, Çetin, Türkan Memiş, Warren J. Eastwood, and Uğur Doğan. 2018a. "Mid-to Late Holocene Paleovegetation Change in Vicinity of Lake Tuzla (Kayseri), Central Anatolia, Turkey." *Quaternary International* 486: 98-106.

Şenkul, Çetin, Aziz Ören, Uğur Doğan, Warren J. Eastwood. 2018b. "Late Holocene Environmental Changes in the Vicinity of Kültepe (Kayseri), Central Anatolia, Turkey." *Quaternary International* 486: 1-9.

Shumilovskikh, Lyudmila, Martin Seeliger, Stefan Feuser, Elena Novenko, Frank Schlütz, Anna Pint, Felix Pirson, and Helmut Brückner. 2016. "The Harbour of Elaia: A Palynological Archive for Human Environmental Interactions during the Last 7500 Years." *Quaternary Science Reviews* 149: 167-87.

Sullivan, D. G. 1989. "Human-Induced Vegetation Change in Western Turkey: Pollen Evidence from Central Lydia." Phd diss., University of California.

van Zeist, Willem, Hendrik Woldring, and Dick Stapert. 1975. "Late Quaternary Vegetation and Climate of Southwestern Turkey." *Palaeohistoria* 17: 55-143.

Veenhof, Klaas R. 2008. "The Old Assyrian Period." In *Mesopotamia: The Old Assyrian Period,* edited by K. R. Veenhof and J. Eidem, 13-263. OBO 160/5. Fribourg: Academic Press; Göttingen: Vandenhoeck & Ruprecht.

Vermoere, Marleen. 2004. "Holocene Vegetation History in the Territory of Sagalassos (Southwest Turkey): A Palynological Approach." In *Studies in Eastern Mediterranean Archaeology-SEMA 6,* edited by M. Waelkens, 1-347. Turnhout: Brepols Publishers.

Vermoere, Marleen, Sytze Bottema, Leo Vanhecke, Marc Waelkens, Etienne Paulissen, and Erik Smets. 2002. "Palynological Evidence for Late Holocene Human Occupation Recorded in Two Wetlands in Southwest Turkey." *Holocene* 12, no. 5: 569-84.

Vermoere, Marleen, Leo Vanhecke, Marc Waelkens, and Erik Smets. 2003. "Modern and Ancient Olive Stands near Sagalassos (South-west Turkey) and Reconstruction of the Ancient Agricultural Landscape in Two Valleys." Global Ecology and Biogeography 12, no. 3: 217-35.

Vermoere, Marleen, Marc Waelkens, Hannelore Vanhaverbeke, Ireen Librecht, Leo Vanhecke, Etienne Paulissen, and Erik Smets. 2000. "Late Holocene Environmental Change and the Record of Human Impact at Gravgaz near Sagalassos, Southwest Turkey." Journal of Archaeological Science 27, no. 7: 571-91.

Weiss, Harvey. 2015. "Megadrought, Collapse, and Resilience in Late 3rd Millennium BC Mesopotamia." In 2200 BC - A Climatic Breakdown as a Cause for the Collapse of the Old World?, edited by H. Meller, H. W. Arz, R. Jung, R. and Risch, 35–52. Halle: Landesmuseum für Vorgeschichte.

———. 2016. "Global Megadrought, Societal Collapse and Resilience at 4.2–3.9 ka BP across the Mediterranean and West Asia." PAGES 24: 62–3.

Weiss, Harvey, Marie A. Courty, Wilma Wetterstorm, François Guichard, L. Senior, Richard Meadow, A. Curnow. 1993. "The Genesis and the Collapse of Third Millennium North Mesopotamian Civilization." Science 261: 995-1004.

Welc, Fabian, and Leszek Marks. 2014. "Climate Change at the End of the Old Kingdom in Egypt around 4200 BP: New Geoarchaeological Evidence." *Quaternary International* 324: 124–133.

Woldring, Hendrik. 2001. "Climate Change and the Onset of Sedentism in Cappadocia." In *The Neolithic of Central Anatolia*, edited by F. Gerard and L. Thissen, 59–66. Ankara: British Institute of Archaeology.

Assessing Continuity and Change in the Sixth to Ninth Century Landscape of North-Central Anatolia

James Newhard

Hugh Elton

John Haldon

Introduction

When studying the past, the amount of attention paid to causation, correlation, and coincidence is voluminous and need not be fully entertained here (Arponen et al. 2019; Ion 2019; Riede 2019). Rather, it is enough to say that determining the relationships between events and conditions and human actions is a well-worn path. It is especially trodden when considering the relationship between environmental shifts and changes in societies (Haldon et al. 2014; Cheddadi et al. 2015; Maley and Vernet 2015; McIntosh, Maley, and Vernet 2015; Izdebski et al. 2016; Kristiansen 2019; Riede 2019). It is tempting to view changes in human structures as a series of quasi-mechanical responses to environmental changes, although the literature is populated with cautionary tales, noting that there is rarely a single triggering element to any major shift in societal structure (e.g., Cline 2015; Knapp and Manning 2016; Lespez et al. 2016). Recent fieldwork at the Middle Byzantine site of Avkat in central Anatolia provides a good case study for the assessment of the notions of continuity and change via the assessment of local spatial shifts in land-use patterns within the larger context of environmental and state-level conditions.

The Adaptive Cycle

Those studying complexity, change, and adaptability have been drawn to the study of resilience—the means by which a system adapts and reconfigures to meet the demands of changing conditions (Redman and Kinzig 2003). A key model within this construct is the Adaptive Cycle (Fig. 1) (Holling 2001; Gunderson and Holling 2002; Redman and

Kinzig 2003; Haldon and Rosen 2018). Basically stated, the Adaptive Cycle suggests a process of systemic evolution. In a human construct, social structures adapt to a series of environmental and social constraints within which the system successfully develops in four phases. As the society develops (r phase), it increases in scale and complexity, with increasing efficiencies in production. These efficiencies, while enhancing the system's capacity to support an increased population and greater integrations of political-economic processes, also lose the flexibility to adapt to external pressure and changes (K phase). At a certain point, the system and its processes become inflexible and unable to adapt to changing conditions. Adverse conditions rise to such an extent that the system becomes unable to accommodate, issuing a release of the structure (often viewed as a collapse or general disintegration of the system—Ω phase), followed by a period of restructuring (α phase). The resulting pattern results conceptually in a repeating process of generation, consolidation, collapse, and reconfiguration. A society's particular arrangement at each stage is dependent upon a host of variables, not excluding environmental constraints, extra-communal ties, and social memory (Redman and Kinzig 2003).

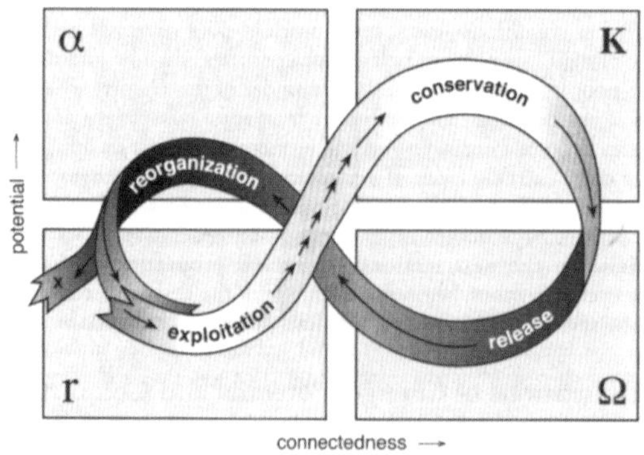

Fig. 1. Conceptual representation of the Adaptive Cycle. From *Panarchy* (Gunderson and Holling 2002, fig. 2-1). Copyright © 2002 Island Press. Reproduced by permission of Island Press, Washington, DC.

In addition to the Adaptive Cycle, the concept of scale requires consideration. Complex systems operate at a series of scales, whether this be a leaf on a tree within a forest or a household in a village of a larger state system (Holling 2001). Each of these units holds the capacity for operating within an Adaptive Cycle, yet the processes within each will be affected by actions taken at larger or smaller scales with which they are associated (Fig. 2). This interdependent, networked system of adaptation was termed "Panarchy" by Gunderson and Holling (2002; Holling 2001) and serves as a means by which one can conceptualize complex adaptive systems and their impacts upon associated systems of organization.

The multi-scalar construct is flexible, such that its application is applicable to systems of varying scale and complexity, yet detailed enough to allow for processes of adaptability and change to be observed and compared across time and space and at differing scales of operation. Applications of the Adaptive Cycle have been applied to a wide variety of contexts to try to explain the transformations of human systems.[1]

1 E.g., modern responses in the German bioenergy sector to changes in climate, as well as economic

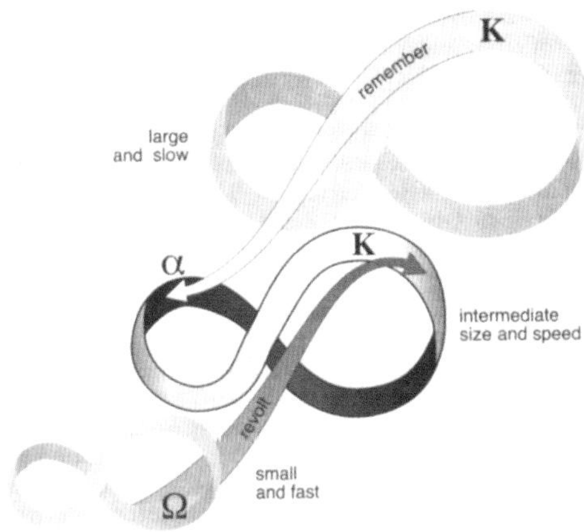

Fig. 2. The construct of Panarchy. In this example, a smaller system, as part of its Ω phase, foments revolution, which impacts a larger system undergoing a hyper-extended phase (K), thereby contributing to an ensuing Ω phase. Alternatively, a larger system in a K phase contributes previously forgotten constructs of organization to a smaller system undergoing restructuring (α phase). From Panarchy (Gunderson and Holling 2002, fig. 3-10). Copyright © 2002 Island Press. Reproduced by permission of Island Press, Washington, DC.

Middle Byzantine Anatolia

As a means to provide a synthesis between environmental, historical, and archaeological data to explain changes to human systems, Newhard, Elton, and Haldon were co-authors among a group that applied the adaptive cycle in reviewing the environmental and human history of Anatolia from the fourth to thirteenth centuries CE (Roberts et al. 2018). Although the study was careful to note the limitations of the palynological and archaeological data (see also Cassis et al. 2018), it argues that recognizable shifts in land use and settlement history correlated to the overextension (K), release (Ω), and recovery (α) phases of the Adaptive Cycle. At an Anatolian scale, this is a sustainable conclusion, but a closer look at one of the datasets involved in the study—that of the Avkat Archaeological Project (under the direction of Newhard, Elton, and Haldon)—sheds light upon the idiosyncratic nature of regional survey data and highlights the construct of panarchy in understanding the interplay between differing scales of a social-ecological system.

Figure 3 presents information from Roberts et al. 2018 that compares the combined trends in cereal crops, tree crops, hydro-climate, and settlement patterns across four regions of Anatolia (Figs. 3-4). When displayed in summary form, the data present similar patterns in ecological and cultural data. Interpreting the Late Roman period as an over-extended K-phase, the period between 700 and 900 CE is identified as an Ω-phase, where an overall decline in rural settlement is observed in correspondence with a decline in cereal and tree crops, as well as overall drier conditions. A subsequent α phase after 900 CE presents a generalized increase in both rural settlement and ecological indices, although the extent to which recovery of conditions varies to some extent. For example, the palynological evidence presents an increase of cereal crops to

----

and political institutions (Grundmann, Ehlers, and Uckert 2012); mid-Holocene climatic impacts on agropastoralism (Marsh 2016); urban landscape design (Luo et al. 2018); and "collapse" scenarios in the Late Bronze Age eastern Mediterranean (Newhard and Cline, forthcoming).

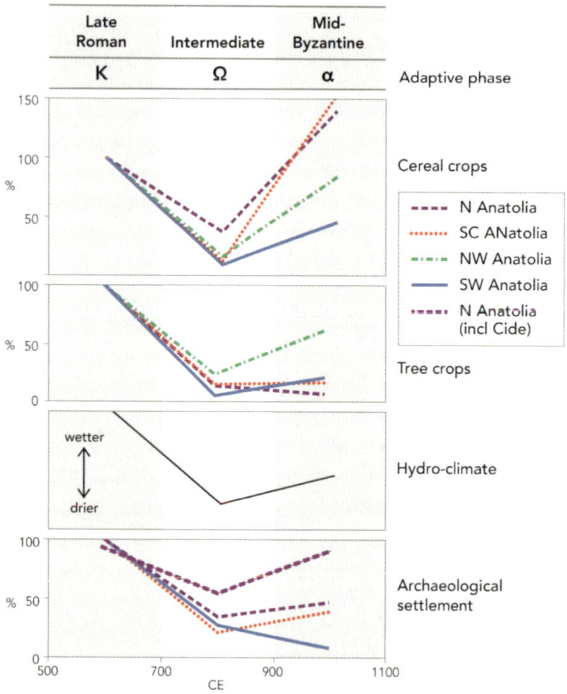

Fig. 3. From Roberts et al. (2018, fig. 6). Summary diagram, showing changes in rural settlement, hydro-climate, cereals, and tree crops during different phases of the Adaptive Cycle in Anatolia (600–1000 CE). All changes are scaled relative to the Late Roman period. Reproduced by permission of Springer Nature BV.

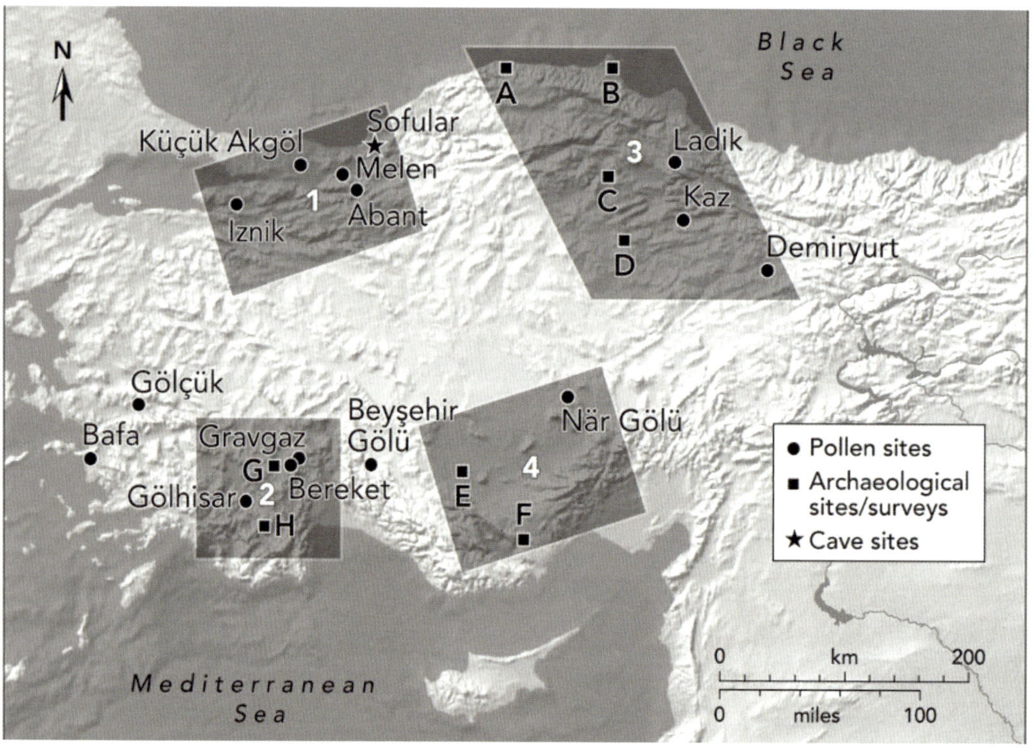

Fig. 4. Map of Anatolia with four sub-regions (1 = northwest, 2 = southwest, 3 = northern, and 4 = south-central) (from Roberts et al. 2018, fig. 1. Reproduced by permission of Springer Nature BV).

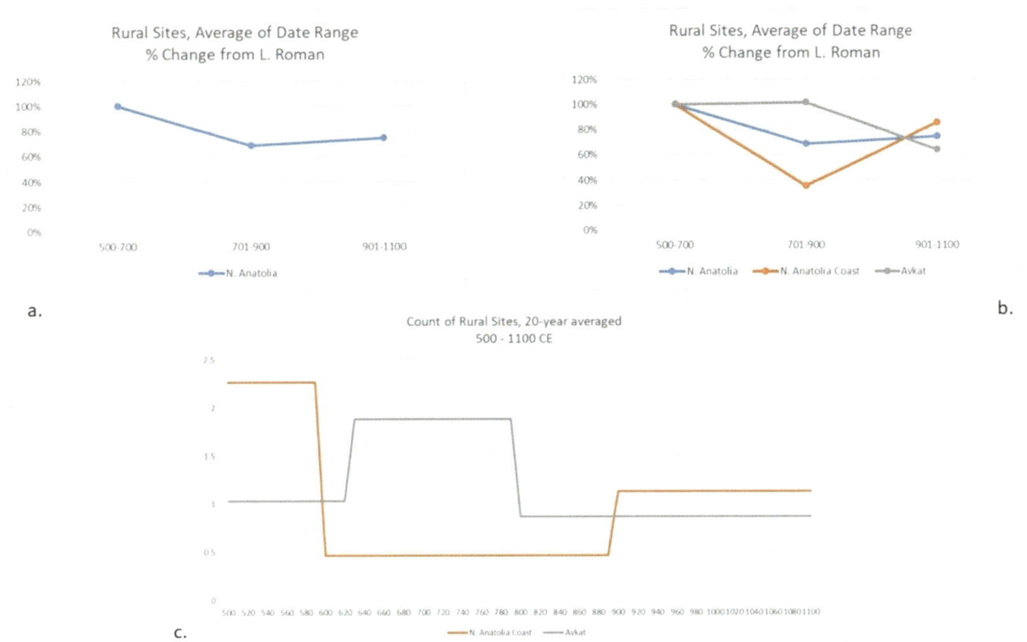

Fig. 5. Combined and disaggregated survey data from northern Anatolia. Northern Anatolia coast data consists of data from both Cide and Sinop.

a level that in most regions matched or exceeded the previous K-phase, while tree crops recovered somewhat (northwest Anatolia), recovered moderately (southwest Anatolia), remained relatively unchanged from the Ω-phase (south-central Anatolia), or declined (northern Anatolia). This pattern would be expected of an α phase within the construct of the Adaptive Cycle, which is characterized by processes of reorganization, some of which configurations mimic previous structures, while others deviate.

When the cultural data from northern Anatolia is disaggregated, divergent patterns emerge (Fig. 5). Figure 5a essentially mimics the pattern reported by Roberts and colleagues (2018, fig. 6; this chapter, Fig. 3). When the data from the Avkat Archaeological Project is disaggregated from Cide and Sinop, however, the pattern diverges (Fig. 5b)—the data from the coastal surveys masks that which is observed from the interior. Reviewing the trends at 10-year intervals (Fig. 5c) for the period captured by Roberts and colleagues, the anomaly from the Avkat survey merits investigation and a broader contextualization of the data. This should not be viewed as a critique of the 2018 paper,[2] but an attempt to show the challenges of synthesizing datasets into a differently scaled interpretation. The present study interrogates the data of

2 It should be noted that this aberration in the data was recognized in the 2018 study, and the local differentiation was taken to indicate the position of Avkat within the larger defensive strategy of the Byzantine state (Roberts et al. 2018, 311).

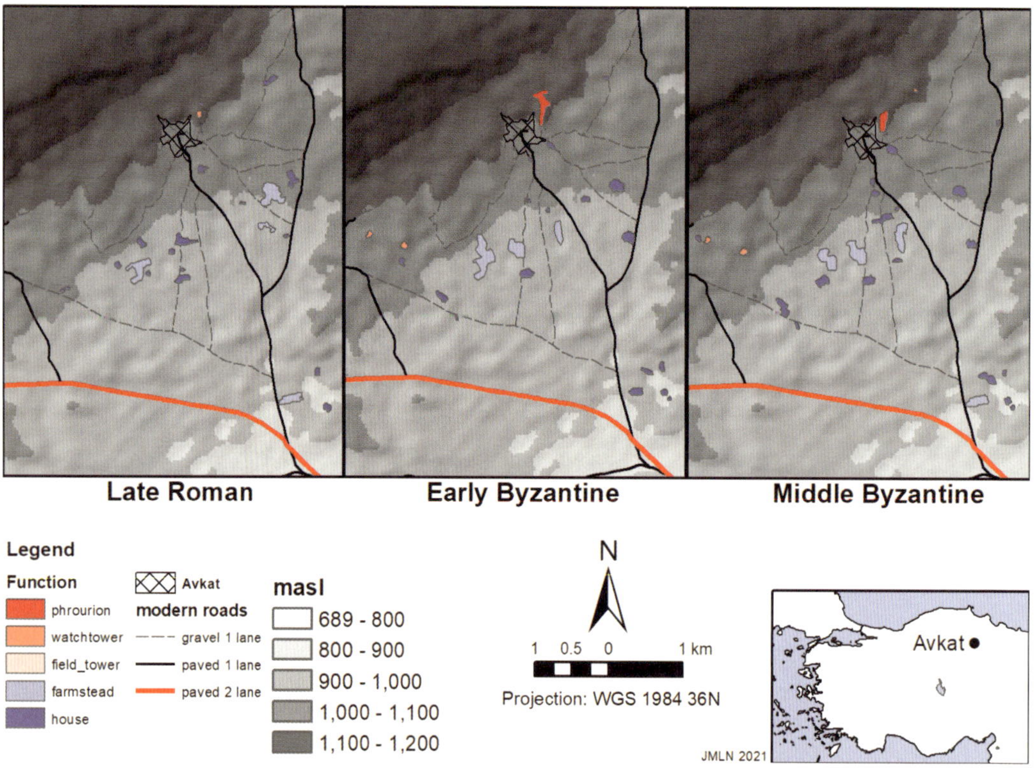

Fig. 6. Archaeological features in the Avkat Archaeological Project dated to the a) Late Roman (284–625 CE), b) Early Byzantine (626–799 CE), and c) Middle Byzantine (800–1261 CE) phases.

this particular region, discussing the extent to which resiliency and adaption can be observed, and contextualizing those patterns within the broader conditions of early seventh-century Anatolia.

Assessing Spatial Continuity and Change at Avkat

If the anticipated results of an Ω phase in medieval Anatolia include decreasing land use and a decrease in the number and size of archaeological features, the area around Euchaïta, the ancient name for Avkat and the settlement at the center of the Avkat project survey area, does not fit the pattern. When compared to prior phases, the seventh-century information shows that there are more features in the landscape, most often associated with previously used space, and these are of larger size (Fig. 6).

Figure 7 notes the number of features identified as agricultural or habitational per period. Between the Late Roman and Early Byzantine phases (see Table 1 for periodization), the count of these sites was unchanged. When normalizing the data to account for the varying length of periods (the Late Roman phase consists of 341 years;

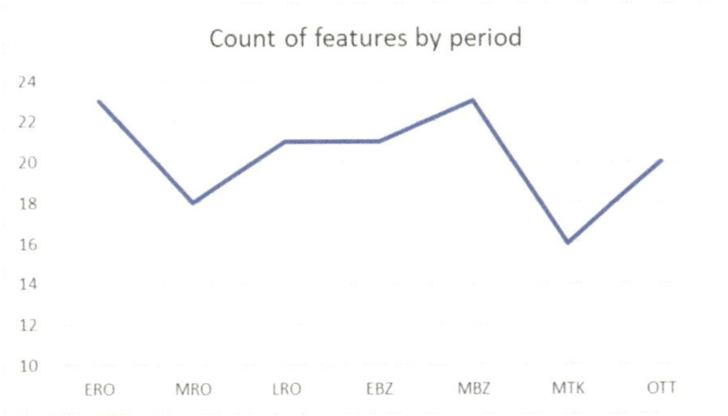

Fig. 7. Count of features by period.

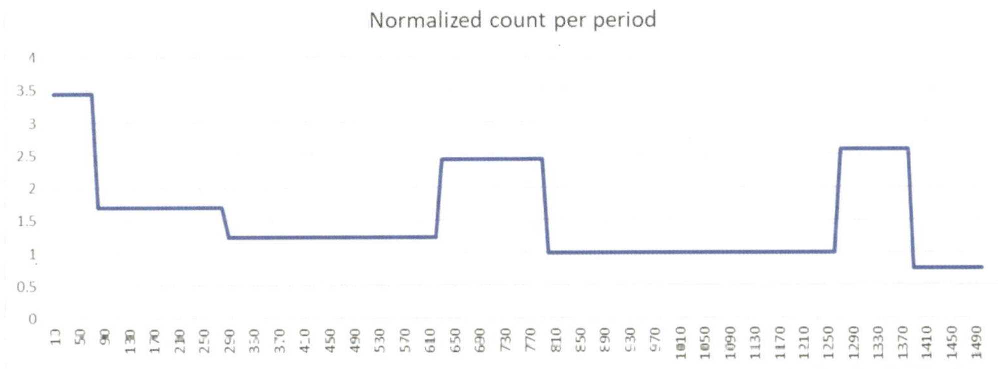

Fig. 8. Count of features, normalized by the number of years per period and multiplied by 20.

Table 1. Periods, abbreviations, and start and end dates used by the Avkat Archaeological Project.

Period	Abbreviation	Start Date	End Date
Early Roman	ERO	63 BCE	71 CE
Middle Roman	MRO	71 CE	283 CE
Late Roman	LRO	284 CE	625 CE
Early Byzantine	EBZ	626 CE	799 CE
Middle Byzantine	MBZ	800 CE	1261 CE
Medieval Turkish	MTK	1262 CE	1386 CE
Ottoman	OTT	1387 CE	1921 CE

the Early Byzantine, 173),[3] the Early Byzantine phase (beginning in 626 CE) shows a distinct increase in comparison with prior and succeeding periods (Fig. 8).

Not only are there more features in the landscape, but they are also larger. Figure 9 presents the total mean area for agricultural and/or habitational features per period (Fig. 9a) and the average area for each feature (Fig. 9b), indicating a modest rise from the Late Roman through the Early and Middle Byzantine phases. Figure 10 presents the same data normalized by dividing the number of years in each period and multiplying by 20. This provides a very rough rate of activity per generation, which suggests an even sharper discrepancy between the Late Roman and Early Byzantine phases.

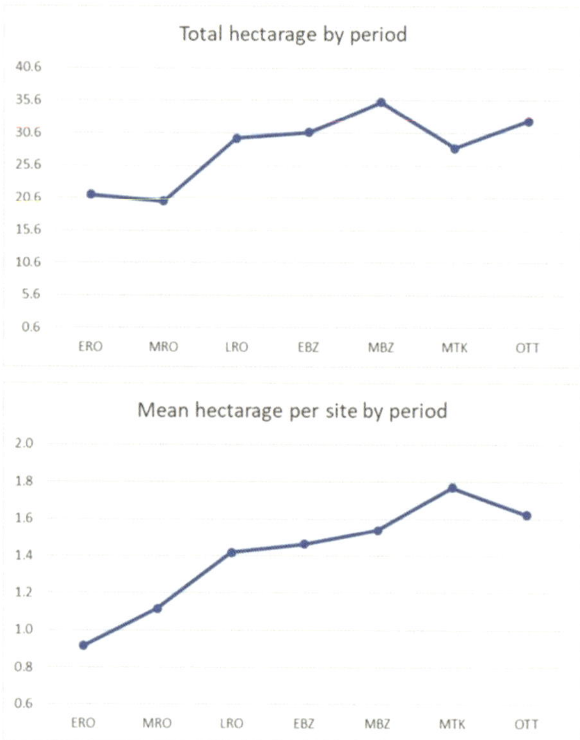

Fig. 9. Average area per period and per feature.

Beyond number and size, the transition between the Late Roman and Early Byzantine periods appears to have included a landscape presence in the Avkat region, reflecting a combination of both continuity and diversity. This dynamic can be reviewed via aspects of persistence and change. Often viewed as negatively correlated, this may not always be the case.

3 This paper employs a normalization formula employed by other studies of the relationship between environment and society in the medieval eastern Mediterranean (e.g., Izdebski et al. 2016; Cassis et al. 2018; Roberts et al. 2018): ([N of sites] / [N of years in period]) x 20.

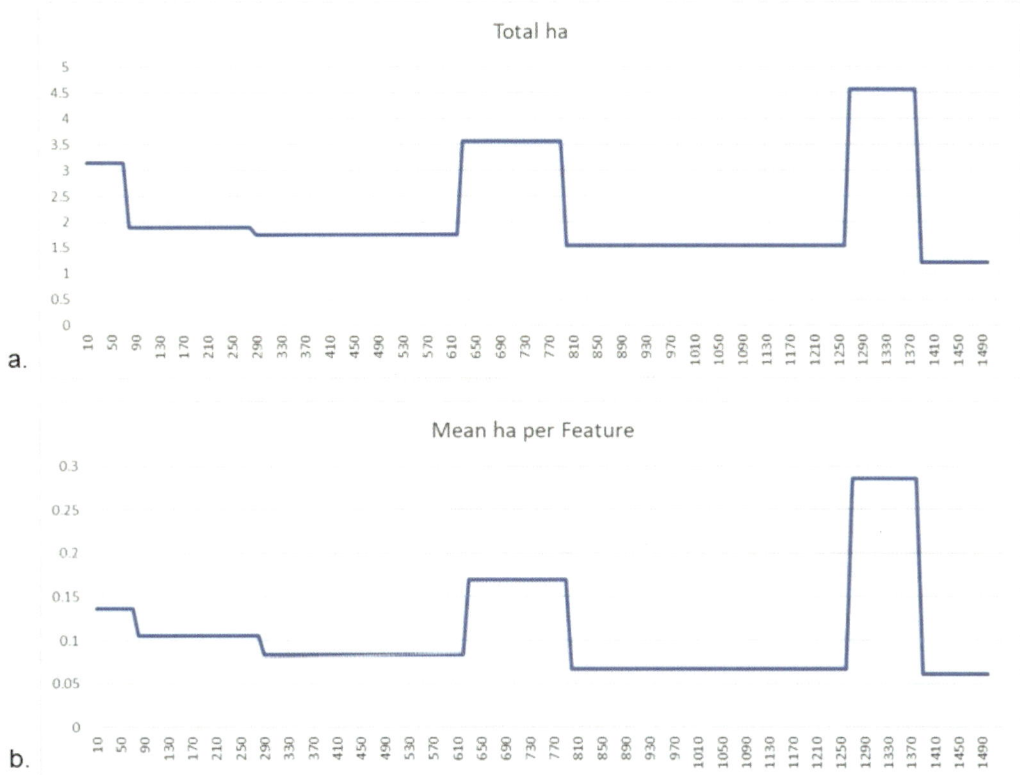

Fig. 10. Average area per period and per feature, normalized by the number of years per period and multiplied by 20.

Persistence is the first aspect to assess. Newhard (2018) suggested that the Avkat agrarian landscape reflected the notion of "persistent places"—spaces within the landscape that exhibited similar functional use over long periods of time (Schlanger 1992: 91). To visualize this (Fig. 11), the number of features in the Avkat region were subjected to a coefficient:

$$\text{persistence coef} = \frac{|A \cap B|}{A}$$

where A is the number of features of the earliest period and B the number for the subsequent period. The resultant coefficient provides a measure of how many features continued—in some fashion—into the next phase. In performing this measure, $A \cap B$ was counted with respect to mere absence or presence. Results of this calculation suggest that the transition from the Late Roman to Early Byzantine phases represents a good amount of consistency: of features dated to the Late Roman phase, 62% exhibited some form of activity in the subsequent period.

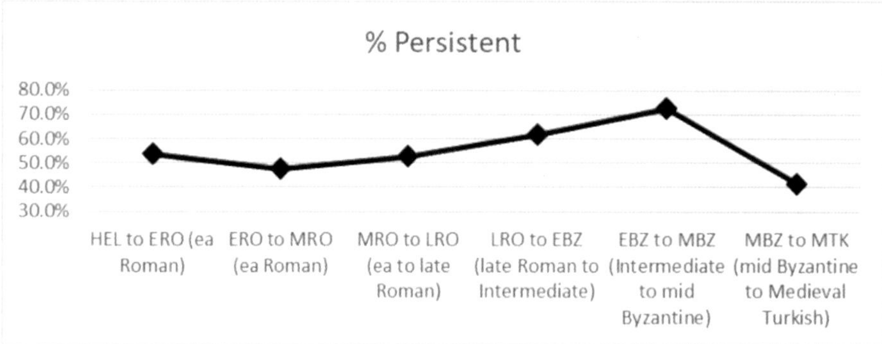

Fig. 11. Measures of persistent spaces in the Avkat region.

A second measure, the Jacccard coefficient (Fig. 12b),[4] was applied to assess the extent of similarity (in terms of size and shape) of features between periods. This measure and its derivations are used often in machine-based assessments of similarity in a range of applications.[5] The Jaccard coefficient was applied geospatially to the agricultural/habitational features within the Avkat region, assessing the area of each phase and its similarity to the subsequent period (Table 2). When comparing the Late Roman to Early Byzantine transition, the Avkat region expressed a low level of similarity; the only transition with a lower coefficient was that between the Middle Byzantine and Medieval Turkish periods. The measure suggests that there is a recognizable difference between the land-use footprint of the Late Roman and Early Byzantine periods. Given that the Jaccard coefficient measures similarity between two entities but does not provide a measure of difference from one to the other, a further assessment was taken. To obtain a measure of how much a subsequent period differed from the prior, a coefficient that evaluated the extent to which activities of a later phase were not a residual of past spaces was obtained (Fig. 12a):

$$\text{change coef} = \frac{B - |A \cap B|}{B}$$

The resulting coefficients present the Early Byzantine footprint as rather different from the Late Roman, with 63% of the area identified as Early Byzantine located in spaces not used in the prior period (see Table 2).

4 Jaccard coef. A,B = $|A \cap B| \ / \ |A \cup B|$. See (Jaccard 1912).

5 E.g., detecting plagiarism (Shuai Wang et al. 2013), data mining and filtering (Plansangket and Gan 2015; Ayub et al. 2020), and imagery analysis (Kobayakawa et al. 2009).

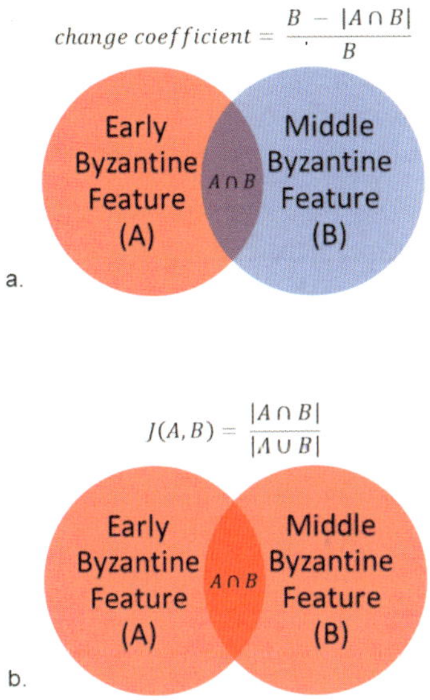

Fig. 12. Conceptual diagrams for the change (a) and Jaccard (b) coefficients. For (a), the area for B that is not intersecting is represented as a percentage of the total area of B. For the Jaccard coefficient (b), the combined area of both A and B is determined, and the amount of intersect is expressed as a percentage of the entire union of A and B.

Table 2. Persistence, Jaccard, and change coefficients for each transition in the Avkat region.

Transition	Persistence coef.	Jaccard	Change coef.
ERO - MRO	0.476	0.317	0.502
MRO - LRO	0.529	0.316	0.596
LRO - EBZ	0.619	0.233	0.627
EBZ - MBZ	0.727	0.513	0.368
MBZ - MTK	0.417	0.142	0.721
MTK - OTT	not recorded	0.538	0.345

When taken together, the Late Roman to Early Byzantine periods in the Avkat region present a land-use pattern that suggests a strong tie to past practices (high persistence) but considerable expansion and alteration (low Jaccard coefficient, high change coefficient, and enlarged archaeological footprint). When viewing these factors within the larger context of the characteristics of the Ω phase of the Adaptive Cycle, the operative words to remember are release, change, and reorganization.

Larger Context: The Late Roman/Byzantine World, 500–1100 CE

For the Byzantine state, the period between 500 and 1100 CE has been described as a process of increasing connectedness, followed by over-connectedness and rigidity, followed by a period of catastrophe and reorganization (Haldon 2016; Haldon and

Rosen 2018). Late Roman/Byzantine history is punctuated in the sixth century by the reign of Justinian I (527–565 CE), who is known for reclaiming much of the previously recognized Roman world under his domain. The period is known for increasing economic and political interconnectedness and prosperity (r phase). Success leads to the continuation of past successful practices, resulting in rigidity, consolidation, and over-connectedness (K phase). Beginning in the 630s and continuing over the course of the next century, the Byzantine state lost nearly 75% of its territory to the Islamic-Arabs and other successful incursions (Haldon and Rosen 2018). Viewed as a period of crisis (Ω phase), palynological data shows a significant decline in agrarian production across Anatolia (Izdebski, Koloch, and Słoczyński 2016; Roberts et al. 2018). By the eighth century (α phase), systemic political and fiscal reorganization brought about a level of stability (Haldon 2016; Haldon and Rosen 2018); agricultural production appears to have increased shortly thereafter (Izdebski, Koloch, and Słoczyński 2016; Roberts et al. 2018; Xoplaki et al. 2018).

Meeting the Islamic-Arab threat in the 630s required the Byzantine state to increase its military presence in central Anatolia significantly, the result of deploying troops from other parts of the empire. As a part of this strategy, Euchaïta was identified as a military outpost (Haldon 2018). This is corroborated by the archaeological and remote sensing survey, which records significant activity dating to this period related to increased defensive elements in the landscape (Elton 2018; Newhard 2018; Watters Wilkes 2018). Despite shifting environmental conditions, the archaeological data points toward a continuation of the scale of agricultural activity consistent with previous phases. Beyond the continuity and persistence of activity within the rural landscape, Euchaïta gains significance within the overall Byzantine defense system. What is recorded as a period of retraction in other regions is a boon (of sorts) to Euchaïta. Within the construct of the Adaptive Cycle, there was in all likelihood a stress to the regional system, given the environmental record and broader social and political conditions. The archaeological and historical evidence, however, suggest an instance of panarchic (top-down) intervention in the case of the Avkat region via the larger Byzantine system, impacting the regional adaptive cycle of Euchaïta.[6]

Conclusion

Within the Adaptive Cycle, the Ω phase, characterized as a period of crisis, collapse, and restructuring, is often assumed to be indicated by abandonment, retrenchment, and simplification within the archaeological and historical data (Strunz, Marselle, and Schröter 2019). We expect declines in demography and land use, or their assumed proxies in counts of archaeological features and/or cereal spore counts. The data from Euchaïta from the sixth–ninth centuries reminds us of not only the interplay between

6 A similar example on a larger scale includes Byzantine policies regarding Antioch in the sixth century CE (Haldon et al. 2020b).

environment and regional human organizations, but of the larger systems in play that can impact local systems and engagements. The environmental data suggests widespread decrease in agricultural production and abandonment of agricultural lands, historical data notes increased fragility and threat to the Byzantine political system, and archaeological data shows broad trends in decreased presence in the extra-urban landscape. Regional systems are not necessarily uniformly impacted, however, and stresses that impact social-ecological systems constructed at higher scales of organization can impact lower systems, presenting us with trends that run counter to those expected. The example from the Avkat region is one such instance. Here, there was no retraction in land use and human activity, as expected under an Ω phase, which would be in parallel with regions to the north (or the south, for that matter; see Cassis et al. 2018; Roberts et al. 2018). Rather, it is an example of how the interconnectedness of adaptive cycles operating at differing levels of scale (panarchy) leads to an *observed* non-linear process of sustainability, adaptation, and change (Walker et al. 2004). Rather than hold to the criticism that such non-linearity potentially reduces the efficacy of the Adaptive Cycle as a helpful framework for discussing transformation and resiliency (Middleton 2017), the Avkat region underlines the principal dynamic of the Ω phase—the release of energy. In a phase of systemic deconstruction, it is inevitable that components of a system will be unequally impacted. Adverse impacts can be asynchronous and spatially non-uniform, based upon the extent to which systems are exposed to the variables predicating release and their internal capacity to adapt, and the extent to which systems hierarchically positioned above and below impact response (Haldon et al. 2020a). The data from Avkat illustrates this point, emphasizing the importance of considering concepts of resiliency and transformation from a multi-scalar perspective.

Works Cited

Arponen, V. P. J., Walter Dörfler, Ingo Feeser, Sonja Grimm, Daniel Groß, Martin Hinz, Daniel Knitter, Nils Müller-Scheeßel, Konrad Ott, and Artur Ribeiro. 2019. "Environmental Determinism and Archaeology. Understanding and Evaluating Determinism in Research Design." *Archaeological Dialogues* 26, no. 1: 1–9. https://doi.org/10.1017/S1380203819000059.

Ayub, Mubbashir, Mustansar Ali Ghazanfar, Tasawer Khan, and Asjad Saleem. 2020. "An Effective Model for Jaccard Coefficient to Increase the Performance of Collaborative Filtering." *Arabian Journal for Science and Engineering* 45, no. 12: 9997–10017. https://doi.org/10.1007/s13369-020-04568-6.

Cassis, Marica, Owen Doonan, Hugh Elton, and James Newhard. 2018. "Evaluating Archaeological Evidence for Demographics, Abandonment, and Recovery in Late Antique and Byzantine Anatolia." *Human Ecology* 46, no. 3: 381–98. https://doi.org/10.1007/s10745-018-0003-1.

Cheddadi, Rachid, Majda Nourelbait, Ouafaa Bouaissa, Jalal Tabel, Ali Rhoujjati, José Antonio López-Sáez, Francisca Alba-Sánchez, et al. 2015. "A History of Human Impact on Moroccan Mountain Landscapes." *African Archaeological Review* 32, no. 2: 233–48. https://doi.org/10.1007/s10437-015-9186-7.

Cline, Eric H. 2015. *1177 B.C.: The Year Civilization Collapsed*. Rev. Princeton, N.J.: Princeton University Press.

Elton, Hugh. 2018. "The Archaeology of the City and Its Hinterland." In *Archaeology and Urban Settlement in Late Roman and Byzantine Anatolia: Investigating the History of Euchaïta-Avkat-Beyözü and Its Environment*, edited by John Haldon, Hugh Elton, and James Newhard, 185–209. Cambridge: Cambridge University Press.

Grundmann, Philipp, Melf-Hinrich Ehlers, and Götz Uckert. 2012. "Responses of Agricultural Bioenergy Sectors in Brandenburg (Germany) to Climate, Economic and Legal Changes: An Application of Holling's Adaptive Cycle." *Energy Policy*, Special Section: Frontiers of Sustainability, 48: 118–29. https://doi.org/10.1016/j.enpol.2012.04.051.

Gunderson, Lance H., and C. S. Holling, eds. 2002. *Panarchy: Understanding Transformations in Human and Natural Systems*. Washington, DC: Island Press.

Haldon, John. 2016. *The Empire That Would Not Die. The Paradox of Eastern Roman Survival 640–740*. Cambridge: Cambridge University Press.

– – –. 2018. "Euchaïta: From Late Roman and Byzantine Town to Ottoman Village." In *Archaeology and Urban Settlement in Late Roman and Byzantine Anatolia: Investigating the History of Euchaïta-Avkat-Beyözü and Its Environment*, edited by John Haldon, Hugh Elton, and James Newhard, 210–54. Cambridge: Cambridge University Press.

Haldon, John, Arlen F. Chase, Warren Eastwood, Martin Medina-Elizalde, Adam Izdebski, Francis Ludlow, Guy Middleton, Lee Mordechai, Jason Nesbitt, and B. L. Turner. 2020a. "Demystifying Collapse: Climate, Environment, and Social Agency in Pre-Modern Societies." *Millennium* 17, no. 1: 1–33. https://doi.org/10.1515/mill-2020-0002.

Haldon, John, Merle Eisenberg, Lee Mordechai, Adam Izdebski, and Sam White. 2020b. "Lessons from the Past, Policies for the Future: Resilience and Sustainability in Past Crises." *Environment Systems and Decisions* 40: 287–97. https://doi.org/10.1007/s10669-020-09778-9.

Haldon, John, Neil Roberts, Adam Izdebski, Dominik Fleitmann, Michael McCormick, Marica Cassis, Owen Doonan, et al. 2014. "The Climate and Environment of Byzantine Anatolia: Integrating Science, History, and Archaeology." *Journal of Interdisciplinary History* 45, no. 2: 113–61. https://doi.org/10.1162/JINH_a_00682.

Haldon, John, and Arlene Rosen. 2018. "Society and Environment in the East Mediterranean ca 300–1800 CE. Problems of Resilience, Adaptation and Transformation. Introductory Essay." *Human Ecology* 46, no. 3: 275–90. https://doi.org/10.1007/s10745-018-9972-3.

Holling, C. S. 2001. "Understanding the Complexity of Economic, Ecological, and Social Systems." *Ecosystems* 4, no. 5: 390–405. https://doi.org/10.1007/s10021-001-0101-5.

Ion, Alexandra. 2019. "Beyond Determinism. A Case for Complex Explanations and Human Scale in Framing Archaeological Causal Explanations." *Archaeological Dialogues* 26, no. 1: 10–12. https://doi.org/10.1017/S1380203819000084.

Izdebski, Adam, Grzegorz Koloch, and Tymon Słoczyński. 2016. "Exploring Byzantine and Ottoman Economic History with the Use of Palynological Data: A Quantitative Approach." *Jahrbuch der Österreichischen Byzantinistik* 65: 67–110. https://doi.org/10.1553/joeb65s67.

Izdebski, Adam, Jordan Pickett, Neil Roberts, and Tomasz Waliszewski. 2016. "The Environmental, Archaeological and Historical Evidence for Regional Climatic Changes and Their Societal Impacts in the Eastern Mediterranean in Late Antiquity." *Quaternary Science Reviews* 136: 189–208. https://doi.org/10.1016/j.quascirev.2015.07.022.

Jaccard, Paul. 1912. "The Distribution of the Flora in the Alpine Zone." *The New Phytologist* 11, no. 2: 37–50.

Knapp, A. Bernard, and Sturt W. Manning. "Crisis in Context: The End of the Late Bronze Age in the Eastern Mediterranean." *American Journal of Archaeology* 120, no. 1: 99–149. https://doi.org/10.3764/aja.120.1.0099.

Kobayakawa, Michihiro, Shigenao Kinjo, Mamoru Hoshi, Tadashi Ohmori, and Atsushi Yamamoto. 2009. "Fast Computation of Similarity Based on Jaccard Coefficient for Composition-Based Image Retrieval." In *Advances in Multimedia Information Processing - PCM 2009*, edited by Paisarn Muneesawang, Feng Wu, Itsuo Kumazawa, Athikom Roeksabutr, Mark Liao, and Xiaoou Tang, 949–55. Lecture Notes in Computer Science. Berlin, Heidelberg: Springer. https://doi.org/10.1007/978-3-642-10467-1_87.

Kristiansen, Kristian. 2019. "Who Is Deterministic? On the Nature of Interdisciplinary Research in Archaeology." *Archaeological Dialogues* 26, no. 1: 12–14. https://doi.org/10.1017/S1380203819000060.

Lespez, Laurent, Arthur Glais, José-Antonio Lopez-Saez, Yann Le Drezen, Zoï Tsirtsoni, Robert Davidson, Laetitia Biree, and Dimitra Malamidou. 2016. "Middle Holocene Rapid Environmental Changes and Human Adaptation in Greece." *Quaternary Research* 85, no. 2: 227–44. https://doi.org/10.1016/j.yqres.2016.02.002.

Luo, Fanghan, Yanxu Liu, Jian Peng, and Jiansheng Wu. 2018. "Assessing Urban Landscape Ecological Risk through an Adaptive Cycle Framework." *Landscape and Urban Planning* 180: 125–34. https://doi.org/10.1016/j.landurbplan.2018.08.014.

Maley, Jean, and Robert Vernet. 2015. "Populations and Climatic Evolution in North Tropical Africa from the End of the Neolithic to the Dawn of the Modern Era." *African Archaeological Review* 32, no. 2: 179–232. https://doi.org/10.1007/s10437-015-9190-y.

Marsh, Erik J. 2016. "The Disappearing Desert and the Emergence of Agropastoralism: An Adaptive Cycle of Rapid Change in the Mid-Holocene Lake Titicaca Basin (Peru-Bolivia)." *Quaternary International*, Advances in Arid Zone Archaeology: The 4th Southern Deserts Conference, 422: 123–34. https://doi.org/10.1016/j.quaint.2015.12.081.

McIntosh, Roderick J., Jean Maley, and Robert Vernet. 2015. "African Palaeoclimate and Human Response: A Special Issue of the African Archaeological Review." *African Archaeological Review* 32, no. 2: 167–77. https://doi.org/10.1007/s10437-015-9187-6.

Middleton, Guy D. 2017. "The Show Must Go on: Collapse, Resilience, and Transformation in 21st-Century Archaeology." *Reviews in Anthropology* 46, no. 2–3: 78–105. https://doi.org/10.1080/00938157.2017.1343025.

Newhard, James. 2018. "The Survey: Methods of Survey, Data Collection, and Management and Artefactual Patterning." In *Archaeology and Urban Settlement in Late Roman and Byzantine Anatolia: Investigating the History of Euchaïta-Avkat-Beyözü and Its Environment*, edited by John Haldon, Hugh Elton, and James Newhard, 49–69. Cambridge: Cambridge University Press.

Newhard, James, and Eric H. Cline. Forthcoming. "Panarchy and the Adaptive Cycle: A Case Study from Mycenaean Greece." In *What Sort of the Past Does the Future Need? Policy Lessons and the Historical Imagination*, edited by Piotr Filipkowski, John Haldon, and Adam Izdebski. Risk, Systems, and Decisions. New York: Springer.

Plansangket, Suthira, and John Q. Gan. 2015. "A Query Suggestion Method Combining TF-IDF and Jaccard Coefficient for Interactive Web Search." *Artificial Intelligence Research* 4 (2). https://doi.org/10.5430/air.v4n2p119.

Redman, Charles L., and Ann P. Kinzig. 2003. "Resilience of Past Landscapes: Resilience Theory, Society, and the Long Durée." *Conservation Ecology* 7, no. 1: 14.

Riede, Felix. 2019. "Environmental Determinism and Archaeology. Red Flag, Red Herring." *Archaeological Dialogues* 26, no. 1: 17–19. https://doi.org/10.1017/S1380203819000072.

Roberts, Neil, Marica Cassis, Owen Doonan, Warren Eastwood, Hugh Elton, John Haldon, Adam Izdebski, and James Newhard. 2018. "Not the End of the World? Post-Classical Decline and Recovery in Rural Anatolia." *Human Ecology* 46, no. 3: 305–22. https://doi.org/10.1007/s10745-018-9973-2.

Schlanger, Sarah H. 1992. "Recognizing Persistent Places in the Anasazi Settlement Systems." In *Space, Time, and Archaeological Landscapes*, edited by J. Rossignol and L. Wandsnider, 91–112. Interdisciplinary Contributions to Archaeology. New York: Plenum.

Shuai Wang, Haoliang Qi, Leilei Kong, and Cuixia Nu. 2013. "Combination of VSM and Jaccard Coefficient for External Plagiarism Detection." *2013 International Conference on Machine Learning and Cybernetics* 4: 1880–85. https://doi.org/10.1109/ICMLC.2013.6890902.

Strunz, Sebastian, Melissa Marselle, and Matthias Schröter. 2019. "Leaving the 'Sustainability or Collapse' Narrative Behind." *Sustainability Science* 14, no. 6: 1717–28. https://doi.org/10.1007/s11625-019-00673-0.

Walker, Brian, C. S. Holling, Stephen R. Carpenter, and Ann Kinzig. 2004. "Resilience, Adaptability and Transformability in Social–Ecological Systems." *Ecology and Society* 9, no. 2. Accessed April 28, 2021. http://www.jstor.org/stable/26267673

Watters Wilkes, Margaret. 2018. "Remote Sensing and Geophysical Prospection." In *Archaeology and Urban Settlement in Late Roman and Byzantine Anatolia: Investigating the History of Euchaïta-Avkat-Beyözü and Its Environment*, edited by John Haldon, Hugh Elton, and James Newhard, 257–68. Cambridge: Cambridge University Press.

Xoplaki, Elena, Jürg Luterbacher, Sebastian Wagner, Eduardo Zorita, Dominik Fleitmann, Johannes Preiser-Kapeller, Abigail M. Sargent, et al. 2018. "Modelling Climate and Societal Resilience in the Eastern Mediterranean in the Last Millennium." *Human Ecology* 46, no. 3: 363–79. https://doi.org/10.1007/s10745-018-9995-9.

II

Events, Landscapes, and People: Change and Response

Re-Thinking Thera: Tree-Rings, Radiocarbon, and Response in the Second Millennium BCE

Charlotte Pearson

Introduction

The Bronze Age eruption of Thera (Santorini) is one of the best-known, intensively studied, and emphatically debated volcanic eruptions of antiquity. Questions about the size of the eruption, the precise sequence of eruptive events, and impacts on the climate, culture, and ecosystems of the ancient Mediterranean all remain interconnected with the need to finally resolve the question of when, precisely, the eruption happened. As has been well described in the literature (e.g., Hardy 1990; Warburton 2009; Manning 2014; Manning et al. 2014; Höflmayer 2012; Wiener and Earle 2014; McAneney and Baillie 2019), prior to the release of the IntCal20 radiocarbon calibration curve, a gap of ca. 50–100 years divided two clusters of dating estimates for the eruption date. The older is based on radiocarbon dating of, primarily, mixed-aged seeds (Bronk Ramsey, Manning, and Galimberti 2004; Manning et al. 2006, 2014) and four segments of an olive tree (Friedrich et al. 2006, 2014) buried by the eruption deposits and dated to the later seventeenth century BCE (Manning 2014; Manning et al. 2014), while the younger, dating to the mid to late sixteenth century BCE, is based on archaeological interconnections with Egypt and dated using a range of evidence including king-lists, texts, monumental records, and ceramic sequences (Warren 2006a, 2006b, 2010; Wiener 2009, 2010; Höflmayer 2012). Pearson and colleagues (2018) measured the radiocarbon content of calendar-dated, single tree rings across the period from 1700–1500 BCE (during which more or less everyone would agree that the Thera eruption took place). This indicated that the courser-resolution radiocarbon calibration data which underlay the IntCal13 radiocarbon calibration curve could be refined and improved. Work to test these findings (Friedrich et al. 2020; Pearson et al. 2020a, 2020b) has resulted in this section of the IntCal20 curve (Reimer et al. 2020) becoming the most highly resolved in the current ca. 55,000 years covered by IntCal, with over 800 single-year, calendar-dated tree-ring measurements

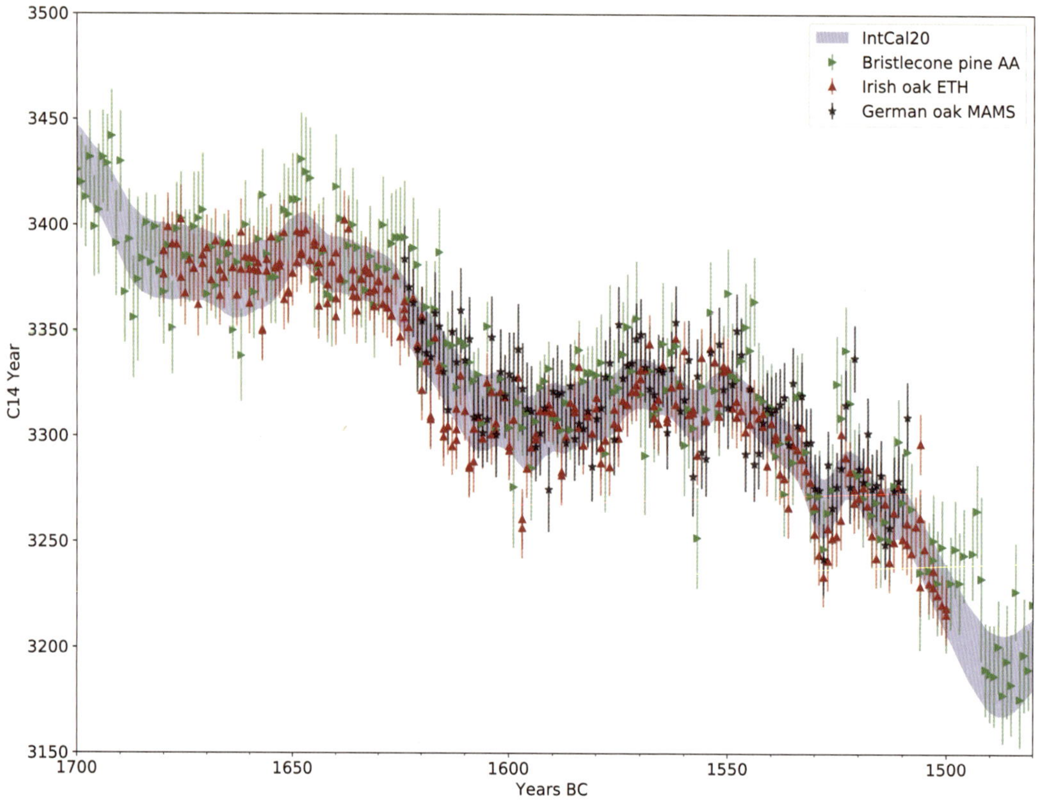

Fig. 1. The IntCal20 curve and three examples of annual measurements of radiocarbon made on contemporary, calendar-dated tree-rings from North America, Ireland, and Germany run at three different radiocarbon laboratories across the period 1700–1500 BCE. The data are all within error ranges of one another and show strong agreement. Note the spread of the annual ^{14}C data for each distinct year using the BP years vertical axis compared with the smoothed, reduced error IntCal20 curve. This provides a visual guide to how much true variation might be expected for a single radiocarbon sample representing a single year at any point across this time period. In our quest for the highest possible precision when aiming to calibrate a radiocarbon date, it is worth keeping this reality in mind. Sometimes the highest accuracy comes with lower precision. The difference that these more accurate data made to curve shape is illustrated in Figure 3. Data used to create this figure can be found in the following publications: Pearson et al. 2018, 2020 (AA,ETH); Friedrich et al. 2020 (MAMS).

produced at multiple radiocarbon laboratories (Fig. 1). This takes care of one part of the necessary investigation of the process of radiocarbon dating in terms of attempting to finally resolve the Thera date debate. This annually based, multi-lab calibration dataset provides the most accurate representation of levels of radiocarbon in the environment currently possible for the Thera period. This is critically important for establishing the exact position and shape of a plateau in radiocarbon across this period which is of primary relevance for radiocarbon dating Thera. Measurement of radiocarbon in contemporary growth from Turkish tree rings has also shown that this new curve (which contains measurements on samples from Germany, Ireland, and North America) provides an applicable resource for the Mediterranean during this period, with only a small regional offset in radiocarbon (Fig. 2; Pearson et al. 2020b).

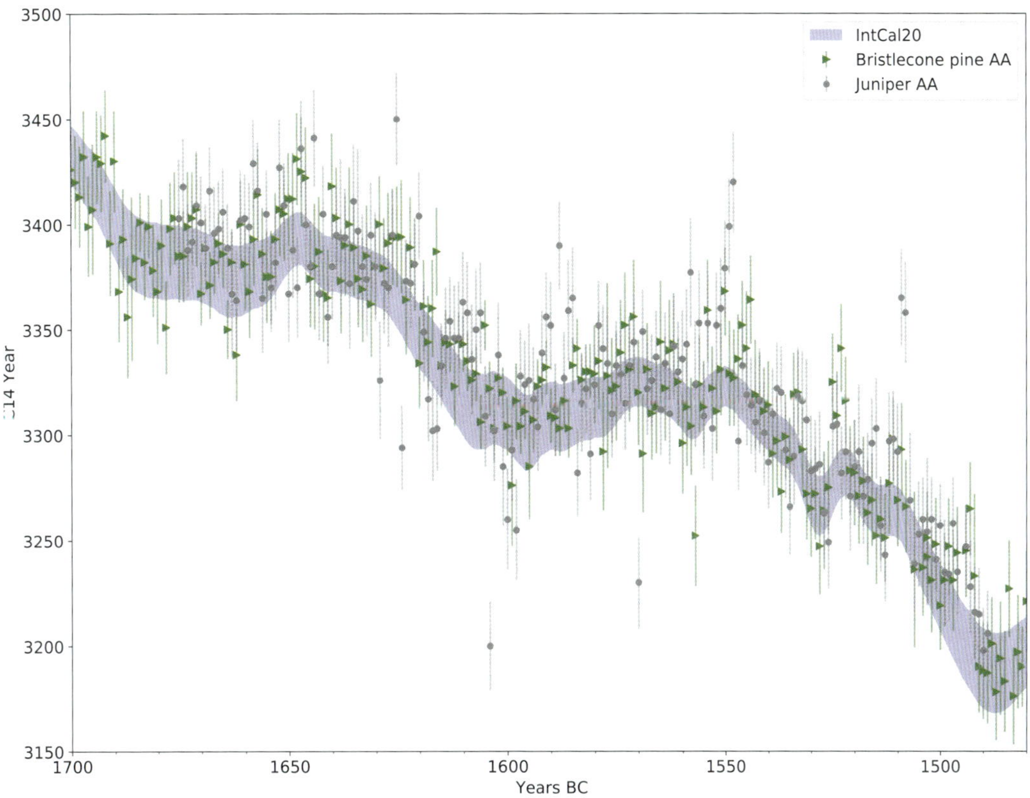

Fig. 2. This figure explores the idea of regional offsets in radiocarbon across the period of greatest relevance to the Thera eruption by showing measurements of radiocarbon in contemporary tree-ring years from Mediterranean (Juniper AA) and North American (Bristlecone pine AA) trees growing at a similar latitude. We see no significant offset; that is, the data overlap well, indicating that growth season or other potential regional ^{14}C factors made little difference on the sample of ^{14}C extracted from the environment at the time of growth in each case. The two datasets were produced at the same laboratory to avoid slight differences that could be introduced by using a range of laboratory pre-treatments or differences in instrumentation. Typically, regional radiocarbon offsets cannot be effectively demonstrated by comparing newly measured series against the calibration curve, as has been a common practice in the past (e.g., Manning and Kromer 2012), because the curve will always include some inter-laboratory variation. For the Thera period, however, the comparison can now be more legitimately made between the Turkish junipers and IntCal20, because it includes the Bristlecone pine AA data shown above, illustrating that Mediterranean samples from this period can now be calibrated to IntCal20 without concern of a substantial regional effect. Pearson and colleagues (2020a) discuss further details. Data used to create this figure can be found in Pearson et al. 2020b.

With this knowledge in place, we can more critically evaluate the impact of radiocarbon calibration curve shape in terms of samples relevant to Thera. For some, it has already been argued that this makes little difference (Van der Plicht et al. 2020; Manning et al. 2020), but the implications for Thera are highly dependent on decisions made by individuals regarding the treatment of existing dates (e.g., averaging of dates or applications of various modeling options [Van der Plicht et al. 2020, 1102–1107; Manning et al. 2020]). Earlier critiques (e.g., Wiener 2012) as well as recent work on annual radiocarbon measurements have underlined that however precise and accurate the radiocarbon calibration data, the calibration accuracy for any given measurement also relies heavily on the accuracy of that individual measurement as well as inherent year to year variability. Accordingly, the quoted error and possibilities for inter-laboratory variability for radiocarbon measurements on samples from archaeological contexts should also be approached critically. In this paper, I explore alternative ways to approach the published radiocarbon data for Thera which would seem to demonstrate that radiocarbon dating and archaeological dating evidence can align convincingly in the mid-late sixteenth century BCE. While this does not wholly resolve the "Thera debate," it offers a concrete framework through which to begin reassessing the implications of the event for climate, environment, and people.

Establishing a "Consensus Date" for Thera in Radiocarbon Years BP

At any point in time in which we attempt to radiocarbon date an event that is close to, or likely within, a radiocarbon plateau period, extra caution and critical awareness of the true limitations of the calibration attempt and the attendant errors of the samples to be calibrated is vital. Subtle inter-laboratory variations on sample measurements or decisions on averages or modeling strategies for assemblages may increase or decrease a given radiocarbon value and associated errors prior to calibration by a few decades. Small differences like these can have very significant effects if one's value is close to a plateau. The slightly higher and more structured shape of the IntCal20 radiocarbon plateau across the critical time period for Thera (Fig. 3) introduces new sixteenth century BCE possibilities for previously obtained radiocarbon dates and complicates things a little by increasing the possibility of multiple calibrated date ranges for any single (averaged or modeled) measurement and its associated error. For example, Van der Plicht and colleagues (2020, 1102) opt to use a ^{14}C date of 3350 ±10 BP to represent the eruption of Thera and highlight around five possible calendar ranges for this date relative to IntCal20, noting that each of these ranges should be reported along with their associated probabilities when discussing the calibrated dating result. Their latter point is important and clearly illustrates the problems of a more finely resolved curve in producing multiple dating possibilities, yet I would respectfully question the use of the averaged 3350 ±10 BP value to represent Thera. Van der Plicht and colleagues (2020, 1102) justify the use of this number because it represents a coinciding result from two separate eruption contexts when groups of values are combined. The first is a modeled average from a large group of short-lived organic material from the volcanic destruction layer at Akrotiri on Thera (Bronk Ramsey, Manning, and Galimberti 2004), a figure for which there have been a number of subsequent iterations involving a range of sample selections and modeling approaches (e.g., Manning et al. 2006, 2020). The second is an average from two animal bones obtained from the Palaikastro tsunami deposits on Crete, which are considered to have been deposited as a direct result of the eruption (Bruins, Van Der Plicht, and MacGillivray 2009, 397). In both cases, the archaeological unit is securely connected to

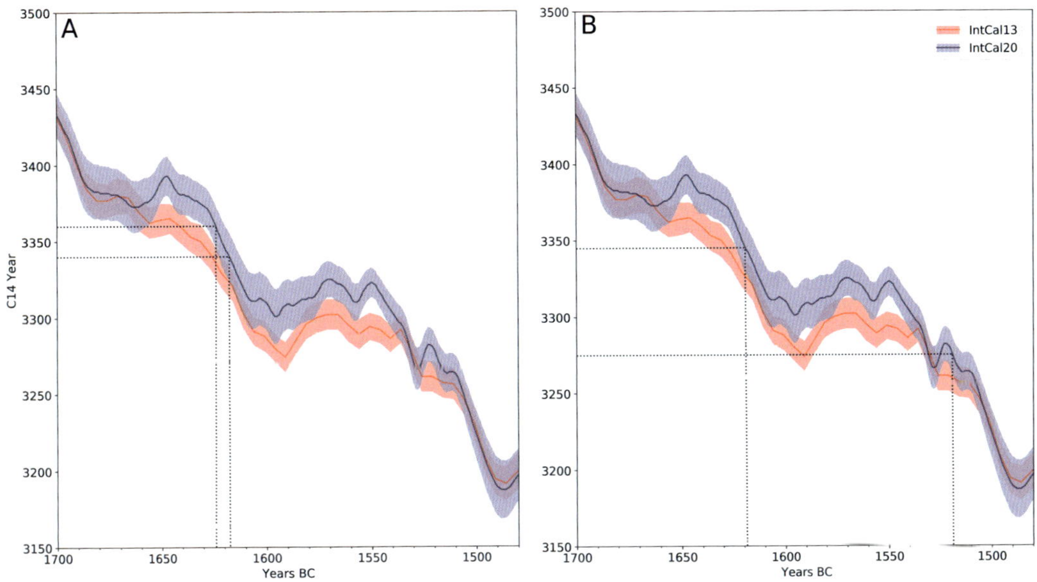

Fig. 3. Examples of how slight differences in quoted error or choices in averaging or modeling groups of mixed-age samples can impact dating results in periods where the calibration curve transitions from slope to plateau. The former IntCal curve is shown in red to illustrate the improvement in the accuracy of curve shape for this period in IntCal20. Note how the level of the radiocarbon plateau has moved ca. 20 years BP higher. This means that samples which might previously have firmly calibrated to the slope region prior to ca. 1620 BCE would now instead intersect with the plateau, losing precision but increasing accuracy, as the new curve section is a truer representation of past radiocarbon levels. In the two examples described in the text, representing older and younger extremes of dating possibilities for Thera-relevant samples, we see (A) how smaller errors (dotted lines) associated with the value of 3350 ± 10 BP mean that even with the revised curve shape, the dating probabilities change very little, as this value is still dominantly intersecting with the slope and not the plateau (with an error of ±20 BP on this same value, it would hit the plateau and the indicated ranges would be much broader). The re-occurring value of ca. 3310 ±35 BP from immediate pre-eruption contexts (B) cannot offer us the precision we need but incorporates a more reasonable measure of error based on the range of annual ^{14}C variability show in Figure 1.

Thera, however in the case of the Akrotiri samples, different laboratories have produced a mixture of dating results run at different times with quite large errors and wide-ranging ages. In the case of the bones, these would seem to represent two animals, one a bit younger than the other, with dates of 3310 ±35 BP (GrA-30336) and 3390 ±35 BP (GrA-30339), respectively (Bruins, Van Der Plicht, and MacGillivray 2009, 404), and both presumably already dead for an unknown period of time before the eruption, as they were apparently disarticulated at the time of the tsunami (imbricated bones rather than whole skeletons are reported). Two further dates (3325 ±40 BP and 3345 ±40 BP) were also reported from a single sheep or goat jaw found in a water-lain drain deposit thought to be contemporary with the tsunami. The two dates from this animal are identical within the given errors and overlap with the youngest of the other two animals. This being the case, an argument can be presented to consider the sample(s) representing the youngest material in the two contemporary deposits to be a *terminus post quem* for the eruption, rather than going with an average of younger and older animals from a single deposit. This case can be built further by comparing other non-calibrated values from pre-Theran contexts. The 3310 ±35 years BP date is within the error ranges of more than ten of the youngest seed determinations found at Akrotiri (Manning et al. 2006) and in similar overlap with dates from a number of other immediate pre/post eruption contexts in the wider region, measured at a variety of laboratories (see published data: Bronk Ramsey, Manning, and Galimberti 2004; Manning et al. 2006; Kutschera et al. 2012; Fischer 2009). For example, at Tell el-Ajjul (Gaza), two BP dates (3310 ±30, 3310±35, VERA 1905, 1906) mark unit H6, which directly underlies an influx of geochemically provenanced pumice from Thera in unit H5 (Fischer 2009), archaeologically dated to the start of the New Kingdom in Egypt.

Similarly, the youngest determination (oldest *terminus post quem*), from pre-eruption context D/2 at Tell el-Dab'a (Kutschera et al. 2012, VERA-3616) in Egypt, is also within the error range of 3310 ±35 years BP at 3337 ±44 BP. This context can be dated archaeologically to 1560–1530 BCE (Kutschera et al. 2012). The determination from context D/1 (VERA-3032), which post-dated or was contemporary with D/2, during which Thera pumice arrives at this site, is 3314 ±36 BP, archaeologically 1530–1500 BCE.

Of course it is not standard (nor necessarily good) practice to single out measurements from a mixed-age assemblage to represent a *terminus post quem* for a context; in this case, however, the consistency of the younger measurements from a variety of immediately pre-Theran contexts—run at different laboratories—is quite striking. While a particular context can logically contain a mixture of material of older, but not younger, age than the event that seals it, erroneous dates do occur as a result of contamination, inter-laboratory factors, or some aspect of stratigraphic mixing; so care is needed. Nevertheless, the 3310 ±35 BP measurement is right on the plateau, as shown on the IntCal20 curve (Fig. 3), suggesting that the Thera eruption occurred during this plateau period and explaining the observed repetition of other measurements overlapping with 3310 ±35 years BP (i.e., a plateau effect). Please note that the 3350 ±10 BP value used by Van der Plicht and colleagues (2020) is also plotted to illustrate that differences of a decade or so, easily introduced by a range of factors, can have a significant impact on calibrated ranges in plateau regions.

The one sample group with some prospects to combat the plateau effect is the much-debated Santorini olive branch (Friedrich et al. 2006, 548; see also *Antiquity* 88, 2014 and multiple references therein). At 3331 ±10 BP, the youngest of the four blocks into which the branch was subdivided is also consistent with the previously listed uncalibrated determinations; in this case, however, the oldest blocks calibrate above the plateau. If the olive had been a juniper, pine, or cedar, it could simply have been calendrically dated via dendrochronology using existing tree-ring sequences from the growth region for this time period (Kuniholm et al. 1996). Had it been oak, a systematic count of the reliably formed annual rings would have made it possible to execute a radiocarbon wigglematch with exact, known spacing between the samples and exact control on the boundaries of each sample block. Unfortunately it was an olive. Olive trees have indistinct growth boundaries which may represent multiple growth phases in a single year or may miss periods of growth altogether (Cherubini et al. 2014; Kuniholm 2014). They have also been shown to return exterior radiocarbon dates that are several decades older than the actual known date (Ehrlich, Regev, and Boaretto 2018). This is clearly a very important finding, as a couple of decades around the mid-sixteenth century BCE could make quite a difference to the Thera dating. Add to this the potential for sub-annual scale fluctuations in old (volcanic) carbon over the lifespan of the tree that could have slightly skewed one or more of the four radiocarbon measurements made on the divided sequence of ca. 70 growth increments, and the need for a cautious and measured approach to interpretation is implicit. Following a series of articles in *Antiquity* (Manning et al. 2014; Bruins and van der Plicht 2014; Bietak 2014; Friedrich et al. 2014; Kuniholm 2014; Cherubini et al. 2014), the authors of the original olive study increased the margins for error on sampling of the measurement blocks of the olive to a simple "ordered sequence" (Friedrich et al. 2006), meaning that the model used, instead of attempting to add defined spacing in terms of "years," simply applied the certain premise that the four samples move from oldest at the inside of the sample toward youngest at the outside of the sample. A recent study by Ehrlich, Regev, and Boaretto (2021) builds on this further to demonstrate potential radiocarbon dating possibilities for the olive, which encompass most of the sixteenth century BCE. The upshot of all this is to conclude that the radiocarbon evidence for Thera to date is not sufficiently strong to rule out archaeological arguments for a mid- (or post mid-) sixteenth century BCE date, and indeed can, through various proposed scenarios, overlap with the archaeological chronology.

Thera and Egypt

This conclusion can be further supported by exploring the agreement between the radiocarbon and archaeological dating for the start of the New Kingdom in Egypt. Strong archaeological arguments and stratigraphic evidence indicate that the Thera eruption took place after the start of the Egyptian New Kingdom (Höflmayer 2012; Wiener 2009; Warren 2010). This includes discussion of the Thera pumice and its first arrival in various New Kingdom contexts (as at Tell el-Ajjul and Tell el-Daba, e.g., Kutschera et al. 2012; Bietak and Höflmayer 2007). The presence of pumice indicates that by the time the associated stratigraphic layer and type artifacts were deposited, the eruption had already taken place. What is not absolutely clear is how much time may have passed between the eruption and the arrival of Thera pumice on site, and this matter has been a point of considerable discussion (e.g., Wiener 2010, 70; Höflmayer 2012, 441; Manning 2014). Whitham and Sparks (1986) showed that certain types of pumice can float on sea water for more than 1.5 years; therefore, Fischer (2009, 254) considers it likely that the sudden, intensive period of Theran pumice deposition in H5 at Tell el-Ajjul seems most likely to have occurred within a year or so of the eruption with transport across the sea via the winds and tides (Fischer 2009, 252), though not completely ruling out the possibility that pumice may exist in earlier unexcavated layers (Fischer 2009, 257) or have arrived anthropogenically somewhat later. Wiener (2010, 70) also advocates for a rapid use of new pumice as and when it became available, due to the wide range of uses for the material in antiquity. While the debate over the timing of pumice use and deposition, and various associated lines of chronological evidence, may continue in terms of a few decades for Thera, that the eruption occurred after the start of the New Kingdom seems in little doubt. Archaeologically, the New Kingdom begins sometime after ca. 1550 BCE (Warren 2010, 72), in good agreement with the results of a systematic radiocarbon study for Egypt (Bronk Ramsey et al. 2010) which places the start between 1566–1552 cal BCE at 68.2% probability or between 1570–1544 cal BCE at 95.4% probability. One could question the logic of challenging radiocarbon evidence for a seventeenth century BCE date for Thera and yet accepting radiocarbon evidence for the Egyptian study, but in fact there is sound basis to do so. Bronk Ramsey and colleagues (2010) use two hundred and eleven radiocarbon measurements from short-lived plants, together with a Bayesian model incorporating secure historical information on reign lengths, to produce a chronology for dynastic Egypt that extends from ca. 2650 to ca. 1100 BCE. This long time period with historically based anchors avoids the "suck-in" effect of any single radiocarbon plateau to achieve a much more robust dating than is possible for the Thera period samples, where the plateau effect is dominant. In fact, I would argue that the agreement between the radiocarbon-based and historically-based start of the New Kingdom (even if the former extends the margins a little earlier than is accepted by some scholars) supports the possibilities for a mid-sixteenth century BCE radiocarbon date for Thera as well, assuming the archaeological associations between Thera and the New Kingdom are sound.

If Thera erupted at the very start of the New Kingdom, the association with the reign of Ahmose (Bronk Ramsey et al. 2010; Warren 2010) and the Tempest Stela (Foster, Ritner, and Foster 1996; Ritner and Moeller 2014) could be revisited. While noting that such an association is not without chronological issues (see Warren 2010, 72; Wiener and Allen 1998; Warren 2010), the Stela famously describes what is considered to be an eyewitness account of an unusual, widespread, and destructive rainstorm, featuring darkness, a common association with volcanic eruptions and storm events (Ritner and Moeller 2014, 7, 14). Ritner and Moeller (2014, 7) also connect this description with another document which they suggest dates to the reign of Ahmose: a docket from year 11 in the Rhind Mathematical Papyrus that relates to the sound of unusual thunder followed by intensive rainfall— of which there is evidence on Crete following the Thera eruption (Athanassas et al. 2018)—again consistent with more recent accounts of the local impacts of documented eruptions (e.g., Krakatoa [Gil Guirado, Olcina-Cantos, and Pérez-Morales 2021]). Trevisanato (2007, 447) goes on to associate six medical papyri from approximately the same period with the Thera eruption (the Ebers Papyrus, Hearst Medical Papyrus, London Medical Papyrus, Carlsberg Papyrus 8, Ramesseum Papyrus III, and Edwin Smith Surgical Papyrus), going as far as to link medical impacts with volcanic acid fallout and the biblical plagues of Egypt, which include a "thick darkness" and poisoning of the Nile (Block 1976, 522, 524).

While many of these hypothesized connections seem unlikely based on what we know of the eruption, and there are some clear chronological issues, a further reference to the docket in the Rhind Mathematical Papyrus (Trevisanato 2007, 448) is worth some discussion. It is argued that the unusual thunder and heavy rain occurred three days before "New Year's Day," which, according to an opening note in the Ebers Papyrus (von Deines, Grapow, and Westendorf 1958), fell on the 309th day of the 365-day Egyptian legal year, that is, around August. This is interesting because it is consistent with what we know about the seasonality of the Thera eruption. Evidence for a summer eruption comes from analysis of a "death assemblage" of bean weevils (*Bruchus rufipes*), buried under the eruption deposits on Thera (Panagiotakopulu et al. 2013) and from other insect remains (MacGillivray and Warburton 2009). A summer eruption was also suggested by Doumas (1983), based on the paucity of food-crops in the storerooms of Akrotiri, and by models of tephra distribution that independently indicate scenarios for a spring summer (Johnston et al. 2012) or summer-autumn eruption (Sewell 2001). So while there is ambiguity in the association with the Rhind Mathematical Papyrus, which was "copied," "in likeness to writings of old" (Chase 1927, 1), the suggestion is intriguing.

Other Possible Cultural Connections

Whether any of the previously described records will ever be confirmed to link with Thera will probably come down to the final dating of the eruption, but it would seem odd that an event of the magnitude of Thera, including the dramatic burial of life at Akrotiri, was not recorded somewhere by the highly literate cultures of the surrounding regions. This seems especially so as it has been speculated that earlier cultures at Çatalhöyük in southern Anatolia may have recorded an eruption of Hasan Dağ (Schmitt et al. 2014, although see Meece 2006 for an alternative interpretation). Athanassas and colleagues (2018, 169) suggest that future translations of Linear A and Cretan hieroglyphics, both of which remain largely undeciphered, may yet hold an account of the eruption. They also speculate about Greek mythology, suggesting that the cataclysmic events described in Hesiod's Titanomachy (War of the Titans, ca. 750 BCE) were in fact formulated long before Hesiod as part of an inherited memory of the people of the region linked back to Thera (Athanassas et al. 2018, 170). While this might sound like a stretch, a growing number of recent studies demonstrate that anthropological memories of volcanic events can be handed down with considerable accuracy over many thousands of years (e.g., Wilkie, Cahir, and Clark 2020). The argument that the text describes an eruption of a volcano seems compelling based on the passage they quote from Evelyn-White's translation (1936, 171):

> "... [the] sea rang terribly around, and the earth crashed loudly: wide Heaven was shaken and groaned .. and a heavy quaking reached dim Tartarus... All the land seethed... The hot vapour lapped round the earthborn Titans: flame unspeakable rose to the bright upper air: the flashing glare of the thunder-stone and lightning... Also the winds brought rumbling earthquake and dust storm, thunder and lightning..."

Similarly, the argument for descriptions of Typhon as a metaphor for the classic volcanic eruption column with associated lightening (see McNutt and Williams 2010) is quite compelling along with speculated volcanic connections with the Hittite Storm God who plunged Anatolia into darkness with dust clouds and smoke, choking crops and pasture, before sending floods to clean off and invigorate the land (Athanassas et al. 2018, 171). It is important to note, however, that the Thera eruption should not be considered to be the only volcanic eruption witnessed by collective cultural memory in the wider region, and indeed it may not have been the only eruption to affect the region in very diverse ways. A part of the legacy of the various "Thera debates" is that this huge, much studied, much publicized event has emerged as the most obvious choice with which to associate a range of cultural references such as these, and also with which to associate a range of paleoenvironmental signals. What we are now learning is that larger volcanic eruptions are not so uncommon and that Thera may not have had the wide-ranging impacts that some have suggested (e.g., LaMoreaux 1995). Indeed Marinatos' (1939) early suggestions that the Thera Eruption led directly to the fall of the Minoan civilization have not gained much weight in a range of studies in the interceding years (Pyle 1997). Ash fall hypothesized to have had crop-destroying potential probably did not exceed a few centimeters on Crete (Johnston et al. 2012), and at Trianda on the island of Rhodes, where there was more significant ash fall, recovery is demonstrated in the archaeological record (Doumas and Papazoglou 1980). It also seems that heavy rains associated with the eruption (Athanassas et al. 2018) may have washed much of the ash away and actually been more damaging to field systems (Vokotopoulos, Plath, and McCoy 2014, 257–258), but again, not catastrophically so. Similarly, regional tsunamis were undoubtedly generated by the eruption (McCoy and Heiken 2000), but there is little evidence that they were catastrophic in nature, especially on Crete (Driessen 2019). Estimates of a maximum run-up height of 10–12 m (e.g., Nomikou et al. 2016; McCoy and Heiken 2000) seem possible, but only the deposits at Palaikastro, on the east coast of Crete, appear to provide any evidence of a more extreme impact (Bruins, Van Der Plicht, and MacGillivray 2009). Even with the possibility for a mid-late sixteenth century BCE radiocarbon date for Thera closing in on the temporal gap between the eruption and the decline of the Minoans on Crete, a gap would still remain, indicating that while the eruption may have weakened Cretan society through a range of lesser impacts (damage to coastal towns or interruptions in maritime trade), the localized effects of the event itself cannot be blamed for the eventual demise of the Minoan civilization.

Environmental and Climatic Records

This brings us to the various "natural archives" in which evidence of the Thera eruption has also been sought, with aims to refine the dating and better understand any more broadly ranging impacts. I'll start with the tree-ring record and my own incorrect hypothesis (Pearson et al. 2009) that a chemical change in juniper trees from Porsuk, Turkey, ca. 400 km downwind of Thera, could be a direct result of the Thera eruption. While I maintain that the chemical data are consistent with a volcanic event (or perhaps a forest fire), I now know that I was incorrect in hypothesizing that this could *only* be Thera, based on both geographic proximity and the radiocarbon ranges presented by Manning and colleagues (2006) and Friedrich and colleagues (2006). We now know much more about a wide range of volcanic events in the seventeenth and sixteenth centuries BCE, and we also know that major volcanic eruptions as far away as Alaska (McConnell et al. 2020), Iceland (Oman et al. 2006), and the tropics (Toohey et al. 2016) can have significant consequences for Mediterranean and Near Eastern societies. In addition, we know that a recent re-analysis of the existing tree-ring and ice-core data (McAneney and Baillie 2019) has demonstrated that the tree-ring date of 1628 BCE, long connected to Thera on the basis of its strong agreement with the debated seventeenth-century radiocarbon possibilities, is in fact now directly linked to the eruption of Aniakchak II in Alaska. Interestingly, McAneney and Baillie (2019, 109) confirm an ice-core sulphate layer at the time of the chemical change in the trees from Porsuk, adding support that it was indeed caused by a volcanic event but arguing on the basis of their re-evaluation of the revised ice-core and tree-ring data that Thera may have occurred later than the seventeenth century BCE. I agree with this conclusion.

For a number of decades, tree-ring records have been explored in tandem with ice-core records for evidence of Thera in the late seventeenth century BCE radiocarbon ranges. The North American bristlecone-pine record is the most robustly correlated with the ice cores. In these trees, layers of frost-damaged cells, formed due to rapid onset, short-term cooling during the growth season (caused by volcanic aerosols in the stratosphere), are now strongly correlated with the records of ice-core sulfate over the last 2500 years (Sigl et al. 2015). The association between 1628/27 BCE and the Thera eruption was made by LaMarche and Hirschboeck (1984, 126) at the very start of this now very well-established and utilized scientific pairing. The frost event reported at 1627 BCE was one of the strongest and best replicated in the whole bristlecone record, clearly resulting from an eruption of considerable magnitude. At the time of publication, the only known option was Thera. Similarly, when Baillie and Munro (1988) compared their Irish tree-ring record and found a significant growth suppression starting at 1628 BCE and lasting for about ten years, the logical assumption, given an absence of any record so striking in the sixteenth century BCE, was that here again was further confirmation of the wide-ranging impact of Thera.

In the 1990s, however, it began to emerge from the ice records that there were indeed other eruptions within the broader time period (e.g., Zielinski and Germani 1998). Revisiting the tree-ring records over the years has produced a number of other possible event years in the sixteenth century that are not as well replicated or widespread as the event at 1628 BCE but are nevertheless indicative of the type of small-scale climate perturbations associated with volcanic eruptions. Unfortunately, the causal connections for the majority of these are not as well established as for bristlecone-pine frost-event / volcanic / ice-core scenarios but relate to lesser climatic fluctuations and, in one case, volcanic aerosols (Helama et al. 2019). Still, we can derive a short list of single year dates from the sixteenth century BCE to explore further against the full range of radiocarbon, archaeological, and paleoenvironmental evidence. One of the more strongly confirmed of these is at 1524 BCE, where Salzer and Hughes (2010) explored comparisons between the high-altitude bristlecone chronology and the Yamal Peninsula chronology (Hantemirov and Shiyatov 2002) to conclude that the correlation of narrow growth in this year provided evidence for a climatically-effective event. This date has been proposed to be in better alignment with the lower estimates for the Egyptian chronology, in particular in relation to the seemingly abrupt arrival of pumice in Tuthmosis III contexts (Wiener 2006; Warren 2010). However, there are also ring-width minima in bristlecone-pine at 1597, 1560, 1546, and 1544 BCE (Salzer and Hughes 2007, 62; Pearson et al. 2018), two of which (1560 BCE and 1546 BCE) correspond with fluctuations in the $\delta^{13}C$ record from pines in Finish Lapland indicating a reduction in irradiance, potentially linked with the effects of a "volcanic acid fog" (Helama et al. 2019, 212). Helama and colleagues (2013; 2019, 212) have also suggested that a severe regional cooling in Finish Lapland at 1584 BCE may be the impact of Thera.

One further event has been noted at 1554 BCE in the trees from the Yamal Peninsula (Hantemirov and Shiyatov 2002), which is not replicated in bristlecone pines. All these possibilities (summarized in Fig. 4) must now be more closely investigated with a view to either exclusion as possible markers for the eruption of Thera, or, ideally, proof of correlation. For those that still prefer modeled radiocarbon date ranges in or close to the seventeenth century BCE (Manning et al. 2020; van der Plicht et al. 2020), 1597 BCE and 1584 BCE might be the most likely sixteenth century candidates, but both also acknowledge increased probabilities around 1560 BCE. Within this scheme, however, 1524 BCE would fall outside the possible range. Likewise for those favoring the 1524 BCE possibility on the basis of archaeological evidence, 1560 BCE, and certainly any date older than that, would be ruled out. There is some middle ground (e.g., Höflmayer 2012), however, and it is of obvious importance given the history of this debate to keep an open mind. The key point archaeologically is that the eruption took place after the start of the New Kingdom, so after 1550/1540 BCE (Warren 2010) or between 1566–1552 cal BCE (68.2% probability) or 1570–1544 cal BCE (95.4% probability) (Bronk Ramsey et al. 2010). Because 1560 BCE can now also work in terms of radiocarbon dating of Theran samples, it seems like we are homing in on a consensus period of a few decades in which numerous lines of dating evidence and paleoenvironmental records broadly agree.

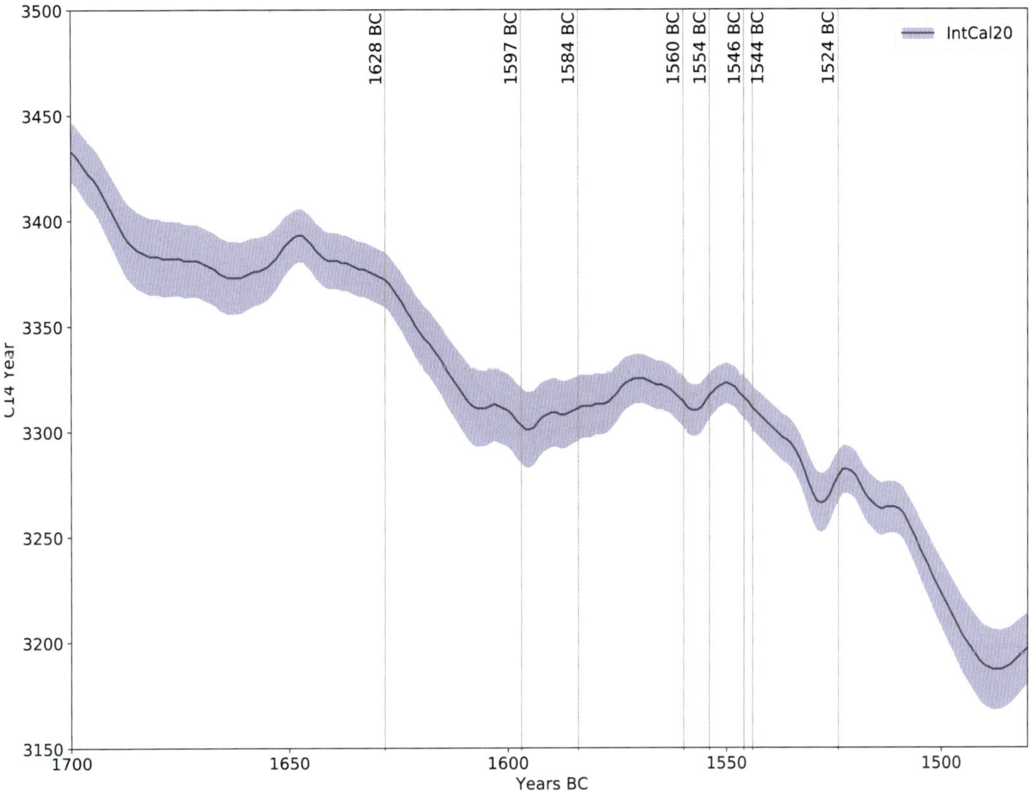

Fig. 4. This figure summarizes possible tree-ring evidence for volcanically forced climatic changes across the period as discussed in the text. The dates shown are not exhaustive or indeed conclusive but are provided here for quick reference with the calibration curve so that the reader can examine the possibilities for themselves. The earliest start date of the Egyptian New Kingdom is 1550 BCE according to the archaeological chronology (e.g., Warren 2010), but a decade or so earlier is also considered possible (e.g., Höflmayer 2012), on the basis of Bronk Ramsey and colleagues (2010).

There is one more tree-ring record to explore, from trees growing on the Anatolian Plateau at the time of the Thera eruption. The Gordion tree-ring chronology (Kuniholm et al. 1996), dated to within one year of an absolute calendar placement (Pearson et al. 2020b), can now be evaluated in terms of marker rings in other chronologies, the main limiting factor being that for the sixteenth century the sample depth is low. Nonetheless, we have uncovered an unusual chemical change in the rings of these trees around ca. 1562–1559 BCE which is clearly indicative of an alteration in the chemical environment at this time. Such a change, as at Porsuk (Pearson et al. 2009), would be consistent with either volcanic acidification or a forest fire. Given the history of dating Thera, we have to be very cautious in remembering that the existence of a correlation does not prove a causal connection. Nevertheless, with marker correlations in the bristlecone and Finland records at this same point in time, and the possible associations with the Athmose Tempest Stela, the volcanic connection becomes far more credible.

Thera as a Less Climatically Effective Event

Previous association of Thera with the 1628/27 BCE event implied a significant and wide-ranging climatic impact, but as mentioned above, none of the climatic forcing evident in tree rings from the sixteenth century is as clear or as well replicated as the 1628/27 BCE event. The implication, then, is that Thera did not have as strong an impact on climate as might be expected given the size of the eruption (Johnston et al. 2014; Newhall, Self, and Robock 2018) and the upper estimates for sulfate production (Cadoux et al. 2015). This does, however, align well with petrological estimates for Thera as a lower sulfate event (Pyle 1997) and with what we find in the few paleoenvironmental contexts in which deposits that can be geochemically tied to Thera are found. No significant climatic impact was recorded in a pollen record from northwest Crete, where the eruption has a definitive pumice marker (Bottema and Sarpaki 2003). Similarly, Eastwood and colleagues (2002) found no change in the terrestrial pollen composition directly following a layer of Theran tephra at Gölhisar, southwest Turkey, indicating a minimal impact on regional vegetation over decadal-to-century timescales. In fact, they report enhanced lake productivity, suggested to be a beneficial side effect of the eruption (making a stark contrast to attempts to connect a hazardous acidic fall out from Thera with depictions in medical papyri [Trevisanato 2007]). Eastwood, Roberts, and Lamb (1998) also rule out any direct correlation between the Thera ash and the initiation of the Beyşehir Occupation Phase, noting that the characteristic pattern of anthropogenic disturbance in the pollen sequences both precedes and follows the ash in a range of stratigraphic sequences. A paleoclimatic record from a sediment core from the Aegean Sea (Rohling et al. 2002) shows a clear decrease in the percentage of warm-water foraminifera species some time after the Thera tephra layer; however, due to a break in the sequence and the resolution of the record, this cannot be connected directly to an impact from Thera (Rohling et al. 2002, 588). It may correlate with a change to cooler and possibly wetter conditions shown as a negative $\delta^{18}O$ signal in Sofular cave, Turkey, dated to around 1510 BCE ±25 (Badertscher et al. 2014, 62), but if so, the interpretation of that sequence (and that of the Trio cave in Hungary [Siklósy et. al 2009]), linked geochemically to Thera, implies that this climatic change was not caused by the eruption (Badertscher et al. 2014, 64). The uranium-series-based dating of the signals at both the caves display some slight disparities, and there are possibilities for both to shift by a few decades. The sulfur record in Sofular is elevated between 1620–1530 ±25 BCE (based on uranium-series dating), peaking ca. 1589 BCE ±25 years. This is currently interpreted to be a lagged response to a single event that the authors at the time connected with the late seventeenth-century radiocarbon placement for Thera. This should now be revisited, because Frisia and colleagues (2008, 25) also showed that geographically distant eruptions were captured in the same record and the "phased signal" could actually represent the multiple different eruptions known across this period. At Trio cave, a marked influx of the rare-earth element lanthanum occurs in the mid-sixteenth century BCE, which could potentially link with the 1589 BCE ±25 years sulphate peak at Sofular and tree-ring anomalies around 1560 or 1554 BCE. Future studies of these records may bring additional important insights, potentially improved by closer integration with ice-core data.

It may be that many of these regional paleoenvironmental records are still too coarse resolution or insufficiently well dated for a short-term climatic forcing event to stand out; however it may also be that there really was very little climatic impact from Thera, as would be predicted if the eruption were at the lower end of sulfate production estimates (Cadoux et al. 2015; Pyle 1997). Alternatively it is perhaps possible that Plinian eruptions in Mediterranean latitudes do not impact the climate of the local regions beyond some more immediate regional weather disturbances like dramatic rainstorms, which appear to have impacted Crete very soon after (or during) the Thera event (Athanassas et al. 2018, 166). Certainly, the famous and first-named "Plinian" eruption of Mt. Vesuvius in 79 CE does not appear to have left a strong signal in paleoclimatic records or ice cores. Sigl and colleagues (2015) demonstrate that for the past 2,500 years, large eruptions in the tropics and high latitudes were the primary drivers of interannual-to-decadal temperature variability in the Northern Hemisphere. The impact of large eruptions at the latitude of Thera seems less clearly defined. We do, however, now have a number of examples of how the Mediterranean region has been significantly impacted by more distant events. For example, in 43 BCE an eruption of Alaska's Okmok volcano—not, as previously thought, an eruption of Mt. Etna (Stothers and Rampino 1983; Forsyth 1988), which does not seem to have impacted climate led to several years of cooler conditions and significant impacts in the Mediterranean region (McConnell et al. 2020). Connections between high-latitude explosive volcanism and the East African monsoon, which controls the majority of the River Nile's floodwater, have also been established (Oman et al. 2006). Finally, a study by Köse and colleagues (2013) explored relationships between volcanism and May–June precipitation in the interior region of southwestern Turkey evidenced in tree rings and showed how volcanic eruptions of known date between 1813 and 2004 CE resulted in wetter than normal years during or immediately after the years with the largest volcanic eruptions. Unfortunately, the Gordion tree-ring record, as mentioned previously, consists of only three trees across the sixteenth century, albeit with very securely dated sequences. While these work for dating and investigating chemical responses, they are not sufficient for climatic reconstruction in a strict sense.

So Where Does All This Leave Us?

We can conclude that the eruption of Thera was a large and impressive event that would surely have been recorded in some way by surrounding cultures, even if there is still debate about exactly how large it was (e.g., Johnston et al. 2014). It seems likely it took place in the summer, and it seems clear that associated phenomena such as earthquakes before and during the event had extensive, although not catastrophic impacts. At Akrotiri, Doumas (1983, 134–35) provides details of repair activities: unsafe walls were torn down, beds taken out into the streets, and tools and re-purposed building stones buried in situ by the main eruption. There is also archaeological evidence to support a similar sequence of events on the surrounding islands of Crete, Kos, and Rhodes, where ash layers from Thera are associated with earthquake damage in a number of contexts (Driessen 2019, 196). There is no doubt that volcanic ash fell widely (although unevenly [e.g., Driessen 2019, 197]) across the region (Guichard et al. 1993; Sulpizio et al. 2013; Johnston et al. 2012; Eastwood, Roberts, and Lamb 1998; Bruins et al. 2019) and would likely have been experienced as a "volcanic darkness" in nearby areas, possibly (though debatably) as far away as Egypt (Stanley and Sheng 1986, 733). The ash or associated run-off events may have had some short term impacts for crops in some regions. So too might damage from tsunamis generated (e.g., Sakellariou et al. 2012) by the event, although the scale of these tsunamis, once described in catastrophic terms (e.g., Marriner and Morhange 2013), now appears to have been fairly modest and related to pyroclastic flows rather than caldera collapse (Driessen 2019, 196). A recent study has hypothesized that these same pyroclastic flows may have incinerated the people of Akrotiri as they tried to escape by boat (Evans and McCoy 2020), explaining why so little evidence of Minoan refugees has been reported; however, no human casualties resulting directly from the eruption have been found to date.

So, at present we can imagine an impressive series of natural phenomena that erased an important Minoan trade center, damaged other harbors, and likely caused a range of shorter term difficulties for local populations. The event marks the LM IA to LM IB transition and even though this now can be argued to fit with both archaeological and radiocarbon evidence for the start of the Egyptian New Kingdom, this dating still fails to establish a direct causal connection with the decline of the Minoan civilization on Crete. It does however narrow the temporal gap, strengthening the case for an eruption impact in the process of decline. Further refinement of the dating should help to resolve remaining questions. The best hope of achieving this is through the ice-core records, which are now well dated for the sixteenth century BCE. These records hold the potential to combine direct evidence of volcanic sulfate with the possibility to find volcanic ash which can be directly linked via geochemistry to Thera. Until this (or other new evidence emerges), however, we have at least an opportunity to narrow the dating possibilities for Thera down to a few calendar dates. Three separate tree-ring records provide lines of evidence consistent in marking a volcanic eruption that occurred at ca. 1560 BCE. This date could potentially fit with references in the Ahmose Tempest Stela and Rhind Mathematical papyrus, as well as speleothem evidence, and be supported by radiocarbon dating for both Theran materials and the start of the New Kingdom. There is a problem with this however, because ca. 1560 BCE is earlier for Thera than would be considered possible according to the archaeological chronology (e.g., Warren 2010). Less well replicated tree-ring markers at 1554, 1548, 1546, 1544, and 1524 BCE also offer exact dates for further exploration which could align more comfortably with the archaeology, but these are currently less well substantiated records. For a topic that has been the focus of so many for so long, and from which we can derive some cautionary tales, it is clear that debate will only be finally settled by direct evidence. Until then, it seems prudent to keep an open mind and focus our search for the final clinching evidence on the full, but limited, range of dating possibilities in the sixteenth century BCE.

Works Cited

Athanassas, Constantin D., Konstantinos Modis, Mehmet Cihat Alçiçek, and Katerina Theodorakopoulou. 2018. "Contouring the Cataclysm: A Geographical Analysis of the Effects of the Minoan Eruption of the Santorini Volcano." *Environmental Archaeology* 23, no. 2: 160–76.

Badertscher, Silvia, Andrea Borsato, Silvia Frisia, Hai Cheng, R. Lawrence Edwards, Okan Tüysüz, and Dominik Fleitmann. 2014. "Speleothems as Sensitive Recorders of Volcanic Eruptions—the Bronze Age Minoan Eruption Recorded in a Stalagmite from Turkey." *Earth and Planetary Science Letters* 392: 58–66.

Baillie, Michael G. L., and Martin Munro. 1988. "Irish Tree Rings, Santorini and Volcanic Dust Veils." *Nature* 332, no. 6162: 344–46.

Bietak, Manfried. 2014. "Radiocarbon and the Date of the Thera Eruption." *Antiquity* 88, no. 339: 277–82.

Bietak, Manfried, and Felix Höflmayer. 2007. "Introduction: High and Low Chronology." In *The Synchronisation of Civilisations in the Eastern Mediterranean in the Second Millennium BCE III: Proceedings of the SCIEM 2000–2nd EuroConference Vienna, 28th of May–1st of June 2003. Vienna: Verlag der Österreichischen Akademie der Wissenschaften*, edited by Manfred Bietak and Ernst Czerny, 13–23. Vienna: Verlag der Österreichischen Akademie der Wissenschaften.

Block, Joel. 1976. "The Ten Plagues of Egypt." *Religious Education* 71, no. 5: 519–26.

Bottema, Sytze, and Anaya Sarpaki. 2003. "Environmental Change in Crete: A 9000-Year Record of Holocene Vegetation History and the Effect of the Santorini Eruption." *The Holocene* 13, no. 5: 733–49.

Bronk Ramsey, Christopher, Michael W. Dee, Joanne M. Rowland, Thomas F. G. Higham, Stephen A. Harris, Fiona Brock, Anita Quiles, Eva M. Wild, Ezra S. Marcus, and Andrew J. Shortland. 2010. "Radiocarbon-Based Chronology for Dynastic Egypt." *Science* 328, no. 5985: 1554–57.

Bronk Ramsey, Christopher, Sturt W. Manning, and Mariagrazia Galimberti. 2004. "Dating the Volcanic Eruption at Thera." *Radiocarbon* 46, no. 1: 325–44.

Bruins, Hendrik J., Jörg Keller, Andreas Klügel, Hanan J. Kisch, Itzhak Katra, and Johannes van der Plicht. 2019. "Tephra in Caves: Distal Deposits of the Minoan Santorini Eruption and the Campanian Super-Eruption." *Quaternary International* 499: 135–47.

Bruins, Hendrik J., and Johannes Van der Plicht. 2014. "The Thera Olive Branch, Akrotiri (Thera) and Palaikastro (Crete): Comparing Radiocarbon Results of the Santorini Eruption." *Antiquity* 88, no. 339: 282–87.

Bruins, Hendrik J., Johannes Van Der Plicht, and J. Alexander MacGillivray. 2009. "The Minoan Santorini Eruption and Tsunami Deposits in Palaikastro (Crete): Dating by Geology, Archaeology, ^{14}C, and Egyptian Chronology." *Radiocarbon* 51, no. 2: 397–411.

Cadoux, Anita, Bruno Scaillet, Slimane Bekki, Clive Oppenheimer, and Timothy H. Druitt. 2015. "Stratospheric Ozone Destruction by the Bronze-Age Minoan Eruption (Santorini Volcano, Greece)." *Scientific Reports* 5, no. 1: 1–12.

Chase, Arnold B. 1927. *The Rhind Mathematical Papyrus. British Museum 10057 and 10058, Volume 1. Free Translation and Commentary.* Oberlin, OH: Mathematical Association of America.

Cherubini, Paolo, Turi Humbel, Hans Beeckman, Holger Gärtner, David Mannes, Charlotte Pearson, Werner Schoch, Roberto Tognetti, and Simcha Lev-Yadun. 2014. "The Olive-Branch Dating of the Santorini Eruption." *Antiquity* 88, no. 339: 267–73.

Doumas, Christos. 1983. *Thera: Pompeii of the Ancient Aegean: Excavations at Akrotiri, 1967–79.* London: Thames and Hudson.

Doumas, Christos, and Lena Papazoglou. 1980. "Santorini Tephra from Rhodes." *Nature* 287, no. 5780: 322–24.

Driessen, Jan. 2019. "The Santorini Eruption: An Archaeological Investigation of Its Distal Impacts on Minoan Crete." *Quaternary International* 499: 195–204.

Eastwood, Warren J., John Tibby, Neil Roberts, Harry John Betteley Birks, and Henry F. Lamb. 2002. "The Environmental Impact of the Minoan Eruption of Santorini (Thera): Statistical Analysis of Palaeoecological Data from Gölhisar, Southwest Turkey." *The Holocene* 12, no. 4: 431–44.

Eastwood, Warren J., Neil Roberts, and Henry F. Lamb. 1998. "Palaeoecological and Archaeological Evidence for Human Occupance in Southwest Turkey: The Beyşehir Occupation Phase." *Anatolian Studies* 48: 69–86.

Ehrlich, Yael, Lior Regev, and Elisabetta Boaretto. 2018. "Radiocarbon Analysis of Modern Olive Wood Raises Doubts Concerning a Crucial Piece of Evidence in Dating the Santorini Eruption." *Scientific Reports* 8, no. 1: 1–8.

———. 2021. "Discovery of Annual Growth in a Modern Olive Branch Based on Carbon Isotopes and Implications for the Bronze Age Volcanic Eruption of Santorini." *Scientific Reports* 11, no. 1: 1–11.

Evans, Krista, and Floyd W. McCoy. 2020. "Precursory Eruptive Activity and Implied Cultural Responses to the Late Bronze Age (LBA) Eruption of Thera (Santorini, Greece)." *Journal of Volcanology and Geothermal Research* 397: 1–8.

Evelyn-White, Hugh G. 1936. *Hesiod, the Homeric Hymns and Homerica.* Cambridge, MA/London: William Heinemann.

Fischer, Peter M. 2009. "The Chronology of Tell el-Ajjul, Gaza: Stratigraphy, Thera, Pumice and Radiocarbon Dating." In *Time's Up! Dating the Minoan Eruption of Santorini. Acts of the Minoan Eruption Chronology Workshop, Sandbjerg, November 2007*, Monographs of the Danish Institute, Vol. 10, edited by David A. Warburton, 253–65. Leuven: Aarhus University Press.

Forsyth, Peter. 1988. "In the Wake of Etna, 44 BCE." *Classical Antiquity* 7, no. 1: 49–57.

Foster, Karen Polinger, Robert K. Ritner, and Benjamin R. Foster. 1996. "Texts, Storms, and the Thera Eruption." *Journal of Near Eastern Studies* 55, no. 1: 1–14.

Friedrich, Ronny, Bernd Kromer, Lukas Wacker, Jesper Olsen, Sabine Remmele, Susanne Lindauer, Alexander Land, and Charlotte Pearson. 2020. "A New Annual ^{14}C Dataset for Calibrating the Thera Eruption." *Radiocarbon* 62, no. 4: 953–61.

Friedrich, Walter L., Bernd Kromer, Michael Friedrich, Jan Heinemeier, Tom Pfeiffer, and Sahra Talamo. 2006. "Santorini Eruption Radiocarbon Dated to 1627–1600 BCE." *Science* 312, no. 5773: 548.

———. 2014. "The Olive Branch Chronology Stands Irrespective of Tree-Ring Counting." *Antiquity* 88, no. 339: 274–77.

Frisia, Silvia, Seraina Badertscher, Andrea Borsato, Jean Susini, Ozan Mert Göktürk, Hai Cheng, R. Lawrence Edwards, Jan Kramers, Okan Tüysüz, and Dominik Fleitmann. 2008. "The Use of Stalagmite Geochemistry to Detect Past Volcanic Eruptions and their Environmental Impacts." *PAGES news* 16, no. 3: 25–26.

Gil Guirado, Salvador, Jorge Olcina-Cantos, and Alfredo Pérez-Morales. 2021. "The Blessing of the 'Year without Summer': Climatic and Socioeconomic Impact of the Krakatoa Eruption (1883) in the South-east of the Iberian Peninsula." *International Journal of Climatology*, 41, no. 4: 2279–300.

Guichard, F., S. Carey, M. A. Arthur, H. Sigurdsson, and M. Arnold. 1993. "Tephra from the Minoan Eruption of Santorini in Sediments of the Black Sea." *Nature* 363, no. 6430: 610–12.

Hantemirov, Rashit M., and Stepan G. Shiyatov. 2002. "A Continuous Multimillennial Ring-Width Chronology in Yamal, Northwestern Siberia." *The Holocene* 12, no. 6: 717–26.

Hardy, David A., ed. 1990. *Thera and the Aegean World III*. Vol. 3. London: Thera Foundation.

Helama, Samuli, Jari Holopainen, Marc Macias-Fauria, Mauri Timonen, and Kari Mielikäinen. 2013. "A Chronology of Climatic Downturns through the Mid- and Late-Holocene: Tracing the Distant Effects of Explosive Eruptions from Palaeoclimatic and Historical Evidence in Northern Europe." *Polar Research* 32, no. 1: 15866.

Helama, Samuli, Pekka Saranpää, Charlotte L. Pearson, Laura Arppe, Jari Holopainen, Harri Mäkinen, Kari Mielikäinen, et al. 2019. "Frost Rings in 1627 BC and AD 536 in Subfossil Pinewood from Finnish Lapland." *Quaternary Science Reviews* 204: 208–15.

Höflmayer, Felix. 2012. "The Date of the Minoan Santorini Eruption: Quantifying the "Offset"." *Radiocarbon* 54, no. 3–4: 435–48.

Johnston, Emma, Jeremy Phillips, Costanza Bonadonna, and Matthew Watson. 2012. "Reconstructing the Tephra Dispersal Pattern from the Bronze Age Eruption of Santorini Using an Advection-Diffusion Model." *Bulletin of Volcanology* 74, no. 6: 1485–1507.

Johnston, Emma, Stephen Sparks, Jeremy Phillips, and Steven Carey. 2014. "Revised Estimates for the Volume of the Late Bronze Age Minoan Eruption, Santorini, Greece." *Journal of the Geological Society* 171, no. 4: 583–90.

Köse, Nesibe, Ünal Akkemik, H. Tuncay Güner, H. Nüzhet Dalfes, Henri D. Grissino-Mayer, M. Sinan Özeren, and Tayfun Kındap. 2013. "An Improved Reconstruction of May-June Precipitation Using Tree-Ring Data from Western Turkey and its Links to Volcanic Eruptions." *International Journal of Biometeorology* 57, no. 5: 691–701.

Kuniholm, Peter Ian. 2014. "The Difficulties of Dating Olive Wood." *Antiquity* 88, no. 339: 287–88.

Kuniholm, Peter Ian, Bernd Kromer, Sturt W. Manning, Maryanne Newton, Christine E. Latini, and Mary Jaye Bruce. 1996. "Anatolian Tree Rings and the Absolute Chronology of the Eastern Mediterranean, 2220–718 BCE." *Nature* 381, no. 6585: 780–83.

Kutschera, Walter, Manfred Bietak, Eva Maria Wild, Christopher Bronk Ramsey, Michael Dee, Robin Golser, Karin Kopetzky, et al. 2012. "The Chronology of Tell el-Daba: A Crucial Meeting Point of ^{14}C Dating, Archaeology, and Egyptology in the 2nd Millennium BCE." *Radiocarbon* 54, no. 3–4: 407–22.

LaMarche, Valmore C., and Katherine K. Hirschboeck. 1984. "Frost Rings in Trees as Records of Major Volcanic Eruptions." *Nature* 307, no. 5947: 121–26.

LaMoreaux, Philip. 1995. "Worldwide Environmental Impacts from the Eruption of Thera." *Environmental Geology* 26, no. 3: 172–81.

MacGillivray, J., Alexander, and David V. Warburton. 2009. "Thera, Hatshepsut, and the Keftiu." In *Time's Up! Dating the Minoan Eruption of Santorini: Acts of the Minoan Eruption Chronology Workshop, Sandbjerg November 2007*, Monographs of the Danish Institute at Athens, Vol. 10, edited by David A. Warburton, 155–70. Leuven: Aarhus University Press.

Manning, Sturt. W. 2014. *A Test of Time and a Test of Time Revisited*. Oxford: Oxbow Books.

Manning, Sturt W., Christopher Bronk Ramsey, Walter Kutschera, Thomas Higham, Bernd Kromer, Peter Steier, and Eva M. Wild. 2006. "Chronology for the Aegean Late Bronze Age 1700-1400 BCE." *Science* 312, no. 5773: 565-69.

Manning, Sturt W., and Bernd Kromer. 2012. "Considerations of the Scale of Radiocarbon Offsets in the East Mediterranean, and Considering a Case for the Latest (Most Recent) Likely Date for the Santorini Eruption." *Radiocarbon* 54 (3-4): 449-74.

Manning, Sturt W., Felix Höflmayer, Nadine Moeller, Michael W. Dee, Christopher Bronk Ramsey, Dominik Fleitmann, Thomas Higham, Walter Kutschera, and Eva Maria Wild. 2014. "Dating the Thera (Santorini) Eruption: Archaeological and Scientific Evidence Supporting a High Chronology." *Antiquity* 88, no. 342: 1164.

Manning, Sturt W., Lukas Wacker, Ulf Büntgen, Christopher Bronk Ramsey, Michael W. Dee, Bernd Kromer, Brita Lorentzen, and Willy Tegel. 2020. "Radiocarbon Offsets and Old World Chronology as Relevant to Mesopotamia, Egypt, Anatolia and Thera (Santorini)." *Scientific Reports* 10, no. 1: 1-14.

Marinatos, Spiridon. 1939. "Volcanic Destruction of Minoan Crete." *Antiquity* 13: 425.

Marriner, Nick, and Christophe Morhange. 2013. "Data Mining the Intellectual Revival of 'Catastrophic' Mother Nature." *Foundations of Science* 18, no. 2: 245-57.

McAneney, Jonny, and Mike Baillie. 2019. "Absolute Tree-Ring Dates for the Late Bronze Age Eruptions of Aniakchak and Thera in Light of a Proposed Revision of Ice-Core Chronologies." *Antiquity* 93, no. 367: 99-112.

McConnell, Joseph R., Michael Sigl, Gill Plunkett, Andrea Burke, Woon Mi Kim, Christoph C. Raible, Andrew I. Wilson, et al. 2020. "Extreme Climate after Massive Eruption of Alaska's Okmok Volcano in 43 BCE and Effects on the Late Roman Republic and Ptolemaic Kingdom." *Proceedings of the National Academy of Sciences* 117, no. 27: 15443-49.

McCoy, Floyd W., and Grant Heiken. 2000. "Tsunami Generated by the Late Bronze Age Eruption of Thera (Santorini), Greece." *Pure and Applied Geophysics* 157, no. 6: 1227-56.

McNutt, Stephen R., and Earle R. Williams. 2010. "Volcanic Lightning: Global Observations and Constraints on Source Mechanisms." *Bulletin of Volcanology* 72, no. 10: 1153-67.

Meece, Stephanie. 2006. "A Bird's Eye View - Of a Leopard's Spots: The Çatalhöyük 'Map' and the Development of Cartographic Representation in Prehistory." *Anatolian Studies* 56: 1-16.

Newhall, Chris, Stephen Self, and Alan Robock. 2018. "Anticipating Future Volcanic Explosivity Index (VEI) 7 Eruptions and their Chilling Impacts." *Geosphere* 14, no. 2: 572-603.

Nomikou, Paraskevi, T. H. Druitt, Christian Hübscher, T. A. Mather, Michele Paulatto, L. M. Kalnins, Karim Kelfoun, et al. 2016. "Post-Eruptive Flooding of Santorini Caldera and Implications for Tsunami Generation." *Nature Communications* 7, no. 1: 1–10.

Oman, Luke, Alan Robock, Georgiy L. Stenchikov, and Thorvaldur Thordarson. 2006. "High-Latitude Eruptions Cast Shadow over the African Monsoon and the Flow of the Nile." *Geophysical Research Letters* 33, no. 18: 1–5.

Panagiotakopulu, Eva, Thomas Higham, Anaya Sarpaki, Paul Buckland, and Christos Doumas. 2013. "Ancient Pests: The Season of the Santorini Minoan Volcanic Eruption and a Date from Insect Chitin." *Naturwissenschaften* 100, no. 7: 683–89.

Pearson, Charlotte L., Peter W. Brewer, David Brown, Timothy J. Heaton, Gregory W. L. Hodgins, A. J. Timothy Jull, Todd Lange, and Matthew W. Salzer. 2018. "Annual Radiocarbon Record Indicates 16th Century BCE Date for the Thera Eruption." *Science Advances* 4, no. 8: eaar8241.

Pearson, Charlotte L., Darren S. Dale, Peter W. Brewer, Peter I. Kuniholm, Jeffrey Lipton, and Sturt W. Manning. 2009. "Dendrochemical Analysis of a Tree-Ring Growth Anomaly Associated with the Late Bronze Age Eruption of Thera." *Journal of Archaeological Science* 36, no. 6: 1206–14.

Pearson, Charlotte L., Lukas Wacker, Alex Bayliss, David Brown, Matthew Salzer, Peter Brewer, Silvia Bollhalder, Gretel Boswijk, and Gregory Hodgins. 2020a. "Annual Variation in Atmospheric ^{14}C between 1700 BCE and 1480 BCE." *Radiocarbon* 62, no. 4: 939–52.

Pearson, Charlotte L., Matthew Salzer, Lukas Wacker, Peter Brewer, Adam Sookdeo, and Peter Kuniholm. 2020b. "Securing Timelines in the Ancient Mediterranean Using Multiproxy Annual Tree-Ring Data." *Proceedings of the National Academy of Sciences* 117, no. 15: 8410–15.

Pyle, David M. 1997. "The Global Impact of the Minoan Eruption of Santorini, Greece." *Environmental Geology* 30, no. 1: 59–61.

Reimer, Paula J., William E. N. Austin, Edouard Bard, Alex Bayliss, Paul G. Blackwell, Christopher B. Ramsey, Martin Butzin, et al. 2020. "The IntCal20 Northern Hemisphere Radiocarbon Age Calibration Curve (0–55 cal kBP)." *Radiocarbon* 62, no. 4: 725–57.

Ritner, Robert K., and Nadine Moeller. 2014. "The Ahmose 'Tempest Stela', Thera and Comparative Chronology." *Journal of Near Eastern Studies* 73, no. 1: 1–19.

Rohling, Eelco, Paul Mayewski, R. Abu-Zied, J. Casford, and A. Hayes. 2002. "Holocene Atmosphere-Ocean Interactions: Records from Greenland and the Aegean Sea." *Climate Dynamics* 18, no. 7: 587–93.

Sakellariou, Dimitris, Grigoris Rousakis, Paraskevi Nomikou, Kathy Croff Bell, Steve Carey, and Haraldur Sigurdsson. 2012. "Tsunami Triggering Mechanisms Associated with the 17th Cent. BC Minoan Eruption of Thera Volcano, Greece." In *The Twenty-Second International Offshore and Polar Engineering Conference*, edited by Jin S. Chung, Ivar Langen, Sa Young Hong, and Simon J. Prinsenberg, 61–67. Cupertino, CA: International Society of Offshore and Polar Engineers (ISOPE).

Salzer, Matthew W., and Malcolm K. Hughes. 2007. "Bristlecone Pine Tree Rings and Volcanic Eruptions over the Last 5000 Yr." *Quaternary Research* 67, no. 1: 57–68.

Salzer, Matthew W., and Malcolm K. Hughes. 2010. "Volcanic Eruptions over the Last 5,000 Years from High Elevation Tree-Ring Widths and Frost Rings." In *Tree Rings and Natural Hazards*, edited by Markus Stoffel, Michelle Bollschweiler, David R. Butler, and Brian H. Luckman, 469–82. Dordrecht: Springer.

Schmitt, Axel K., Martin Danišík, Erkan Aydar, Erdal Şen, İnan Ulusoy, and Oscar M. Lovera. 2014. "Identifying the Volcanic Eruption Depicted in a Neolithic Painting at Çatalhöyük, Central Anatolia, Turkey." *PLoS One* 9, no. 1: e84711.

Sewell, David Alexander. 2001. "Earth, Air, Fire and Water: An Elemental Analysis of the Minoan Eruption of Santorini Volcano in the Late Bronze Age." PhD diss., University of Reading.

Sigl, Michael, Mai Winstrup, Joseph R. McConnell, Kees C. Welten, Gill Plunkett, Francis Ludlow, Ulf Büntgen, et al. 2015. "Timing and Climate Forcing of Volcanic Eruptions for the Past 2,500 Years." *Nature* 523, no. 7562: 543–49.

Siklósy, Zoltan, Attila Demeny, Torsten W. Vennemann, Sebastien Pilet, Jan Kramers, Szabolcs Leél-Őssy, Mária Bondár, Chuan-Chou Shen, and Ernst Hegner. 2009. "Bronze Age Volcanic Event Recorded in Stalagmites by Combined Isotope and Trace Element Studies." *Rapid Communications in Mass Spectrometry: An International Journal Devoted to the Rapid Dissemination of Up-to-the-Minute Research in Mass Spectrometry* 23, no. 6: 801–08.

Stanley, Daniel Jean, and Harrison Sheng. 1986. "Volcanic Shards from Santorini (Upper Minoan Ash) in the Nile Delta, Egypt." *Nature* 320, no. 6064: 733–35.

Stothers, Richard B., and Michael R. Rampino. 1983. "Volcanic Eruptions in the Mediterranean Before AD 630 from Written and Archaeological Sources." *Journal of Geophysical Research: Solid Earth* 88, no. B8: 6357–71.

Sulpizio, R., Mehmet Cihat Alçiçek, Giovanni Zanchetta, and Luigi Solari. 2013. "Recognition of the Minoan Tephra in the Acigöl Basin, Western Turkey: Implications for Inter-Archive Correlations and Fine Ash Dispersal." *Journal of Quaternary Science* 28, no. 4: 329–35.

Toohey, Matthew, Kirstin Krüger, Michael Sigl, Frode Stordal, and Henrik Svensen. 2016. "Climatic and Societal Impacts of a Volcanic Double Event at the Dawn of the Middle Ages." *Climatic Change* 136, no. 3: 401–12.

Trevisanato, Siro Igino. 2007. "Medical Papyri Describe the Effects of the Santorini Eruption on Human Health, and Date the Eruption to August 1603–March 1601 BCE." *Medical Hypotheses* 68, no. 2: 446–49.

Van der Plicht, Johannes, Christopher Bronk Ramsey, Timothy J. Heaton, E. Marian Scott, and Sahra Talamo. 2020. "Recent Developments in Calibration for Archaeological and Environmental Samples." *Radiocarbon* 62, no. 4: 1095–117.

Vokotopoulos, Leonardis, Gerhard Plath, and Floyd W. McCoy. 2014. "The Yield of the Land: Soil Conservation and the Exploitation of Arable Land at Choiromandres, Zakros in the New Palace Period." In *Physis. L'environnement naturel et la relation home-milieu dans le monde égéen protohistorique*, Actes de la 14e Rencontre égéenne internationale, Paris, Institut National d'Histoire de l'Art (INHA), 11–14 December 2012, Aegaeum 37, edited by G. Touchais, R. Laffineur, and F. Rougemont, 251–62. Liège: Belgium.

von Deines, Hildegard, Hermann Grapow, and Wolfhart Westendorf. 1958. *Grundniss der Medizin der alten Ägypter IV: Überset-zung der medizinischen Texte.* Berlin: Akademie Verlag.

Warburton, David A., ed. 2009. *Time's Up! Dating the Minoan Eruption of Santorini: Acts of the Minoan Eruption Chronology Workshop, Sandbjerg, November 2007.* Monographs of the Danish Institute, Vol. 10. Leuven: Aarhus University Press.

Warren, Peter M. 2006a. "The Date of the Thera Eruption in Relation to Aegean-Egyptian Interconnections and the Egyptian Historical Chronology." In *Timelines: Studies in Honour of Manfred Bietak* 2, edited by E. Czerny, I. Hein, H. Hunger, D. Melman, and A. Schwab, 305–21. Leuven: Peeters.

———. 2006b. "The Absolute Chronology of the Aegean Circa 2000 BCE–1400 BCE A Summary." In *Die Bedeutung der minoischen und mykenischen Glyptik: VI. Internationales Siegel-Symposium aus Anlass des*, Vol. 50, edited by W. Muller, 383–94. Mainz am Rhein: Phillip von Zabern.

———. 2010. "The Date of the Late Bronze Age Eruption of Santorini on the Basis of the Historical Chronology." In *The Date of the Late Bronze Age Eruption of Santorini on the Basis of the Historical Chronology*, Pasiphae 4, edited by Louis Godart, 67–72. Rome: Fabrizio Serra Editore.

Whitham, Andrew G., and Robert Stephen J. Sparks. 1986. "Pumice." *Bulletin of Volcanology* 48, no. 4: 209–23.

Wiener, Malcolm H. 2006. “Chronology Going Forward (with a Query about 1525/4 B.C.).” In *Timelines. Studies in Honour of Manfred Bietak*, edited by E. Czerny, I. Hein, H. Hunger, D. Melman, A. Schwab, 317–28. Leuven: Peeters.

———. 2009. “The State of the Debate about the Date of the Theran Eruption.” In *Time's Up! Dating the Minoan Eruption of Santorini: Acts of the Minoan Eruption Chronology Workshop, Sandbjerg, November 2007*, Monographs of the Danish Institute, Vol. 10, edited by David A. Warburton, 197–206. Leuven: Aarhus University Press.

———. 2010. “A Point in Time.” In *Cretan Offerings: Festschrift, Studies in Honour of Peter Warren*, British School at Athens Studies 18, edited by Olga Krzyszkowska, 367–94. London: British School at Athens.

———.. 2012. “Problems in the Measurement, Calibration, Analysis, and Communication of Radiocarbon Dates (with Special Reference to the Prehistory of the Aegean World).” *Radiocarbon* 54, no. 3–4: 423–34.

Wiener, Malcolm H., and James P. Allen. 1998. “Separate Lives: The Ahmose Tempest Stela and the Theran Eruption.” *Journal of Near Eastern Studies* 57, no. 1: 1–28.

Wiener, Malcolm H., and Jason W. Earle. 2014. “Radiocarbon Dating of the Theran Eruption.” *Open Journal of Archaeometry* 2, 5265: 60–64.

Wilkie, Benjamin, Fred Cahir, and Ian D. Clark. 2020. “Volcanism in Aboriginal Australian Oral Traditions: Ethnographic Evidence from the Newer Volcanics Province.” *Journal of Volcanology and Geothermal Research* 403: 106999.

Zielinski, Gregory A., and Mark S. Germani. 1998. “New Ice-Core Evidence Challenges the 1620s BCE Age for the Santorini (Minoan) Eruption.” *Journal of Archaeological Science* 25, no. 3: 279–89.

The Justinianic Plague: Pressing Questions, Interdisciplinary Answers?

Lee Mordechai

Introduction

The first plague pandemic, also known colloquially as the Justinianic Plague (hereafter JP), has drawn significant scholarly attention since the beginning of the twenty-first century. Research has accelerated as well: in the last four years alone, over a couple dozen papers have either covered the pandemic directly or touched upon topics relevant to it—from ancient DNA (aDNA) to studies of pre-JP plague. A substantial amount of additional non-traditional publications such as blog posts, interviews, podcast episodes, and even YouTube videos have similarly addressed the pandemic (e.g., Invicta 2021). Covid has stimulated interest in the pandemic further, introducing it to public discourse more frequently than ever before (e.g., LePan 2020).

Some structural factors have hindered this rapidly-developing conversation. One recurring issue is its sprawling publication landscape: an agglomeration of English and non-English language publications, historical and scientific journals, peer-reviewed articles, non-peer-reviewed contributions, book chapters, and books. Some publications are indexed in repositories such as Web of Science or Google Scholar, while others are barely noticed. Online discourse—in the form of blog posts or social media—serves as an informal forum to hash out ideas and make arguments. Moreover, the considerable lag between submission and publication has meant that some papers are published much after they are written and do not appear to take into consideration research that is already published.

As research on the JP has accumulated, several reviews of the current state of scholarship have appeared at varying resolutions. The first, by Dionysios Stathakopoulos (2000), covered research until the year of its publication. Merle Eisenberg and Lee Mordechai (2019) published a follow-up review that covered the period from 2000–2018. Other publications include shorter bibliographies of the Late Antique western (Eisenberg et al. 2018) and eastern Mediterranean evidence (McMahon and Sargent 2018), generally until 2018. This paper follows these earlier reviews to examine the current state of the JP field, specifically covering publications between 2017 and 2021. After providing some background through 2017, it focuses on the discussion surrounding the mortality and effects of plague, which has drawn much attention and led to the hottest scholarly debate. Subsequent sections include topics related to the JP that have seen considerable advances over the past few years—such as volcanism/climate and the Byzantine cities in the Negev—as well as a more disparate collection of studies that examine the JP or parts of it but are yet to coalesce into or inspire a broader trend.

Background

The origins of the most recent trend of JP interest can be traced to a specialized conference on the topic at Rome in 2001. Most of the talks in the conference were eventually published in a volume edited by Lester Little (2007): *Plague and the End of Antiquity: The Pandemic of 541–750*. To date this remains the main—arguably the only—scholarly monograph on the topic in book format. Contributions to the book contain diverse methodological approaches, some of which remain relevant, while others have become obsolete. Notably, several chapters examined non-literary sources, most obviously in discussions of archaeology and molecular biology, a trend that has grown in subsequent years (e.g., Sarris 2007; McCormick 2007; Sallares 2007). A second book on the topic, William Rosen's (2007) *Justinian's Flea*, has been poorly received by historians, who pointed out the author's lack of expertise in the discipline or understanding of the source material or debate. However, scientists were far more positive towards this account and cited it in some of their analyses of the JP. The discrepancy in the book's reception between both audiences likely contributed to subsequent divergent readings of the JP that became apparent in the mid-2010s (Eisenberg and Mordechai 2019, 161–62).

For most of the first two decades of the twenty-first century, scholars attempted to understand the JP better, particularly by mapping the contours of the pandemic as a phenomenon. After earlier scholarship expanded the chronological and spatial range of the JP, more recent scholarship focused its efforts on expanding the JP's source base. Since the amount of written evidence has been (or was widely perceived to have been) exhausted, scholars moved to examine types of evidence that could be considered as factors that enabled the JP (e.g., climate, rats), or potential evidence for its effects. Scholars worked on non-literary evidence (e.g., mass burials), complemented by additional work on literary evidence for the JP's effects (e.g., liturgification).

Worthy of mention here is Dionysios Stathakopoulos' (2004) book, *Famine and Pestilence in the Late Roman and Early Byzantine Empire: A Systematic Survey of Subsistence Crises and Epidemics*. Although this book opted for a broader approach, surveying food shortages, famines, outbreaks, and epidemics between 284–750 CE, the JP featured prominently in its structure. Stathakopoulos' catalog of relevant events organized and summarized the sources, including an emphasis on the Greek literary accounts, in a running list of famines and disease outbreaks. The result introduced many sources to scholars who might not have the required language skills to investigate them for themselves and thus remains indispensable for JP research. On the other hand, the decision not to include the primary sources themselves in the original text or translation, as in catalogs of other events such as earthquakes (e.g., Guidoboni 1994; Ambraseys 2009), has de facto distanced subsequent scholars from the primary sources, with important detrimental effects (discussed below).

This early twenty-first century research trend culminated with Kyle Harper's (2017) *The Fate of Rome*, which had several features that cemented its place as a key publication on the JP. First, it situated the JP within a broader context of environmental stresses that strongly influenced Roman history. The JP disease event, Harper argued, must be viewed in the context of earlier disease events, such as the Antonine Plague and Cyprianic Plague, and contemporaneous climatic events such as the Late Antique Little Ice Age (LALIA). Harper's book drew considerable attention. As a book directed at a broad, non-specialist audience, it was also picked up by media outlets. A review in the Guardian (P. Smith 2019) noted its "remarkable erudition and synthesis," while the Los Angeles Review of Books stated that it was "an ambitious and convincing reappraisal" of Rome's collapse (Robins 2017). Academic reviewers were similarly positive. As one reviewer put it, "Harper's *Fate of Rome* is an intellectual *tour de force*, combining prodigious scientific evidence with historical insight, and presented with superb prose" (Wijnendaele 2019). Another reviewer praised the book as "intellectually compelling as a study and also an enjoyable read" (Sarris 2018). A third simply described the book as a "marvelous contribution" (Reilly 2020).

Within the long chapter he dedicated to the JP, Harper synthesized much earlier research. Using engaging prose, the chapter blended historical discussions of sixth-century Constantinople with epidemiology and paleogenetics. The chapter was accompanied by an appendix that cataloged plague amplification events (i.e., recurrences) in the two centuries after the first outbreak. Although the appendix drew heavily on Stathakopoulos' catalogue and the average length of entries was considerably shorter, Harper added a substantial number of primary sources in Latin, Syriac, and Arabic.

The JP evidence Harper laid out in his chapter was not groundbreaking—most of the different pieces were well-known before his work—but it was the most persuasively written. According to Harper's forceful conclusion, the JP was an unparalleled demographic catastrophe. In the course of a few years, the population of the Eastern Roman Empire dropped by half. Subsequent demographic disasters—the result of later plague amplification events—periodically curbed population growth over the next couple of centuries. Harper's chapter ultimately emphasized the role of demography in modulating the JP's effects. To put it simply, the JP had such a catastrophic effect because it killed so many people (Harper 2017, 199–245, 304–15).

Plague Mortality

As *The Fate of Rome* galvanized the debate surrounding the JP, scholarly discourse began to incorporate specific terminology for the divergent positions that developed. The so-called maximalist position came to be used to describe those scholars who interpret the JP as having had significant societal-scale effects on the Late Antique Mediterranean. These effects often include some combination of apocalyptic mortality (e.g., of 25–50% of the population, or counted in the range of 15–100 million deaths), drastic transformation (e.g., economic collapse, unexpected cultural changes), and societal decline or collapse (e.g., the fall of the Roman Empire). Within the maximalist position there is much variance—some scholars are more explicit than others or suggest more catastrophic interpretations. Maximalists tend to accept a verbatim reading of the primary sources, particularly the few numbers that appear in Procopius and John of Ephesus. In recent years, many frequently incorporate finds from the natural sciences, often with little to no reflection on the biases and methodological issues that shape the results of such research. Theoretical and critical approaches are often absent from their accounts (see discussion of Sessa 2019 below). Notably, some scholars have argued against the use of the term "maximalist position" at all.[1]

The high mortality associated with the maximalist position continued to feature prominently in subsequent accounts of plague. Stephen Mitchell, for example, discussed "colossal mortality rates," accepted both Procopius and John of Ephesus' numbers on mortality in Constantinople, and argued that "the figures are under-estimates," because according to John it became too difficult to keep track of the dead after the authorities recorded 230,000 fatalities (Mitchell 2018, 28–29). Michael Bonner, who wrote a history of the Sassanian Empire, followed Harper's analysis. Most of his account of the JP focuses on the Roman Empire and on Roman sources. He appears to accept John of Ephesus' account that as a result of plague "most of the population of Egypt died, the death toll in Palestine was yet greater, and entire villages were annihilated in Mesopotamia" (Bonner 2020, 192). Bonner further uses the catastrophic reading of plague's effects in the Eastern Roman Empire as an analogy to its effects in the Sassanian Empire; the effects on Constantinople are seen as equivalent to plague's effects on the Persian capital at Seleucia-Ctesiphon. The concrete evidence for plague in Persia is almost non-existent (Bonner 2020, 194).[2]

1 Meier (2020a, 174–75): "anyone who ignores all these internal differentiations within research and instead speaks of a single 'maximalist position' as 'consensus' is making rough simplifications"; McCormick (2021, 47) arguing that Meier "forcefully rejects as an inaccurate strawman the arbitrary classification as 'maximalist' of the range of opinions of all scholars adhering to views that the pandemic had at least some important impact in some times and some places, including himself." Notably, Meier himself (2020a, 174, footnote 8) considers "plausible" a 25% reduction in population of the eastern Mediterranean between 541 and 544 (compared to Harper's 50%).

2 Note that this is a case of smearing based on little evidence (the term is discussed below), as Bonner (2020, 194) admits that "it is difficult to reconstruct the effect of the Justinianic plague upon the empire of Khusro." Only two sources for epidemic disease (i.e., not necessarily plague) in Persia are cited—that Khusro was

The maximalist position is contrasted with an alternative interpretation. The terminology surrounding this group is more fluid and has included terms such as minimalists (or minimalism), skeptics (or plague skeptics), and the "critical position" (i.e., on the JP), which will be used below. The critical position—despite the variance of opinion—offers a more nuanced interpretation of plague. For example, some scholars in this group accept that a particular outbreak of plague could have a significant impact in a specific location, but this would be a local rather than a societal effect that would cause neither a drastic transformation nor a massive demographic collapse. Until recently, scholars who dissented from the maximalist position tended to draw on a paper written by Jean Durliat (1989). Employing several types of sources, such as inscriptions and coins, Durliat asserted that the evidence for the JP was insufficient to justify the effects claimed by the maximalists. Maximalist scholars have since explicitly argued against Durliat's account (Sarris 2007; Meier 2016). Other scholars, however, were more receptive to his message, and some even cited him as evidence against the arguments of plague's supposed devastation (e.g., Wickham 2016, 43–44, footnote 3).

Things changed in 2018. The first pushback against Harper's (2017) *The Fate of Rome*—and the formation of the contemporary critical position—was a multi-authored three-piece review led by John Haldon (Haldon et al. 2018a, 2018b, 2018c). Although earlier scholars had been skeptical regarding maximalist plague accounts (e.g., Durliat 1989; Wickham 2005, 548–49; Whittow 1996, 66–68), with the exception of Durliat they had not formulated complete arguments against the maximalist account. Each piece of the review focused on a chapter or two of Harper's book, dismantling its arguments. The authors criticized Harper's tendency to accept the literary sources at face value, pointed out various errors and discrepancies in Harper's sources and quotes, called attention to his misrepresentative visualizations, and raised strong doubts that the evidence was enough to back Harper's dramatic statements (e.g., that "the rise of *Y. pestis* [i.e. the bacterium that causes plague] was a landmark event in the history of the human species"; Harper 2017, 244). The group concluded by praising Harper's efforts to synthesize such a large amount of material but emphasized that it required integrated and interdisciplinary teams of historians working together with scientists on specific questions (Haldon et al. 2018c).

A publication in *Past & Present* in mid-2019 authored by Lee Mordechai and Merle Eisenberg made a more comprehensive argument against the maximalist position. In their contribution, titled "Rejecting Catastrophe: The Case of the Justinianic Plague," the authors contextualized plague and espoused "a minimalist perspective" on its

----

infected while on campaign against the Romans in 542 and retreated to Azerbaijan, and that the patriarch of Seleucia-Ctesiphon in the 550s had to bury corpses that littered the streets of the city. Bonner interprets the latter point as evidence that plague remained a problem for the Iranian state for a long time without further evidence. He interprets a few additional anecdotes to reflect the aftermath of a large mortality event in Persia, but these seem unconvincing (e.g., Khusro attempted to augment the strength of his armed forces by recruiting more troops).

consequences (Mordechai and Eisenberg 2019, 7). They surveyed some of the most important literary sources for plague and called attention to the silence in many others, which strongly contrasted with the maximalist argument of high mortality. The article examined some of the non-literary sources (e.g., legislation, archaeology) while also revisiting and correcting some of the old evidence cited for plague's supposed effects. It added new evidence, such as the almost complete absence of known individuals who died of plague. In this context, the authors argued against the automatic association between the JP and the Black Death, suggesting that the Third Pandemic—which killed a far smaller percentage of the population—was just as viable a model.

In 2019, Mordechai and Eisenberg complemented their argument with a subsequent paper titled "The Justinianic Plague: An Inconsequential Pandemic?", this time with a team of scholars. The authors published in a top general science journal, *Proceedings of the National Academy of Sciences*, emphasizing the quantitative evidence and opening the discussion to a broader audience of scientists. After quantitatively discussing the literary evidence for plague, the article introduced to the JP debate large datasets of tens of thousands of inscriptions and papyri, which it collected and analyzed for the first time. Additional quantitative analyses included a survey of over 8,000 burials from pre-JP Britain, which was compared to all the burials in the cemeteries where *Yersinia pestis* was found in Late Antique human remains, showing that the onset of plague did not appear to result in an increase in the ratio of multiple burials—a finding that contrasted with maximalist claims. Furthermore, a survey of pollen from lakebed sediments across the northeast Mediterranean suggested that the onset of plague did not result in any significant change in land use, contrary to clear changes during the Black Death.[3] The authors concluded by pointing out that all the high-resolution (i.e., precise data at an annual level) and most of the low-resolution (i.e., data on the scale of several decades) quantitative measures they knew "show no observable effect of plague on Late Antique demographics" (Mordechai et al. 2019).

As of June 2021, there have been no comprehensive attempts to answer the arguments laid out by the critical position. The closest has been a response by Mischa Meier who pushed back on the quantitative account published in the scientific journal *PNAS* (Meier 2020a; for other attempts see Greatrex 2020; Bresson 2020; and McCormick 2021; perfunctory references are in Green 2019; all are discussed below). Meier acknowledged the qualitative *Past & Present* article was "more profound in its arguments," but chose not to focus on it since it "has not attracted the same amount of media attention and contains qualitative discussions of the material." Instead, he intended his review to be "a critical examination of the general recent trend to market

3 These changes included a decrease in cereal pollen and an increase in pine pollen. The interpretation of the Black Death evidence was that fewer people required less food and therefore grew less cereals, resulting in less pollen. In parallel, some of the former fields were colonized by pine trees, a fast growing species, a phenomenon that resulted in the increase in pine pollen.

research in the humanities more strongly in the media by clothing it in a scientific garment and placing it in scientific publication organs." In a series of subsections corresponding to the subsections of the paper by Mordechai and colleagues, Meier attempted to argue against its thesis, stating that he still thought plausible "a reduction of the population of the eastern Mediterranean by about 25% between 541 and and 544" (Meier 2020a, 173–74). The review triggered a response by Mordechai and colleagues who pointed out that "Meier's review ultimately offers few new ideas and no new evidence. Old literature is reiterated without acknowledging the problems we raised..." and that he "rehearses his own argument" which remains "unrejectable and unprovable." This was followed by a counter-response by Meier and a final response by Mordechai and colleagues (Mordechai et al. 2020a; Meier 2020b; Mordechai et al. 2020b). The debate highlighted a broad gap between both interlocutors on the question of engaging with scientific audiences on interdisciplinary topics such as the JP outside the fora of the historical profession and the necessity to move the discussion forward beyond "the same old sources and refuted arguments" (Mordechai et al. 2020b). Meier would further develop his positions on paleogenetics in a subsequent book.[4]

Broader Research Trends

In parallel to the debate on JP demography and effects, research has followed several threads. The sharply contrasting maximalist and critical positions on plague's effects have stimulated discussion among historians, who have attempted to review the current state of the field. In a second thread, interdisciplinary groups of scholars—historians, archaeologists, and scientists—have been working on certain specific questions that touch upon the JP debate. These include the nexus of volcanic eruptions, climate change, and the JP; the paleogenetics of plague; and, the discussions of the rich evidence coming out of the Byzantine cities in the Negev. In a third thread, a group of scholars—mostly historians—have been at work in outlining potential ways forward through specific methodological and topical case studies. This paper will examine the different threads below.

Thread 1: Reviewing the Debate

The radically different positions held by the maximalist and critical positions regarding the demographic effects of plague require some resolution. Yet, bridging this gap is hardly straightforward. As one commentator put it, the hypotheses of both groups are "built on large bodies of data, well argued and absolutely contradictory" (Bresson 2020, 240). The rapid pace of the debate between both groups further hinders a resolution, since new research can lead to significant revisions while making earlier reviews obsolete.

4 Meier and Patzold 2021. The author has not been able to consult this recent book by the submission date of this paper.

A review by Geoffrey Greatrex (2020) called attention to the debate, framing it within the juxtaposition of Procopius' description of life in Constantinople with life during the early months of the Covid pandemic. Reflecting on Harper's account and the work of Mordechai and colleagues, Greatrex argued that claiming that the early medieval plague (i.e., the JP) "was 'an inconsequential pandemic' is surely needlessly provocative," but added that Harper's attribution of almost every change to the pandemic is similarly exaggerated. The entire debate, he reflected, seemed to generate more heat than light, and Greatrex concluded that the truth "surely lies somewhere in the (less exciting) middle." Despite this approach, his review ultimately opted for the maximalist position, following his apparent suggestion that scholars should accept the numbers in the primary sources at face value. He concluded with a prediction that "Harper's book is more likely to weather the present storm more successfully than the thesis put forward by his opponents." The review does not explain this conclusion based on the evidence or arguments laid out on both sides, but it does seem to criticize researchers' interests in exaggerating their claims to gain attention "in an age where 'hits' on websites count" and reflecting upon the then-current Covid situation.[5]

Another review by Alain Bresson examined the debate through the arguments of Harper's *The Fate of Rome*. Bresson surveyed the known paleogenetic identifications of plague, which he interpreted as enough to show that "the presence of *Y. pestis* is in fact surprisingly high" (Bresson 2020, 242). Yet "the most difficult question" that remained was the lethality of plague. He chose to interpret the lethality through a biological paradigm, that is, whether the virulence of the *Y. pestis* bacterium changed between pandemics and within the first pandemic. Bresson ultimately acknowledged that the results remain speculative on both sides, because the bacterium's virulence is not understood well even in present-day plague. The demographic impact of plague, another major challenge, similarly remained unclear and varied geographically. Even if population levels decreased, he reflected, Late Antique societies had an unusual capacity for demographic recovery, but this capacity was perhaps limited. Bresson concluded by rejecting the minimalist view while also admitting that the maximalist model had no basis in the archaeology and is therefore "unlikely." The results seemed, again, to be somewhere in the middle. Future research, he asserted, should examine the capacity of Late Antique societies to absorb and recover from shocks (Bresson 2020, 242–43).

5 Greatrex (2020, 9) argues that as of his writing (April 2020) "just over 125,000 [people] have died from it [i.e. Covid]. More than that number died in Constantinople alone during the first outbreak of the Early Medieval Pandemic [i.e., JP], as Procopius and other sources indicate; even sceptics would accept a figure well over this." It is worth mentioning that the only other source who provides a numerical estimate of mortality in Constantinople is John of Ephesus, who argues for over 300,000 deaths in the capital (up to 16,000 deaths a day); elsewhere, John claims that only one of a thousand people survived the plague. For exact citation references: "was an inconsequential pandemic'" (p. 9); "surely lies" (p. 10), "even sceptics" (p. 9); "Harper's book" (p. 8); "in an age" (p. 10).

An alternative take has been suggested by Monica Green, who attempted to sidestep the debate, arguing "that the question 'how many died [in the JP]?' is not the only, nor the most important, question to ask about the plague." Drawing on analogies with the 1918-19 Flu Pandemic and the HIV Pandemic, she claimed that "many of the dead from both pandemics were young, previously healthy adults," and therefore "the economic and cultural impact was profound." In other words, plague did not have to kill a high percentage of the population to have been "profoundly influential as an infectious disease in late antiquity" (Green 2019; no references given). Although Green refrains from tackling the mortality question itself, it appears that her account acknowledges that total JP mortality was lower than estimated by the maximalist side, even as she remains open to maximalist arguments about the JP's social and cultural effects.

Altogether, then, the reviews of the debate surrounding the effects of the JP may have assisted in distilling the arguments in their current form and by collecting at least some of the relevant literature, thus orienting other scholars within the sprawling discussion. On the other hand, these reviews did not engage with the important contribution by Sessa (2019). Moreover, they have not helped answer the underlying key question about the JP's demographic impact and other effects, opting instead towards suggesting a vague solution in the "less exciting" middle or simply skirting the debate.

Thread 2: Interdisciplinary Questions

Volcanic eruptions, climate change, and plague

Perhaps the fastest-developing research topic associated with the JP is the investigation of the mid-sixth century cooling event often called the Late Antique Little Ice Age (LALIA) and its onset, the 536 atmospheric event sometimes described as a "mystery cloud" or a "Dust Veil" (henceforth Dust Veil). Early studies speculated about the origin of the Dust Veil and about the connection between this Dust Veil event and the JP (survey in Newfield 2018a). The absence of hard evidence in this discussion did not prevent scholars from conjecturing about the link between both variables in the past few decades (e.g., Toohey et al. 2016; see also Moreland 2018).

A study by Büntgen and colleagues called attention to significant cooling in late antiquity in 2011. This position entered the consensus view after the publication of their 2016 paper in *Nature Geoscience* (Büntgen et al. 2011, 2016). The latter paper established a temporal correlation between the onset of the LALIA—with two major volcanic eruptions in 536 (which caused the Dust Veil) and 540—and the onset of the JP, visualizing the connection through a controversial figure that heavily implied that the LALIA caused the plague (fig. 4 in Büntgen et al. 2016; for a critique see Moreland 2018, 95). The figure depicted both the LALIA and the JP as co-existent for over a century before the LALIA slowly faded away. The paper did not cover plague at length but suggested that the climate was "an additional environmental factor contributing to the establishment of the Justinianic Plague" and stating that the JP "followed widespread food shortages right after the onset of the LALIA." The authors also admitted that the role climate might have played in the onset of plague was "unclear" and "requires caution" (Büntgen et al. 2016, 231, 235). More recently, Büntgen appeared to signal more caution (Büntgen et al. 2020). In parallel, other research has questioned the temporal extent of the mid-sixth century cooling. One team, for example, used model simulations that suggested that cooling would persist for only about twenty years instead of the more than a century originally suggested by Büntgen and colleagues (2016), which may shorten any long-term effects the LALIA might have had (van Dijk et al. 2021).

Regardless of these more recent reservations, Büntgen and colleagues' (2016) paper opened a path for continued rumination. Several of the scholars who discussed the 536 eruption, its location, or its traces, referred to the JP. These references were brief, superficial, and speculative, often covering the JP in a line or two.[6] While never explicitly stated, it appears that better understanding the 536 eruption would in turn improve our understanding of a major Late Antique demographic event. Moreover, most scientists accepted the maximalist position on the JP, which remained more attractive to them as an event they could try to explain using their own sources and methodologies, while showing no awareness of critical JP interpretations.

There have been several more detailed attempts to outline a causal link between volcanism, climate, and plague outbreaks. The JP is an attractive case in these studies because of its temporal proximity to the 536 and 540 eruptions. Although recent studies increasingly use scientific terminology, this in itself is hardly a new approach. The roots of this connection can be dated back at least to the mid-nineteenth century, even if it has expanded rapidly in the last couple of decades (Newfield 2018b, 472). Tim Newfield (2018a, 97–98) has sorted out the different approaches in the literature to the supposed connection between volcanic forcing (i.e., cooling) and the JP (see also Newfield 2016; 2018b, 472–73). The discussion below will adopt Newfield's numbering for ease of reference.

One interpretation (linkage 1 in Newfield's division) argued that the mid-530s climatic change disrupted a plague reservoir, enabling the pathogen to spread to new areas in the Mediterranean leading to the JP. Earlier scholars speculated that the hypothesized "plague reservoir" was somewhere "east of Lake Victoria in Kenya and Tanzania" in Africa (linkage 1, subsection 1) or perhaps somewhere in Asia—the Himalayan foothills in India or western China (1.2). Plague then moved towards the

6 A few examples will suffice: Dull and colleagues (2019) discuss a concurrent "major outbreak of bubonic plague" and "demographic contraction," and Helama and colleagues (2019) associate a "sustained climatic depression with potential human impacts in the form of famine, the Justinianic plague and human migrations over the following years." The 2019 study by Dull and colleagues pinpointed the 536 eruption to Ilopango (in modern El Salvador), but a recent study by Smith and colleagues (2020) has pushed back the Ilopango eruption by a century to 431 CE while also arguing that its major effects were local in scope. Webb (2018, 359) argued that the eruption caused a failure of harvest, subsequent famine, and thus a heighted biological vulnerability to pathogens, and Bonner (2020, 192–93) stated that the "huge volcanic explosion [sic]... would have favoured either the outbreak or the spread of the Justinianic Plague."

empire either through maritime trade connections with South Asia (1.2.1) or overland through Asia (1.2.2). A second interpretation (2) linked food shortages to the JP: the presumed result of the Dust Veil event and LALIA. The food shortages may have led to malnutrition that made populations more vulnerable to diseases, such as plague, increasing mortality (2.1). Alternatively, environmental stress may have pressured Late Antique populations to migrate, and this migration would facilitate the spread of plague (2.2). After surveying the different linkages and the evidence for them, Newfield concluded that linkages 1.1 and 2.1 are doubtful, though "not so easily dismissed" (Newfield 2018a, 99–100). The evidence for dearth (the assumption underlying linkage 2) appears patchy, even after the largest eruption of 536—for example in Cassiodorus' letters. There is no known evidence for dearth after 538 or for general famine in the Mediterranean world after the 539/540 eruption (Newfield 2018a, 100–102).

Although a significant step forward, Newfield's careful analysis did not receive the attention it deserved. Subsequent research tended to follow one or two of his linkages, albeit with less evidence and more speculation. In a chapter aptly described "Climate and Disease," Peter Sarris attempted to examine the connection between the 536 volcanic event, the ensuing cooling, and the JP. He argued that "climate change and instability may also have played a crucial part" in the bubonic plague of the sixth to eighth centuries (Sarris 2020, 521). The exact mechanism, however, followed Keys' (1999) account: a vague "change in climatic conditions" may have disrupted a plague reservoir in East Africa (linkage 1.1 in Newfield's division) (Sarris 2020, 522ff). Sarris appears to have been unaware that Newfield had argued against this linkage.

The authors of more recent papers—published during the Covid pandemic—highlighted "the urgent need for more collaborative multi-disciplinary work to deepen our understanding of global plague pandemics" (Fell et al. 2020, 37; see also Luterbacher et al. 2020). Fell and colleagues highlighted the cooling associated with the LALIA, which would have caused crop failure and food shortages (linkage 2). These would have weakened people's immune systems and perhaps increased mortality, potentially also contributing to the spread of plague through migration or grain hoarding, which would in turn benefit rodent populations (linkages 2.1 and 2.2). As this brief survey reveals, however, these connections remain conjectural rather than based on hard evidence.[7] Luterbacher and colleagues (2020) collected several paleoclimate proxies of different temporal resolutions. Here the authors chose to align the results of these proxies with JP recurrences but were not able to find a clear connection. They therefore concluded that "any simple correlation between plague reemergence and European summer temperatures or continental European summer precipitation is unapparent."

7 As discussed above, the cooling does have some evidence. However, it is limited spatially, while its effects on crops remains conjectural. The article also includes some apparent contradictions, such as "each volcanic eruption caused intense short-term cooling leading to crop failure, severe famine and large scale depopulation across Eurasia" but also "the impact of LALIA cooling on Mediterranean and Eurasian societies may have... [caused] only a minor regional epidemic with a limited death toll" (Fell et al. 2020, 39).

Bresson's review similarly reiterates the existing hypothetical linkages that Newfield mapped but does not further the discussion or build on Newfield's insights. It is claimed that "less well-fed and less prosperous populations are weaker and less able to resist an epidemic" (linkage 2.1) (Bresson 2020, 238). Yet the review cites no concrete evidence for this truism and does not indicate the spatial and temporal patchiness of dearth reports.[8] Later Bresson speculates that higher precipitation in the semi-arid regions of Asia, the reservoir of the *Y. pestis* bacterium, would cause a proliferation of rodents and accelerate the dissemination of the disease (linkage 1.2).

Discussions about climate and volcanism tend to speculate particularly on pertinent details for historians, leading to conflicting interpretations and narratives. Mid-sixth century Arabia is a case in point. According to Büntgen and colleagues (2016, 235), cooling in the Arabian Peninsula would have led to increased precipitation. This in turn would be beneficial to local populations and contribute "to the rise of the Islamic Empire." Yet, more focused studies reveal a series of droughts in Arabia in the early to mid-sixth century. The most recent account locates the most severe decadal drought around 520 CE, while the 540s feature average rather than increased precipitation. Notably, the research group—which did not focus on the JP—conjectured that the transition to the Islamic era in the seventh century may have been triggered by recurring severe droughts throughout the sixth century, again in contrast to Büntgen and colleagues' conclusions (e.g., 2009; Fleitmann et al., under review).

Although the discussion linking climate and plague is increasingly diverse in the tools it uses, it remains severely restricted. There have been almost no attempts to specify the potential effects of cooling on agricultural species or crop yield. Similarly, there has been almost no involvement by ecologists, who might study the effects of cooling on relevant rodent or flea populations even in the present (for earlier discrepant research, see Stenseth et al. 2006; Schmid et al. 2015). Methodologies such as population modeling, extremely common in contemporary ecological research, are all but absent from discussions surrounding the JP, which therefore remain largely speculative and based on little evidence beyond the environmental trigger. At the core of the effort are paleoscientists who attempt to answer their own questions, namely to determine the extent and effects of the volcanic eruptions and climatic cooling in the mid-sixth century, in a manner reminiscent of paleogeneticists' contribution to discussions surrounding changes in the *Y. pestis* pathogen itself (below). The volcanic/climatic debate has therefore focused on producing quantitative data, interpreting graphs and model results of environmental changes in temperature or precipitation.

8 Bresson (2020, 238, 241) cites Newfield's (2018a) "Mysterious and Mortiferous Clouds" article twice (referring to him as a source for the link between climate and plague, and on the number of JP outbreaks).

Paleogenetics and Plague

A series of studies that has examined the evolution of the *Y. pestis* bacterium, much of which has been published in the past decade, has drawn much recent attention. The mid-2010s paleogenetic studies demonstrated that *Y. pestis* was almost certainly the pathogen behind the JP (Harbeck et al. 2013; Wagner et al. 2014; Feldman et al. 2016). It has since become increasingly common for historians to refer to paleogenetic research; some have similarly begun to collaborate directly with paleogeneticists. At the same time, paleogeneticists have been attending and presenting their work in historical conferences. Several earlier reviews of the paleogenetic evidence have already been published (Little 2011; Eisenberg and Mordechai 2019), so this survey will examine only the material from the past few years to track the latest developments and their implications.

Following earlier trends, scholars continued to discuss the origins of the JP. The debate splintered into a few questions. One focused on the proximate origin of plague: how did plague reach the Mediterranean world? The identification of *Y. pestis* in the remains of an individual from the second-third centuries in the Tien Shan Mountains, together with the finding that the strain (DA101) was basal to the JP strain, suggested to some that plague was brought to the Mediterranean through the southern steppe rather than through the Red Sea (Damgaard et al. 2018). The Tien Shan strain may have been a direct ancestor of the JP strains, but this remains ultimately uncertain and unanswerable considering the current state of evidence (debate in pre-prints: Keller et al. 2019a; Rasmussen et al. 2019).

A continued attractive hypothesis was the possibility of a proximate JP origin in East Africa. Monica Green examined the elusive evidence for the JP in Africa and proposed several cursory hypotheses, for example pointing to third century epidemics in "Ethiopia" as potential episodes of plague without any supporting evidence (discussed below). She similarly noted that the Aksumite Kingdom collapsed by the seventh century, while Egypt was "also hit hard [but] was able to survive and soon thrive again." The divergent fates of both polities was potentially related to plague effects, the only reasoning being that "the well-documented affinity of *Y. pestis* for high elevations may have been a factor" (Green 2018, para. 33). In other words, Green used the already-hypothesized presence of the pathogen in the area to subsequently explain regional political developments.

More radically, Green's article also argued "that the evolution of organisms—in this case, a pathogenic bacterium—is a kind of history and therefore can be used by historians for historical argument" (Green 2018, para. 8). Green then suggested an outline of how to use genetics as a source of geographical information—a relatively secure process based on comparing genomes from different areas on a phylogenetic tree—and the use of genetics as a source for chronological information. The latter is more important for historians who aim to reconstruct plague epidemiology but is unfortunately far less precise: the unit of measure, the SNP (a single nucleotide polymorphism), does not correspond to a fixed measure of time even within the same species.[9]

Peter Sarris, who had suggested a couple of decades earlier that the JP originated in Africa, maintained his theory in a recent chapter (Sarris 2002; 2020). As the Tien Shan *Y. pestis* find posed some difficulties with this interpretation, Sarris' most recent account hypothesized that (1) plague reached the Mediterranean from Central Asia a few centuries before the JP; (2) the disease reached East Africa, where it established a reservoir; and, (3) after "re-emerging in late antiquity," the East Africa reservoir spread back to the Mediterranean world, causing the JP (Sarris 2020, 521–24). The evidence supporting this theory, however, remained extremely weak.[10]

The most important recent contribution to paleogenetic research of the JP was Keller and colleagues' (2019b) study. The study included three findings relevant to historical discussions: it identified human remains with *Y. pestis* in new locations; it presented new complete genomes of *Y. pestis* from Late Antiquity; and, it then used these changes to postulate changes in the virulence of the *Y. pestis* bacterium over time.

The identification of *Y. pestis* in a total of 32 teeth, coming from new locations, was a straightforward finding. If earlier human remains with *Y. pestis* came from Germany (securely identified) and France (controversial), Keller and colleagues added a secure identification of *Y. pestis* in human remains in France and added new locations to the list, including Britain and Spain. Earlier scholars had conjectured that plague reached both areas, and Keller and colleagues seemed to resolve these questions (e.g., Maddicott 1997; Kulikowski 2007). Scholars differed on the importance of these findings. Some maximalists, such as Harper, saw plague presence in a specific location as indicative of its affecting a much larger area (Harper 2017, 230; also Bresson 2020, 241–42, using a different methodology). Others, such as Green, asserted that these findings had already been anticipated by those "historians who had been paying attention," and that these results were not paradigm-shifting but paradigm-confirming (Green 2019, citing the entire book of Little 2007). Supporters of the critical position pointed out that the presence of plague said little about its effects; after all, plague exists in the United States today but is barely acknowledged by most of its population (as noted, for example, in Eisenberg and Mordechai 2020, 1663).

9 Green 2018, paras. 25–29. The examples provided by Green feature the evolution of 17 SNPs in 800 years; 40 SNPs in 1700 years; more than 3 times as many SNPs in half the time (i.e., >120 SNPs in 850 years); and, 100 SNPs in 500 years. A broad survey found a total of 2326 SNPs in 133 *Y. pestis* strains, but some 99% of these SNPs were mutations that happened only once in the genealogy that were distributed essentially at random. Only seven genes had a significant excess of function-changing SNPs. Green also pointed out the methodological issues in trying to track epidemics through single genetic specimens—and suggested we should instead track reservoirs, where genetic changes become fixed (stable) in the *Y. pestis* lineage.

10 For example, Sarris followed Green in hypothesizing an early/mid-third century plague precursor in Africa. Working through the citation trail, alongside multiple misreferences, the primary source evidence for this claim is the epidemic Harper describes as the Cyprianic Plague (in his book, which both Sarris and Green cite for other purposes), which was clearly not the *Y. pestis*-caused plague. See Sarris 2020, 523; Green 2018, para. 33; Hatke 2013, 46; Eide 1998, 960–62, 996–97; Banchich and Lane 2009, 101–102).

The reconstruction of new complete genomes of *Y. pestis* opened another set of questions, particularly because Keller and colleagues added eight genomes to the one that had been known. Scholars could now place all nine genomes on a phylogenetic tree and cross-reference that information with spatial data about where each genome was found (Zhou et al. 2020 has the most recent visualization of the *Y. pestis* phylogenetic tree). Somewhat unexpectedly, they found that the *Y. pestis* from Britain was earlier in evolutionary terms than the other Late Antique *Y. pestis* genomes. This raised the possibility that plague reached Britain before other locations, which had not been considered in earlier scholarship. More broadly, some scholars argued, Keller and colleagues' new genomes were an early step in a process that would eventually allow researchers to track the spread of *Y. pestis* across space and time and thus reconstruct JP epidemiology at a high resolution (i.e., annually or even more finely).[11] While such an approach had been suggested over a decade ago (McCormick 2007, 306–307), scholars who continue to support it do not explain how they might get around two key challenges: the low number of *Y. pestis* positive identifications (regardless of the number of false negatives), and perhaps more disturbing, the fact that the *Y. pestis* genome is "genetically monomorphic," that is, stable changes in its genetic lineages are relatively rare (Green 2018, para. 28; following Achtman 2012).

Keller and colleagues also commented on the pathogen's virulence. The team found that a certain area in the *Y. pestis* DNA was deleted and that this area included two "virulence factors involved in host cell invasion and intracellular growth (*mgtB* and *mgtC*)." The same area was also deleted in Second Pandemic genomes from the seventeenth and eighteenth centuries, but not in any other ancient or modern genome. Keller and colleagues admitted that the results were ambiguous. The deletion may have resulted in an "attenuation," that is, the pathogen becoming less virulent, or alternatively, it could have also made the pathogen more virulent by facilitating immune evasion, that is, evading the host's immune response (Keller et al. 2019b). Subsequent studies referred to this finding but, perhaps understandably, could not reach any clear conclusion based on it.[12]

Subsequent paleogenetic research on the JP would have to overcome substantial hurdles to be able to answer the type of questions historians seem most interested in (Eisenberg and Mordechai 2020, 1658–64). Moreover, the type of specialized knowledge

11 Green (2019): "The new genetics, for the first time ever, allows us to ask some of the questions real epidemiologists do: Why is this specific disease presenting here? Why in these populations? Where did it come from in the first place, and by what pathways did it (will it) spread?" As a model for the potential of this approach, Green shared a video reconstructing the movements of 1,610 Ebola genomes at a weekly resolution over less than two years in total. See also McCormick 2021.

12 Mordechai et al. (2019) discussed the attenuation possibility and concluded that the genetic data "leave open the possibility that the First Pandemic was less prevalent than Second Pandemic plague." Bresson's review (2020, 241) noted that the deletion "might either have increased or decreased the lethality [sic] of the virus [sic]."

required to read and evaluate the scientific studies remains rare among historians focusing on the JP. Finally, no JP paleogenetics papers with a historian as their first author have been published in a scientific venue so far. On the other hand, scientism, the "excessive belief in the methods [and data] of the natural sciences" to explain humanistic knowledge, appears common among historians conversant in this debate. In interdisciplinary collaborations, it appears, historians believe they could learn much more from the paleogeneticists than vice versa, while history is often relegated to the role of validating the scientific information, interpreting it and embedding it within an attractive narrative, often following the maximalist position (definition and discussion in Sessa 2019, 223–24). This narrative can then be used to justify further (mostly paleogenetic) research.

The Byzantine Cities in the Negev

The Byzantine cities in the Negev (present-day southern Israel) have long attracted scholarly attention: a handful of desert cities that flourished in Late Antiquity before they declined and were eventually abandoned. The supposed temporal correlation between the end of the cities and the JP was an attractive potential causal link for those seeking evidence for plague's drastic effect. Over the past couple of years, several studies of these cities were published in prestigious journals, drawing scholarly and public attention to what was at best a marginal area for the Late Antique Roman Empire. Perhaps the most significant contribution of these studies, part of a European Research Council grant, was to introduce new high-resolution archaeological data in an attempt to understand better the end of these cities, which were well-preserved because of their remoteness.

The study of Bar-Oz and colleagues (2019) attempted to use well-preserved trash mounds in the urban center of Elusa to establish the end date of organized trash management. Although Elusa survived into the seventh century, trash collection and deposit in the mounds stopped at some point between 400 and 550 CE. This was understood as a signal for "the decline in the city's high municipal standards of living introduced during the Roman and Byzantine periods, indicating an overall downgrading in the regional level of socio-political organization." The authors suggested that perhaps the cessation of trash removal correlates temporally with the LALIA and the first JP outbreak in 541 but explicitly rejected stating a direct causal or mechanistic link with these events, at least partially because their archaeological evidence was not of sufficient temporal resolution. The authors added that present cause-and-effect equations are unbalanced, because there exists substantial evidence for major natural hazards such as climate change, volcanism, or plague; but, the evidence for the societal response to these events remained "remarkably weak" (Bar-Oz et al. 2019) These careful reservations were not shared by all commentators, who preferred to imply a causal connection between the cessation of trash removal and plague, despite the lack of evidence and the broad interval during which the trash removal may have stopped.[13]

A subsequent study attempted to tackle the issue of causality more carefully, this time through the rise and fall of local viticulture in the fourth–sixth centuries. The authors tied the decline of viticulture, represented by the proxy luxury "Gaza wine," to market contraction. This market contraction was perhaps triggered by climate change, plague, or wider sociopolitical developments, rather than the seventh century Islamic conquest. In a section focusing on the connection to plague, the authors acknowledged that the current uncertainties in JP research precluded reaching a concrete conclusion. They conjectured that a demographic decline of some 20% in Constantinople and Alexandria, two cities that were hit hard by plague, could have reduced demand for imported luxuries, such as wine from Gaza. Moreover, the authors conjectured, "if the plague reached the Negev, it could also have harmed the local production capacity and supply of agricultural products in general by inducing a shortage of agricultural laborers." Yet, the authors had no way to evaluate whether plague—rather than other viable explanations such as climate change or a range of sociopolitical developments—was the trigger of reduced demand (Fuks et al. 2020). Another study by the same group, however, used isotopic indicators of animal dietary and mobility patterns to assess the region's vegetative cover. Since both patterns indicated continuity through the sixth and seventh centuries, the authors moved away from a climatic explanation and admitted that "the end of the Byzantine presence in the Negev Desert was not as dramatic as previously suggested, with a catastrophic climatic shift leading to a societal collapse [but] was more likely the result of reorganization of economic and/or territorial priorities within the wide Byzantine empire" (Vaiglova et al. 2020).

The case of the Byzantine cities in the Negev demonstrated the major challenges scholars would have to overcome in order to link archaeology and plague. Even in the context of a mostly undisturbed and well-preserved site, with rich excavation data collected over years, the authors were unable either to establish a robust connection with plague or to reject such a connection with confidence.

13 McCormick (2019) stated that we can now unveil "the extent of what look like exogenous shocks to the resilient Roman system, in the form of rapid climate change and unprecedented disease outbreaks" and that "the new dating [i.e. of the cessation of trash removal] overlaps with those two major exogenous shocks on the Roman economy: rapid climate change starting in 536, and the Justinian Plague pandemic beginning in 541 to 542"; Green (2019) claimed that "we have new methods to document population contractions [i.e. as a result of plague] by, remarkably, noting when trash collections ceased." The 536–542 dates of the Dust Veil and the JP "suck-in" the loosely dated evidence from the trash mounds (see below and for sucking-in Baillie 1991).

Thread 3: Case Studies

Over the last few years, several scholars chose to investigate specific questions associated with the JP. Unlike the previous thread, which featured predominantly large interdisciplinary groups and multiple publications on each research question, this trend includes studies focusing on close reading of a limited number of texts, almost all of which were written by single authors from the humanities. Notably, four of the studies below were published in the same journal issue, and three others originated in a Harvard graduate seminar.

While scholars have long been aware of the existence of literary sources that referred to earlier plague accounts, there had been no concentrated work on this issue (Eisenberg and Mordechai 2019). John Mulhall attempted to rectify this situation by surveying the Greek medical evidence for plague before the sixth century. He pointed out a couple of medical sources from around the second century CE (Rufus of Ephesus and Aretaeus of Cappadocia) who discussed plague, although also noting that Galen did not seem to be aware of the disease, and that the Hippocratic Corpus contained no evidence for plague as early as the fifth or fourth centuries BCE. Mulhall suggested that Rufus and Aretaeus were discussing an earlier outbreak (by the first century CE) but noted that plague, or a similar disease, subsequently disappeared from the sources until the sixth century pandemic (Mulhall 2019).

While Mulhall surveyed medical literary sources, Michael McCormick conducted an in-depth examination of references to plague and other diseases in a single author, Gregory of Tours, while offering a broader context about the JP. McCormick rejected earlier claims that cast doubt on Gregory's account of plague, although he acknowledged that most of Gregory's descriptions were brief and undetailed. After surveying the 22 mentions of plague in Gregory—corresponding to six epidemics in the mid- and late sixth century—he concluded that Gregory's information is robust and reliable, "his authority is unimpeachable," particularly where he offers relatively detailed information. Ultimately, however, Gregory's perspective remains local in scope (McCormick 2021, 88–91). According to McCormick, Gregory's descriptions also fit scientific understandings of *Y. pestis* with regard to symptoms, epidemiology, and mortality (McCormick 2021, 38, 84, 91–92; following Pollitzer 1954). His interpretation of the potential of aDNA, which featured heavily in the article, seemed overly optimistic while overlooking the serious limitations of aDNA discussed above.[14]

14 McCormick avoids discussion of the major issues regarding genetics raised by Green 2018 which were noted above; McCormick cites the paper as part of the special journal issue of *Afriques* on p. 38), preferring instead to attempt to link specific aDNA finds to plague "waves" or amplifications (McCormick 2021, 48, 68–69, 75, 93). Technological optimism is apparent elsewhere in the paper. A few examples: p. 40 ("new and accumulating evidence"); p. 41 ("the new archaeoscience is changing what we know..."); and, p. 43 ("In just seven years, the stunning advances of biomolecular archaeological research have changed the history of plague"). For earlier examples, see e.g., Eisenberg and Mordechai 2019, 160–61.

A recent contribution to the discussion by Eric Faure, however, contradicts parts of McCormick's account. Faure agrees that Gregory had "a good knowledge of diseases," and accepts that his account of the 571 CE outbreak was "very credible" (Faure 2021, 430, 439). However, Faure argues, the first episodes of plague in Frankish Europe (i.e., 543-47 CE) were not even of low intensity as suggested by Mordechai and colleagues (2019). Rather, he asserts, "they did not occur at all" (Faure 2021, 460). His analysis is based on a thorough investigation of Gregory's texts, alongside attempts to date the plague episodes precisely, and reading in parallel other contemporary and near-contemporary authors. Since scholars generally accept that the first outbreak was the most significant outbreak of plague by a large margin, Faure's argument implies that mortality in Western Europe was considerably lower than hitherto estimated.

A third approach to the literary sources was adopted by Henry Gruber, who attempted to use indirect evidence to trace some of the social effects of plague in mid-sixth century Hispania. The heart of his argument focused on a canon of the Council of Valencia (546 CE) which referred to burying bishops who died "a sudden death" in a coffin apart from others. Gruber used several canons from the same council to argue that "the social order was under stress." Although the evidence is indirect, he believed it was suggestive of the effects of plague and further recommended looking elsewhere for new types of evidence for plague, namely "texts that do not mention the plague but that under examination may show signs of the strains it caused" (Gruber 2018, 210-11).

A paper by Lauren White and Lee Mordechai adopted a different approach. The authors used an infectious disease model to examine the effects of the JP in Constantinople in attempt to discern whether the effects reported by Procopius and accepted by maximalist scholars were even feasible. The resulting paper drew upon contemporary plague research to estimate parameters such as the incubation period of bubonic plague or the percentage of cases of bubonic plague that develop secondary pneumonic plague. Historical scholarship and Procopius' account (accepted verbatim in the article as a hypothesis), supplied additional variables such as the population of Constantinople and the length of the outbreak. The model results indicated that the outbreak in Constantinople was almost certainly not pneumonic plague and implied that plague would have varied effects in different large cities (White and Mordechai 2020).

While the aforementioned papers focused on the presence and effects of plague, several scholars chose a different approach—one that emphasizes how the surviving primary sources experienced and wrote about disease in Late Antiquity. Most of the resulting papers do not engage with the JP literature and are not interested in the JP's actual effects. Moreover, they do not differentiate between plague and other disease outbreaks. Notably, three of the papers were published shortly before submission of this review in the same special journal issue of *Vox Patrum*, in the context of the Covid pandemic.

Sally Shockro conducted an early investigation along these lines. Her study focused on Bede and a few other authors who wrote about sickness in England and Ireland in the seventh century. Shockro de-emphasized the identity of the disease and avoided the discussion of almost all plague literature, preferring instead to focus on the texts she analyzed after noting that the sickness her sources were concerned with was "most likely" bubonic plague (Shockro 2018, 320). In her analysis, she pointed out that the biblical interpretation of plague was as a manifestation of divine wrath, reflected also in other early Christian authors. Yet Bede, writing in the late seventh and early eighth centuries, made a considerable effort to emphasize in his *Historia ecclesiastica gentis Anglorum* that the seventh-century plague affecting the English people was *not* a sign of guilt and punishment. Rather, the plague affected the exemplary English people who, according to Bede, enjoyed a special relationship with God. It occurred in the holiest places, sickened the most pious inhabitants, and reinforced survivors' faith. (Shockro 2018, 323–24 and passim).

Alberto Ferreiro (2021) adopted a similar methodological approach, avoiding discussion of the identity of the specific disease and not engaging with any of the JP literature. His paper selected several sources from Late Antique-Visigothic Hispania (fifth-seventh centuries) and examined the manner in which they understood and portrayed plagues, famines and other abnormal phenomena. A common interpretation was to causally link these events with barbarian incursions and to accept that God ended both plagues and invasions directly or through a holy person. Abnormal events such as plagues, therefore, fulfilled God's will, even if observers and commentators did not always immediately understand it.

Along the same lines, Michal Ludewicz (2021) examined how the authors of the *Liber Pontificalis* (lit. the Book of Popes) perceived plagues and illnesses. The paper did not focus on identifying these diseases or tracing their effects. Contrary to Ferreiro's conclusion, Ludewicz pointed out that most diseases were presented in the *Liber Pontificalis* as caused by natural factors, but in some cases they were signs of divine wrath. As might be expected, the individuals most commonly afflicted with disease of any sort in the source were the popes themselves.

Geoffrey Dunn (2021) was similarly interested in Christian responses to Late Antique and early medieval pandemics, writing explicitly in the context of the Covid pandemic. Dunn (2021, 394) was very much aware of the discussions surrounding the JP, but pointed out that "the precise nature of the ailment is not clear nor is it important for our purposes." The subject of his research was "pandemic as a rhetorical reality rather than an epidemiological one." Surveying the writings of Gregory the Great, Dunn argued that the bishop felt the pandemic was so overwhelming that there was no point in attending to physical concerns, and therefore focused his efforts on tending to spiritual care—by exhorting Christians to repent of sin and seek forgiveness and mercy from God.

Paradigm Shifts

Although both the maximalist and critical positions have stated their arguments, as of writing it is clear that the debate has not been resolved and does not appear to be moving in that direction. In this situation, non-specialists would understandably have difficulties keeping up with the sprawling discussion. Two papers appear to offer a path out of the conundrum. Although from a broader perspective of the historical discipline both papers adopt well-established positions, they are radical within the existing discourse on plague. Both focus not on Late Antique plague but push back against the very notion of plague as a seemingly self-explanatory term, while reflecting upon what we think plague does. Kristina Sessa's (2019) paper frames JP research within the recent increase in attention to and scholarship on environmental history, specifically in the Late Antique context. Eisenberg and Mordechai's (2020) paper, on the other hand, frames JP research within broader intellectual trends over the past century and a half.

Sessa suggested a different methodological approach in a key piece that commented on recent developments in environmental history in the context of Late Antiquity in particular. While Sessa, a cultural historian, acknowledged the importance and agency of non-human factors in human history, she offered a forceful critique of existing scholarship, focusing in particular on Harper's *Fate of Rome* and Michael McCormick's scholarship (citing the *Fate of Rome* in 31 of 167 footnotes, and McCormick's scholarship in 35 footnotes in total). Sessa pointed out that many environmental historians have moved away from textual interpretation, preferring instead to attempt to draw connections between scientific sources and historical sources. These links, however, were often superficial and oversimplified. As Sessa put it (2019, 237), "historians should reconsider accepting so simple (though undeniably elegant) a solution to the 'end' of Rome." Existing scholarship on plague, according to Sessa, tends to implement two fallacies, which she summarized as "sucking-in" and "smearing."[15] She expressed skepticism regarding the purported maximalist effects of plague, arguing that we "should resist concluding that the first plague pandemic... constituted *the* catastrophic event that brought about 'the end of Antiquity' when our physical evidentiary base remains as small as it is" (Sessa 2019, 235–36).

15 Sessa adopts both terms from Baillie (1991). In Sessa's explanation (2019, 236–37): "Because the best witnessed, and purportedly most extreme, outbreak of the plague occurred during the reign of Justinian (from 527 to 565), scholars link a wide range of different evidence to this specific time and event. Developments, for example, in the built environment that archaeologists date broadly to the second half of the sixth century get "sucked in" to the explanatory vortex of the "Justinianic Plague," which functions not only as a dating anchor but also as the cause of any change. Moreover, the first plague pandemic was not a single event; rather, it was a series of events that took place in varying degrees and in different locations, between roughly 541 and 750 CE. In this respect, scholars "smear" these different outbreaks into a single event when they characterize the entire first plague pandemic as the "Justinianic Plague," even though Justinian died in 565 CE, and not all outbreaks occurred during his reign."

The significant contributions to the debate outside the critical position—all written by historians—overlook or ignore Sessa's methodological and theoretical arguments, de facto refusing to engage with or even acknowledge her critique. This stands in stark contrast with some of these authors' careful reading of Keller and colleagues' (2019b) paper, published a few weeks later in a scientific journal.[16] This lack of engagement, particularly with an overdue careful reflection on the swift developments in the field, has done much to stifle the discussion and keep it within familiar bounds.

Eisenberg and Mordechai's most recent paper (2020), "The Justinianic Plague and Global Pandemics: The Making of the Plague Concept," drew on some of the ideas and critique explored by Sessa and examined JP-related research over the past century and a half. It deconstructed some of the long-standing assumptions in the discourse on historical plague, including the tripartite division of plague history into three supposedly distinct "pandemics." By focusing on the development of plague features in scholarship over time—its chronological scope, spatial extent, and mortality—the paper argued that JP research had been strongly influenced by much broader trends in historiography and public discourse on infectious diseases. In other words, scholarly arguments on the longevity, scale, and lethality of plague were founded on contemporary twentieth and twenty-first century expectations of what plague *should* have done rather than on Late Antique evidence for plague's effects.

The paper proceeded to demonstrate how such expectations influenced research by focusing on three "truisms:" rats, climate, and aDNA. Rats have been associated with plague since the late nineteenth century, although evidence for rats' presence in the Late Antique Mediterranean has been underwhelming both archaeologically and in literary sources. Climate has long been hypothesized as a factor that determined plague outbreaks and spread; yet, the assumed model had always been the Third Pandemic case study in Bombay (modern Mumbai), with warm weather. This paradigm completely collapsed in 2011, when scientists argued that the mid-sixth century was a cooler period. Instead of reflecting on the reasons for the previous erroneous consensus, historians simply changed the paradigm and have since associated plague with cooling, even if the exact reasons for this connection remain unclear. The third truism has been the entrance of paleogenetics into the historical debate, most importantly in using the aDNA evidence for plague to argue for drastic effects for plague. While some historians continued to hope that science would answer some of the gaping questions on plague, Eisenberg and Mordechai argued that it faced a series of difficult challenges and doubted that it would provide solutions to the most pressing historical questions in the foreseeable future.

16 McCormick (2021), and Sarris (2020) do not cite Sessa at all; Bresson (2020, 233) notes her review is important but never returns to it; Greatrex (2020) mentions her in the context of a tangential point and in his bibliography; Meier (2020a) refers to her in a footnote in the penultimate sentence of his review. In contrast, McCormick and Bresson make much use of Keller and colleagues (2019b), published a few weeks after Sessa's paper. Meier and Sarris cite Keller, while Greatrex does not.

As the Eisenberg and Mordechai paper has been published only recently, it remains to be seen whether subsequent studies will engage with its most significant contributions to the debate. A reflection upon the construction of plague as a rhetorical and imaginary concept, as well as a greater awareness of changing influences on modern historiography that have shaped our understanding of the JP, have a potential to move the discussion forward.

Concluding Remarks

This review has surveyed recent developments in the study of the Justinianic Plague over the past four years, beginning with the publication of *The Fate of Rome*. The topic that has stimulated the liveliest debate concerns the effects and impact of the JP on Late Antique societies, leading to the crystallization of the critical position on the JP. Several interdisciplinary teams examined adjacent fields and topics, such as the potential connections between the JP and climate change and between the JP and the decline of the Byzantine cities in the Negev, as well as a survey of the JP's paleogenetics. Additional research initiatives offered their own questions, methodologies, and source bases, potentially to be followed by subsequent investigations.

Continuing earlier trends, JP research will continue to draw upon evidence and methodologies from multiple disciplines in the future. Yet, as additional papers are published, the field faces two challenges. First, it is increasingly apparent that interdisciplinary research that draws insights from non-historical methodologies cannot and will not be able to answer historians' most pressing questions, which have the broadest implications for demographic, political, social, cultural, and economic issues. Although some studies answer existing historical questions, they raise new sets of questions, of which many seem intractable. One potential way forward for historians would be a careful discussion of the merits and weaknesses of the known evidence—from literary sources and paleogenetics to pollen studies—keeping in mind the theoretical and methodological issues introduced by Sessa and by Eisenberg and Mordechai.

Second, the debate has spread widely at an exceptionally fast pace. This would likely continue to facilitate centrifugal tendencies that have already been observed above—namely the fact that some papers appear to be unaware of key literature, obstructing the broader discussion, and the increasing notion that the study of the JP is splitting into organic or artificial subfields. These subfields continue to refer to the "Justinianic Plague" but the meaning of this concept is increasingly unclear as it is often used as a shorthand, for example to imply a catastrophic event, even though the actual effects of plague remain contested.[17]

17 This is the case, for example, in much of the scientific literature on climate that refers to plague (e.g., Büntgen et al. 2016) and even the archaeological literature that is aware of the divergent interpretations of plague but still uses it as a potential explanatory factor without determining its effects (e.g., Fuks et al. 2020).

Following the aforementioned survey of the state of the field, it is perhaps surprising that the interventions with the highest potential to have a radical impact on the discussion were written by historians, using tools and methodologies from within the discipline: carefully reflecting on past literature, offering key context through a nuanced frame, deconstructing the orthodoxies within the field, and suggesting a new approach forward. The next few years will reveal whether this approach might offer a way out of the current entanglement.

Acknowledgments

Merle Eisenberg offered insightful extensive suggestions to this paper and has my thanks.

Works Cited

Achtman, Mark. 2012. "Insights from Genomic Comparisons of Genetically Monomorphic Bacterial Pathogens." *Philosophical Transactions of the Royal Society B: Biological Sciences* 367, no. 1590: 860–67. https://doi.org/10.1098/rstb.2011.0303.

Ambraseys, Nicholas. 2009. *Earthquakes in the Mediterranean and Middle East: A Multidisciplinary Study of Seismicity up to 1900*. Cambridge: Cambridge University Press.

Baillie, M. 1991. "Suck in and Smear: Two Related Chronological Problems for the 90s." *Journal of Theoretical Archaeology* 2: 12–16.

Banchich, Thomas, and Eugene Lane. 2009. *The History of Zonaras: From Alexander Severus to the Death of Theodosius the Great*. Routledge Classical Translations. London and New York: Routledge.

Bar-Oz, Guy, Lior Weissbrod, Tali Erickson-Gini, Yotam Tepper, Dan Malkinson, Mordechay Benzaquen, Dafna Langgut, Zachary C. Dunseth, Don H. Butler, and Ruth Shahack-Gross. 2019. "Ancient Trash Mounds Unravel Urban Collapse a Century before the End of Byzantine Hegemony in the Southern Levant." *Proceedings of the National Academy of Sciences* 116, no. 17: 8239–48.

Bonner, Michael Richard Jackson. 2020. *The Last Empire of Iran*. Piscataway, NJ: Gorgias Press LLC.

Bresson, Alain. 2020. "Fates of Romes." *The Journal of Roman Studies* 110: 233–46.

Büntgen, Ulf, Dominique Arseneault, Étienne Boucher, Olga V. Churakova (Sidorova), Fabio Gennaretti, Alan Crivellaro, Malcolm K. Hughes, et al. 2020. "Prominent Role of Volcanism in Common Era Climate Variability and Human History." *Dendrochronologia* 64 (December): 125757. https://doi.org/10.1016/j.dendro.2020.125757.

Büntgen, Ulf, Vladimir S. Myglan, Fredrik Charpentier Ljungqvist, Michael McCormick, Nicola Di Cosmo, Michael Sigl, Johann Jungclaus, et al. 2016. "Cooling and Societal Change during the Late Antique Little Ice Age from 536 to around 660 AD." *Nature Geoscience* 9, no. 3: 231–36.

Büntgen, Ulf, Willy Tegel, Kurt Nicolussi, Michael McCormick, David Frank, Valerie Trouet, Jed O. Kaplan, et al. 2011. "2500 Years of European Climate Variability and Human Susceptibility." *Science* 331, no. 6017: 578–82. https://doi.org/10.1126/science.1197175.

Damgaard, Peter de Barros, Nina Marchi, Simon Rasmussen, Michaël Peyrot, Gabriel Renaud, Thorfinn Korneliussen, J. Víctor Moreno-Mayar, et al. 2018. "137 Ancient Human Genomes from across the Eurasian Steppes." *Nature* 557, no. 7705: 369–74. https://doi.org/10.1038/s41586-018-0094-2.

Dull, Robert A., John R. Southon, Steffen Kutterolf, Kevin J. Anchukaitis, Armin Freundt, David B. Wahl, Payson Sheets, et al. 2019. "Radiocarbon and Geologic Evidence Reveal Ilopango Volcano as Source of the Colossal 'Mystery' Eruption of 539/40 CE." *Quaternary Science Reviews* 222 (October): 105855. https://doi.org/10.1016/j.quascirev.2019.07.037.

Durliat, Jean. 1989. "La peste du VIe siècle: pour un nouvel examen des sources byzantines." In *Hommes et richesses dans l'Empire byzantin*, edited by Catherine Abadie-Reynal, Cécile Morrisson, and Jacques Lefort, 107–19. Réalités byzantines 1. Paris: Lethielleux.

Dunn, Geoffrey David. 2021. "'For It Is Improper to Be Addicted to the Tedium of Affliction': Christian Responses to Pandemic in Late Antiquity and the Early Middle Ages." *Vox Patrum* 78 (June): 389–426. https://doi.org/10.31743/vp.11896.

Eide, Tormod. 1998. *Fontes Historiae Nubiorum: From the First to the Sixth Century AD*. PhD diss., University of Bergen.

Eisenberg, Merle, and Lee Mordechai. 2019. "The Justinianic Plague: An Interdisciplinary Review." *Byzantine and Modern Greek Studies* 43, no. 2: 156–80. https://doi.org/10.1017/byz.2019.10.

———. 2020. "The Justinianic Plague and Global Pandemics: The Making of the Plague Concept." *The American Historical Review* 125, no. 5: 1632–67. https://doi.org/10.1093/ahr/rhaa510.

Eisenberg, Merle, David J. Patterson, Jamie Kreiner, Ellen F. Arnold, and Timothy P. Newfield. 2018. "The Environmental History of the Late Antique West." *Late Antique Archaeology* 11, no: 1: 31–50.

Faure, Eric. 2021. "Did the Justinianic Plague Truly Reach Frankish Europe around 543 AD?" *Vox Patrum* 78 (June): 427–66. https://doi.org/10.31743/vp.12278.

Feldman, Michal, Michaela Harbeck, Marcel Keller, Maria A. Spyrou, Andreas Rott, Bernd Trautmann, Holger C. Scholz, et al. 2016. "A High-Coverage Yersinia Pestis Genome from a Sixth-Century Justinianic Plague Victim." *Molecular Biology and Evolution* 33, no. 11: 2911–23. https://doi.org/10.1093/molbev/msw170.

Fell, Henry G., James U. L. Baldini, Ben Dodds, and Gary J. Sharples. 2020. "Volcanism and Global Plague Pandemics: Towards an Interdisciplinary Synthesis." *Journal of Historical Geography* 70 (October): 36–46. https://doi.org/10.1016/j.jhg.2020.10.001.

Ferreiro, Alberto. 2021. "Plagues and Their Function in Sources of Late Antique-Visigothic Hispania and Gregory of Tours." *Vox Patrum* 78 (June): 365–88. https://czasopisma.kul.pl/vp/article/view/11582

Fleitmann, D., M. Mudelsee, R. S. Bradley, P. Pickering, J. Kramers, S. J. Burns, A. Mangini, and A. Matter. 2009. "Megadroughts at the Dawn of Islam Recorded in a 2600-Year Long Stalagmite from Northern Oman" 11 (April): 8174.

Fleitmann, Dominik, John Haldon, Raymond S. Bradley, Stephen J. Burns, Hai Cheng, R. Lawrence Edwards, Christoph C. Raible, Matthew Jacobson, and Albert Matter. Under Review. "Droughts and Societal Change: The Environmental Context for the Emergence of Islam in Late Antique Arabia." *Science*.

Fuks, Daniel, Guy Bar-Oz, Yotam Tepper, Tali Erickson-Gini, Dafna Langgut, Lior Weissbrod, and Ehud Weiss. 2020. "The Rise and Fall of Viticulture in the Late Antique Negev Highlands Reconstructed from Archaeobotanical and Ceramic Data." *Proceedings of the National Academy of Sciences* 117, no. 33: 19780–91. https://doi.org/10.1073/pnas.1922200117.

Greatrex, Geoffrey. 2020. "Procopius and the Plague in 2020." *Boletín de la Sociedad Española de Bizantinística* 35: 5–13. https://doi.org/10.5281/zenodo.3813380.

Green, Monica H. 2018. "Putting Africa on the Black Death Map: Narratives from Genetics and History." In *Sillages de la peste noire en Afrique subsaharienne: une exploration critique du silence / Black Death and its Aftermaths in Sub-Saharan Africa: A Critical Exploration of Silence*, edited by Gérard Chouin. *Afriques. Débats, Méthodes et Terrains d'histoire* 9. https://doi.org/10.4000/afriques.2125.

———. 2019. "When Numbers Don't Count." Accessed November 18, 2019. https://eidolon.pub/when-numbers-dont-count-56a2b3c3d07.

Gruber, Henry. 2018. "Indirect Evidence for the Social Impact of the Justinianic Pandemic: Episcopal Burial and Conciliar Legislation in Visigothic Hispania." *Journal of Late Antiquity* 11, no. 1: 193–215.

Guidoboni, Emanuela. 1994. *Catalogue of Ancient Earthquakes in the Mediterranean Area Up to the 10th Century*. Rome: Istituto Nazionale di Geofisica.

Haldon, John, Hugh Elton, Sabine R. Huebner, Adam Izdebski, Lee Mordechai, and Timothy P. Newfield. 2018a. "Plagues, Climate Change, and the End of an Empire: A Response to Kyle Harper's The Fate of Rome (1): Climate." *History Compass* 16, no. 12: e12508.

———. 2018b. "Plagues, Climate Change, and the End of an Empire. A Response to Kyle Harper's The Fate of Rome (2): Plagues and a Crisis of Empire." *History Compass* 16, no. 12: e12506.

———. 2018c. "Plagues, Climate Change, and the End of an Empire: A Response to Kyle Harper's The Fate of Rome (3): Disease, Agency, and Collapse." *History Compass* 16, no. 12: e12507.

Harbeck, Michaela, Lisa Seifert, Stephanie Hänsch, David M. Wagner, Dawn Birdsell, Katy L. Parise, Ingrid Wiechmann, et al. 2013. "Yersinia Pestis DNA from Skeletal Remains from the 6th Century AD Reveals Insights into Justinianic Plague." *PLOS Pathogens* 9, no. 5: e1003349. https://doi.org/10.1371/journal.ppat.1003349.

Harper, Kyle. 2017. *The Fate of Rome: Climate, Disease, and the End of an Empire*. Princeton: Princeton University Press.

Hatke, George. 2013. *Aksum and Nubia: Warfare, Commerce, and Political Fictions in Ancient Northeast Africa*. New York: NYU Press.

Helama, Samuli, Pekka Saranpää, Charlotte L. Pearson, Laura Arppe, Jari Holopainen, Harri Mäkinen, Kari Mielikäinen, et al. 2019. "Frost Rings in 1627 BC and AD 536 in Subfossil Pinewood from Finnish Lapland." *Quaternary Science Reviews* 204 (January): 208–15. https://doi.org/10.1016/j.quascirev.2018.11.031.

Invicta. 2021. *How the Black Death Killed Rome - The Plague of Justinian DOCUMENTARY*. https://www.youtube.com/watch?v=ByvUZVp3epY.

Keller, Marcel, Maria A. Spyrou, Michael McCormick, Kirsten I. Bos, Alexander Herbig, and Johannes Krause. 2019a. "Ancient Yersinia Pestis Genomes Provide No Evidence for the Origins or Spread of the Justinianic Plague." *BioRxiv* (November): 819698. https://doi.org/10.1101/819698.

Keller, Marcel, Maria A. Spyrou, Christiana L. Scheib, Gunnar U. Neumann, Andreas Kröpelin, Brigitte Haas-Gebhard, Bernd Päffgen, et al. 2019b. "Ancient Yersinia Pestis Genomes from across Western Europe Reveal Early Diversification during the First Pandemic (541–750)." *Proceedings of the National Academy of Sciences* 116, no. 25: 12363–72. https://doi.org/10.1073/pnas.1820447116.

Keys, David. 1999. *Catastrophe: An Investigation into the Origins of the Modern World*. London: Century.

Kulikowski, Michael. 2007. "Plague in Spanish Late Antiquity." In *Plague and the End of Antiquity: The Pandemic of 541-750*, edited by Lester K. Little, 150–70. Cambridge: Cambridge University Press.

LePan, Nicholas. "Visualizing the History of Pandemics." *Visual Capitalist* (blog), March 14, 2020. https://www.visualcapitalist.com/history-of-pandemics-deadliest/.

Little, Lester K., ed. 2007. *Plague and the End of Antiquity: The Pandemic of 541–750*. Cambridge University Press.

———. 2011. "Plague Historians in Lab Coats." *Past & Present* 213, no. 1: 267–90. https://doi.org/10.1093/pastj/gtr014.

Ludewicz, Michal Jan. 2021. "'There Was a Great Mortality in Rome, More Serious than Is Recalled in the Time of Any Other Pontiff'. Plagues and Diseases in the 'Liber Pontificalis.'" *Vox Patrum* 78 (June): 347–64. https://doi.org/10.31743/vp.12190.

Luterbacher, J., T. P. Newfield, E. Xoplaki, E. Nowatzki, N. Luther, M. Zhang, and N. Khelifi. 2020. "Past Pandemics and Climate Variability across the Mediterranean." *Euro-Mediterranean Journal for Environmental Integration* 5, no. 2: 46. https://doi.org/10.1007/s41207-020-00197-5.

Maddicott, J. R. 1997. "Plague in Seventh-Century England." *Past & Present* 156, no. 1: 7–54.

McCormick, Michael. 2007. "Toward a Molecular History of the Justinianic Pandemic." In *Plague and the End of Antiquity: The Pandemic of 541-750*, edited by Lester K. Little, 290–312. Cambridge: Cambridge University Press.

———. 2019. "Radiocarbon Dating the End of Urban Services in a Late Roman Town." *Proceedings of the National Academy of Sciences* 116, no. 17: 8096–98.

———. 2021. "Gregory of Tours on Sixth-Century Plague and Other Epidemics." *Speculum* 96, no. 1: 38–96. https://doi.org/10.1086/711721.

McMahon, Lucas, and Abigail Sargent. 2018. "Environmental History of the Late Antique Eastern Mediterranean." *Late Antique Archaeology* 11, no: 1: 17–30.

Meier, Mischa. 2016. "The 'Justinianic Plague': The Economic Consequences of the Pandemic in the Eastern Roman Empire and Its Cultural and Religious Effects." *Early Medieval Europe* 24, no. 3: 267–92. https://doi.org/10.1111/emed.12152.

———. 2020a. "The 'Justinianic Plague': An 'Inconsequential Pandemic'? A Reply." *Medizinhistorisches Journal* 55, no. 2: 172–99. https://doi.org/10.25162/mhj-2020-0006.

———. 2020b. "What Historians Are Doing – A Final Reply to Mordechai et Al." *Medizinhistorisches Journal* 55, no. 3: 294–96. https://doi.org/10.25162/mhj-2020-0012.

Meier, Mischa, and Steffen Patzold. 2021. *Gene Und Geschichte: Was Die Archäogenetik Zur Geschichtsforschung Beitragen Kann*. Stuttgart: Anton Hiersemann Verlag.

Mitchell, Stephen. 2018. "The Great Plague of Late Antiquity in Asia Minor." In *Geç Antik Çağ'da Lykos Vadisi ve Çevresi / The Lykos Valley and Neighbourhood in Late Antiquity - Laodikeia Calismalari Ek Yayin Dizisi 1*, edited by Celal Şimşek and Turhan Kaçar, 27–35. Istanbul: Ege Yayınları.

Mordechai, Lee, and Merle Eisenberg. 2019. "Rejecting Catastrophe: The Case of the Justinianic Plague." *Past & Present* 244, no. 1: 3–50. https://doi.org/10.1093/pastj/gtz009.

Mordechai, Lee, Merle Eisenberg, Timothy Newfield, Adam Izdebski, and Janet Kay. 2020a. "Doing History: Plague Past and Future – A Second Response to Mischa Meier." *Medizinhistorisches Journal* 55, no. 3: 297–98. https://doi.org/10.25162/mhj-2020-0013.

———. 2020b. "Quantitative Analysis and Plagued Assumptions: A Response to Mischa Meier." *Medizinhistorisches Journal* 55, no. 3: 290–93. https://doi.org/10.25162/mhj-2020-0011.

Mordechai, Lee, Merle Eisenberg, Timothy P. Newfield, Adam Izdebski, Janet E. Kay, and Hendrik Poinar. 2019. "The Justinianic Plague: An Inconsequential Pandemic?" *Proceedings of the National Academy of Sciences* 116, no. 51: 25546–54. https://doi.org/10.1073/pnas.1903797116.

Moreland, John. 2018. "AD 536 - Back to Nature?" *Acta Archaeologica* 89, no. 1: 91–111. https://doi.org/10.1111/j.1600-0390.2018.12194.x.

Mulhall, John. 2019. "Plague before the Pandemics: The Greek Medical Evidence for Bubonic Plague before the Sixth Century." *Bulletin of the History of Medicine* 93, no. 2: 151–79. https://doi.org/10.1353/bhm.2019.0025.

Newfield, Tim. 2016. "The Global Cooling Event of the Sixth Century. Mystery No Longer?" Accessed June 16, 2021. http://www.historicalclimatology.com/1/post/2016/05/something-cooled-the-world-in-the-sixth-century-what-was-it.html.

Newfield, Timothy P. 2018a. "Mysterious and Mortiferous Clouds: The Climate Cooling and Disease Burden of Late Antiquity." *Late Antique Archaeology* 11, no. 1: 271–97.

———. 2018b. "The Climate Downturn of 536–50." In *The Palgrave Handbook of Climate History*, edited by Sam White, Christian Pfister, and Franz Mauelshagen, 447–94. London: Palgrave Macmillan.

Pollitzer, R. 1954. *Plague*. Geneva: World Health Organization.

Rasmussen, Simon, Martin Sikora, Peter de Barros Damgaard, Rasmus Nielsen, and Eske Willerslev. 2019. "Response to Keller Et Al on Justinianic Plague". *figshare* (Preprint). https://doi.org/10.6084/m9.figshare.10290275.v1.

Reilly, Benjamin. 2020. "The Fate of Rome: Climate, Disease, and the End of an Empire by Kyle Harper (Review)." *Journal of World History* 31, no. 2: 450–52. https://doi.org/10.1353/jwh.2020.0026.

Robins, Ellie. 2017. "How the Environment Topples Empires: On Kyle Harper's 'The Fate of Rome.'" Los Angeles Review of Books. December 7, 2017. https://lareviewofbooks.org/article/how-the-environment-topples-empires-on-kyle-harpers-the-fate-of-rome/.

Rosen, William. 2007. *Justinian's Flea: Plague, Empire, and the Birth of Europe*. New York: Penguin.

Sallares, Robert. 2007. "Ecology, Evolution, and Epidemiology of Plague." In *Plague and the End of Antiquity: The Pandemic of 541–750*, edited by Lester K. Little, 231–89. Cambridge: Cambridge University Press.

Sarris, Peter. 2002. "The Justinianic Plague: Origins and Effects." *Continuity and Change* 17, no. 2: 169–82.

———. 2007. "Bubonic Plague in Byzantium: The Evidence of Non-Literary Sources." In *Plague and the End of Antiquity: The Pandemic of 541-750*, edited by Lester K. Little, 119–32. Cambridge: Cambridge University Press.

———. 2018. "Kyle Harper. The Fate of Rome: Climate, Disease, and the End of an Empire." *The American Historical Review* 123, no. 5: 1627–29. https://doi.org/10.1093/ahr/rhy205.

———. 2020. "Climate and Disease." In *A Companion to the Global Early Middle Ages*, edited by Erik Hermans, 511–38. Leeds: Arc Humanities Press.

Schmid, Boris V., Ulf Büntgen, W. Ryan Easterday, Christian Ginzler, Lars Walløe, Barbara Bramanti, and Nils Chr Stenseth. 2015. "Climate-Driven Introduction of the Black Death and Successive Plague Reintroductions into Europe." *Proceedings of the National Academy of Sciences* 112, no. 10: 3020–25. https://doi.org/10.1073/pnas.1412887112.

Sessa, Kristina. 2019. "The New Environmental Fall of Rome: A Methodological Consideration." *Journal of Late Antiquity* 12, no. 1: 211–55. https://doi.org/10.1353/jla.2019.0008.

Shockro, Sally. 2018. "Apocalyptic Disease and the Seventh-Century Plague." In *Trauma in Medieval Society*, edited by Wendy J. Turner and Christina Lee, 320–40. Explorations in Medieval Culture 7. Leiden: Brill.

Smith, P. D. 2019. "The Fate of Rome by Kyle Harper Review - Climate, Disease and the End of an Empire." *The Guardian*, April 11, 2019. Accessed June 24, 2021. http://www.theguardian.com/books/2019/apr/11/fate-of-rome-kyle-harper-review.

Smith, Victoria C., Antonio Costa, Gerardo Aguirre-Díaz, Dario Pedrazzi, Andrea Scifo, Gill Plunkett, Mattieu Poret, et al. 2020. "The Magnitude and Impact of the 431 CE Tierra Blanca Joven Eruption of Ilopango, El Salvador." *Proceedings of the National Academy of Sciences* 117, no. 42: 26061–68.

Stathakopoulos, Dionysios. 2000. "The Justinianic Plague Revisited." *Byzantine and Modern Greek Studies* 24: 256–76. https://doi.org/10.1179/byz.2000.24.1.255.

Stathakopoulos, Dionysios Ch. 2004. *Famine and Pestilence in the Late Roman and Early Byzantine Empire: A Systematic Survey of Subsistence Crises and Epidemics*. Aldershot: Ashgate.

Stenseth, Nils Chr., Noelle I. Samia, Hildegunn Viljugrein, Kyrre Linné Kausrud, Mike Begon, Stephen Davis, Herwig Leirs, et al. 2006. "Plague Dynamics Are Driven by Climate Variation." *Proceedings of the National Academy of Sciences* 103, no. 35: 13110–15. https://doi.org/10.1073/pnas.0602447103.

Toohey, Matthew, Kirstin Krüger, Michael Sigl, Frode Stordal, and Henrik Svensen. 2016. "Climatic and Societal Impacts of a Volcanic Double Event at the Dawn of the Middle Ages." *Climatic Change* 136, no. 3: 401–12. https://doi.org/10.1007/s10584-016-1648-7.

Vaiglova, Petra, Gideon Hartman, Nimrod Marom, Avner Ayalon, Miryam Bar-Matthews, Tami Zilberman, Gal Yasur, et al. 2020. "Climate Stability and Societal Decline on the Margins of the Byzantine Empire in the Negev Desert." *Scientific Reports* 10, no. 1: 1512. https://doi.org/10.1038/s41598-020-58360-5.

van Dijk, Evelien van, Johann Jungclaus, Stephan Lorenz, Claudia Timmreck, and Kirstin Krüger. 2021. "Was There a Volcanic Induced Long Lasting Cooling over the Northern Hemisphere in the Mid 6th–7th Century?" *Climate of the Past Discussions* (May Preprint): 1–33. https://doi.org/10.5194/cp-2021-49.

Wagner, David M., Jennifer Klunk, Michaela Harbeck, Alison Devault, Nicholas Waglechner, Jason W. Sahl, Jacob Enk, et al. 2014. "Yersinia Pestis and the Plague of Justinian 541–543 AD: A Genomic Analysis." *The Lancet Infectious Diseases* 14, no. 4: 319–26. https://doi.org/10.1016/S1473-3099(13)70323-2.

Webb, James L. A. 2018. "Climate, Ecology, and Infectious Human Disease." In *The Palgrave Handbook of Climate History*, edited by Sam White, Christian Pfister, and Franz Mauelshagen, 355–66. London: Palgrave Macmillan.

White, Lauren A., and Lee Mordechai. 2020. "Modeling the Justinianic Plague: Comparing Hypothesized Transmission Routes." *PLOS ONE* 15, no. 4: e0231256. https://doi.org/10.1371/journal.pone.0231256.

Whittow, Mark. 1996. *The Making of Byzantium, 600–1025*. Berkeley: University of California Press.

Wickham, Chris. 2005. *Framing the Early Middle Ages: Europe and the Mediterranean 400–800*. Oxford: Oxford University Press.

———. 2016. *Medieval Europe*. New Haven: Yale University Press.

Wijnendaele, Jeroen W. P. 2019. "The Climate and the History of Rome – (K.) Harper The Fate of Rome. Climate, Disease, and the End of an Empire. Pp. Xvi + 417, Figs, Ills, Maps. Princeton and Oxford: Princeton University Press, 2017. Cased, £27.95, US$35. ISBN: 978-0-691-16683-4." *The Classical Review* 69, no. 2: 560–62. https://doi.org/10.1017/S0009840X19000192.

Zhou, Zhemin, Nabil-Fareed Alikhan, Khaled Mohamed, Yulei Fan, and Mark Achtman. 2020. "The EnteroBase User's Guide, with Case Studies on Salmonella Transmissions, Yersinia Pestis Phylogeny, and Escherichia Core Genomic Diversity." *Genome Research* 30, no. 1: 138–52. https://doi.org/10.1101/gr.251678.119.

Olive Cultivation at High Altitudes in Anatolia: Exploiting Micro-Localities in Ancient Asia Minor

Josh England

Hugh Elton

Anneley Hadland

Çetin Şenkul

Patrick Free

Warren J. Eastwood

Introduction

Pollen data, archaeological evidence, and historical sources attest to the wide extent of olive cultivation, production, and consumption in the ancient Mediterranean world (Aydınoğlu 2009; Aydınoğlu and Şenol 2010; Diler, Şenol, and Aydınoğlu 2015). However, the scope and nature of olive cultivation in high elevation regions during the period ca. 700 BCE–700 CE is the subject of much debate. Although significant amounts of olive pollen are found in sediment cores from inland lake sites, these sites lie at or above the present-day altitudinal limit for successful olive cultivation (ca. 800 masl), determined by the olive's vulnerability to frost. Debate surrounds the extent to which these differing patterns of high-altitude olive cultivation may be the result of climate change (Fig. 1). This paper briefly discusses the importance of olive cultivation and examines the three hypotheses that can be advanced to explain the evidence of olive cultivation at high elevation sites in southwestern Asia Minor.

Discussion of Evidence of High Elevation Olive Cultivation

Olive cultivation is an integral component of ancient Mediterranean lifestyles and the Beyşehir Occupation Phase (BOP; see Table 1), a pronounced period of agrarian land use in Anatolia that intensified around ca. 800 BCE (Bottema, Woldring, and Aytuğ 1986;

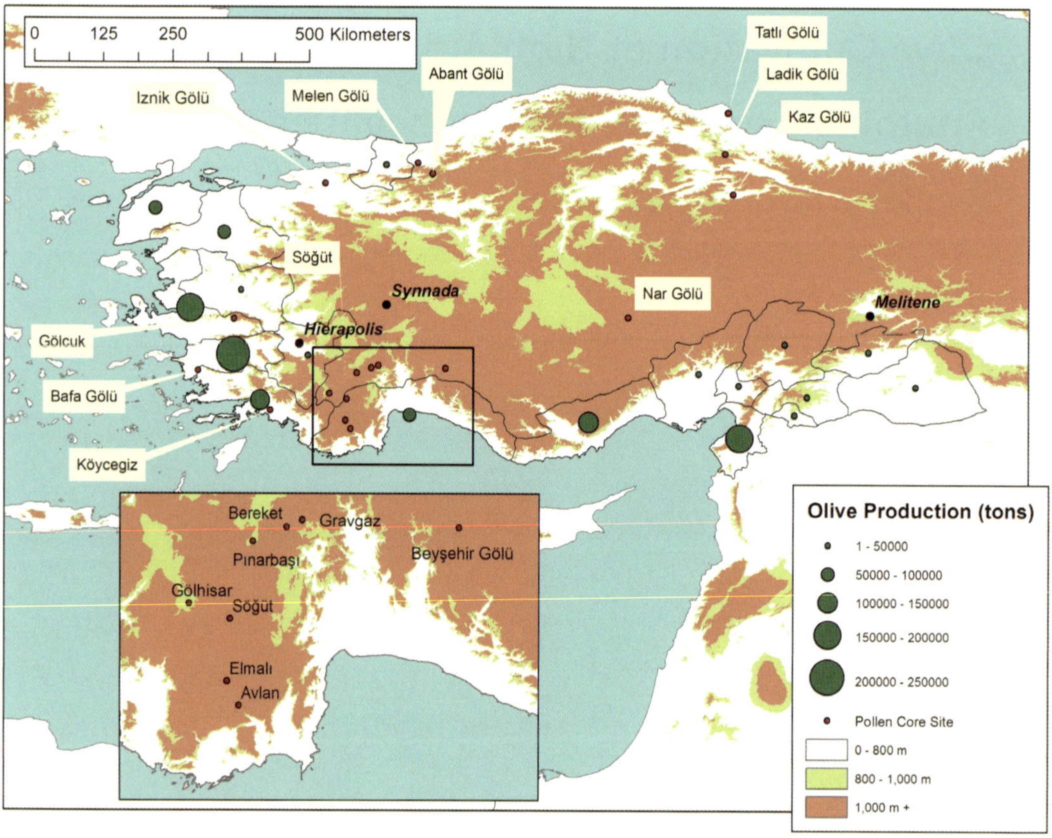

Fig. 1. Selected pollen sites in Asia Minor, and places mentioned in the text and table. Present distribution of olive cultivation (2015 production figures by modern provinces having more than 100 tons of olives).

Eastwood, Roberts, and Lamb 1998; Eastwood et al. 1999; Vermoere et al. 2002; Roberts 2018; England 2021). The BOP was based on agropastoralism, incorporating arboricultural elements to make it a classical example of early Mediterranean polyculture. The crop plants at BOP sites in southwestern Turkey varied but always included the three Mediterranean staples: cereals, vines, and olives. This strategy markedly reduced risk, as each component was vulnerable to different environmental hazards at different times of the year (Butzer 2005).

Pollen data for many high-altitude (over 1000 masl) inland BOP sites record significant amounts of olive pollen during antiquity where olives are not grown now. Although it is generally accepted that values of 5% or greater indicate that olives grew extensively within the catchment of a site (Huntley and Birks 1983; Florenzano et al. 2017; England 2021), scholars have previously explained the presence of olive pollen recorded at these sites as the result of long-distance wind transport. While olive pollen is well-represented at most high-altitude inland sites, elevated olive pollen values are not, however, recorded in pollen diagrams from high altitude sites closer to the Mediterranean,

Table 1. Elevation and maximum percentage pollen values for olive in Anatolian sites (see Fig. 1 for site locations).

Site	Elevation (m)	Max *Olea* Pollen (%)	Reference
Western Anatolia			
Bafa Gölü (S1)	0	5	Knipping, Müllenhoff, and Brückner 2008
Bafa Gölü (S6)	0	7	Knipping, Müllenhoff, and Brückner 2008
Gölcük (Sardis)	1050	18	Sullivan 1989
Southwestern Anatolia			
Beyşehir-I	1120	10	Van Zeist, Woldring, and Stapert 1975
Beyşehir-II	1120	3	Bottema and Woldring 1984
Söğüt	1393	5	Van Zeist, Woldring, and Stapert 1975
Pınarbaşı	980	5	Bottema and Woldring 1984
Gölhisar I	930	19	Bottema and Woldring 1984
Gölhisar II (GHA)	930	7	Eastwood et al. 1999
Gölhisar III (GHC)	930	6	Eastwood et al. 1999
Köyceğiz	0	5	Van Zeist, Woldring, and Stapert 1975
Elmalı	930	<1	Bottema and Woldring 1984
Avlan	1043	<1	Bottema and Woldring 1984
Bereket	1420	7	Kaniewski et al. 2007
Gravgaz	1215	23	Vermoere et al. 2000
Central Anatolia			
Nar	1363	8	England et al. 2008
Northern Anatolia			
Iznik Gölü	85	25	Miebach et al. 2016
Tatlı Gölü	0	3	Bottema, Woldring, and Aytuğ 1993
Melen Gölü	125	2	Bottema, Woldring, and Aytuğ 1993
Kaz Gölü	500	3	Bottema, Woldring, and Aytuğ 1993
Abant Gölü	1300	3	Bottema, Woldring, and Aytuğ 1993
Ladık Gölü	800	2	Bottema, Woldring, and Aytuğ 1993
Yenicaga Gölü	976	<1	Bottema, Woldring, and Aytuğ 1993

casting doubt on the wind transport explanation (Fig. 2). Moreover, present-day high-altitude sites consistently show low olive pollen quantities, despite large-scale olive production in lowland coastal areas of southern and western Anatolia (Eastwood 1997; Bottema and Woldring 1990; Eastwood et al. 2011, 2012).

This combination of high percentage values of olive pollen (>5%) during the BOP phase together with archaeological evidence for olive cultivation and processing above 1000 masl led Roberts to suggest that winter-spring temperatures during the BOP were higher than today by 2–3°C (Roberts 1990). On the other hand, Mitchell argued that growing olives in interior upland Anatolia may have been a cultural decision (Mitchell 2005, 84). His analysis is based on Strabo's account from the first century CE—during

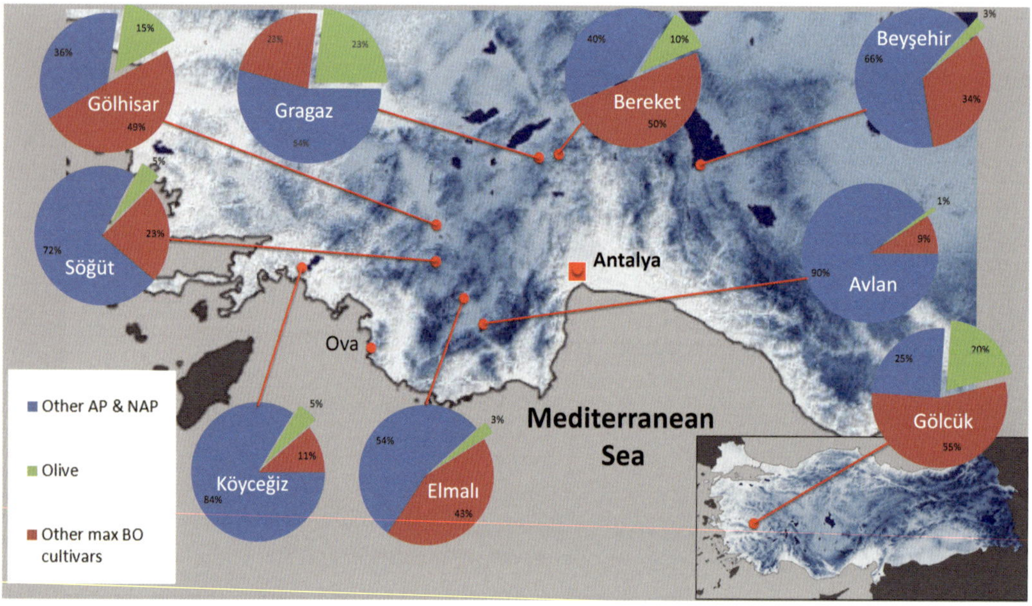

Fig. 2. Selected pollen sites in southwestern Asia Minor, showing maximum percentage values for olive pollen. Note how interior sites show elevated percentage values of olive pollen, while coastal and near coastal (intermediate) sites show low olive pollen percentage values, thus falsifying the long-distance hypothesis.

the Roman Climate Optimum, that is—which mentioned olive cultivation near higher-altitude settlements such as Synnada (near present-day Şuhut) and Melitene (Malatya), where there is no significant present-day cultivation of olives (Mitchell 2005, 88–93).

The Roman Climate Optimum did allow limited olive cultivation at these higher altitudes, but expanding the zone of cultivation brought a higher risk of crop failure, as well as smaller yields and more bitter fruit. This high-altitude cultivation was mostly "domestic" olive cultivation undertaken for local consumption associated with the institution of the gymnasium (Thonemann 2012). Its nature was thus very different from "industrial" cultivation at lower elevations, where large olive groves produced high-quality olives and olive oil for export to large urban markets (Mitchell 2005, 98–99; England et al. 2008).

The extent to which climate change may have shaped olive cultivation patterns for an interior region of southwestern Turkey can be modelled using GIS. Figure 3 depicts the area near the present-day city of Denizli and shows the location of press weights across the territory of the classical city of Hierapolis. Hierapolis itself lies at 360 masl in a valley surrounded by mountains that exceed 1500 m. The press weights can be divided into those that have been positively identified as olive presses (triangles in Fig. 3) and those that may have been used for olive, grape, or walnut crushing (squares in Fig. 3). Those positively identified as olive press weights tend to be located at lower elevations, whereas unclassified press weights tend to be found at higher elevations.

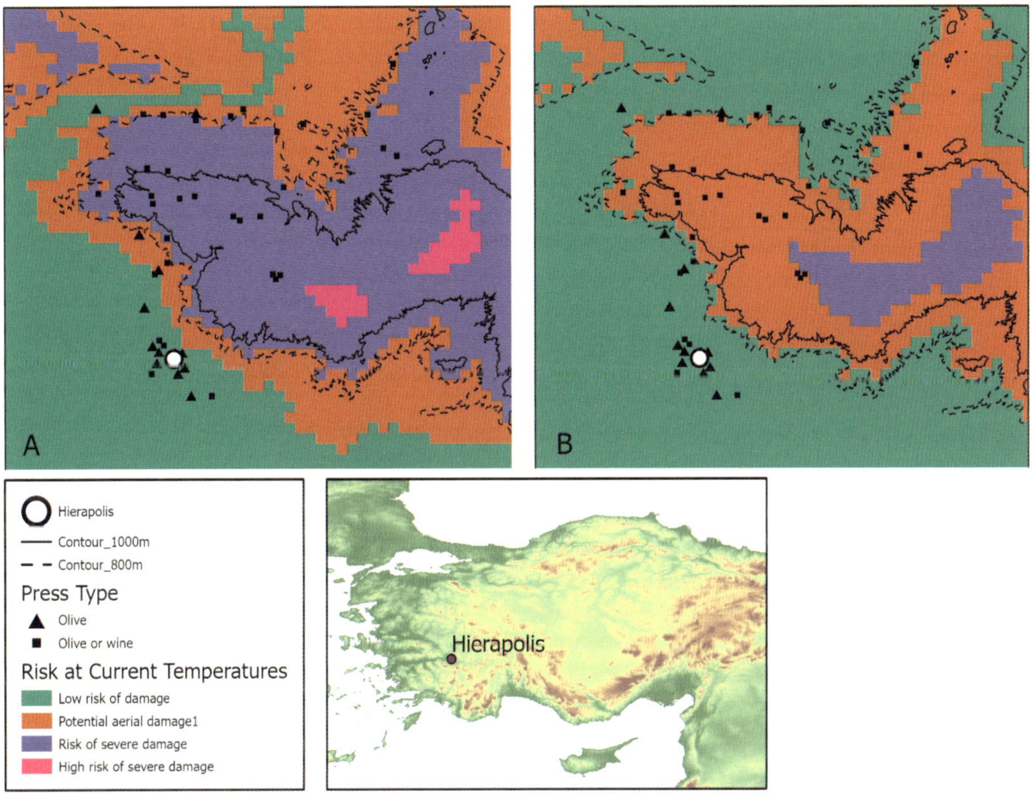

Fig. 3. A: Territory of the classical city of Hierapolis near modern Denizli (southwestern Turkey), showing locations of olive press weights on the hillside or bottom of the valley and unclassified press weights at higher elevations at the top of the plateau. B: When a 2°C increase is modelled, all press weights are found outside areas of risk of severe damage (GIS model from McMillan 2013; press weight data from Scardozzi 2010).

Overlaid onto these findings is a classification of potential damage to olive plants based on recent climatic records (over a 30-year reference period). The result demonstrates that under present-day conditions, some press weights are found in areas that carry the risk of potential aerial damage to trees, but no press weights are found in areas of very high risk. Modeling a scenario of a 2°C increase as suggested by Roberts (1990), the press weights all appear to have been located within areas of low risk of damage (McMillan 2013).

The combination of archaeological and palynological data, as well as GIS modeling and recent comparative data, suggests that a 2°C increase in winter temperatures during parts of the BOP (i.e., the Roman Climate Optimum) could make land between 800–1000 masl suitable for limited olive cultivation as part of an intensive agricultural regime. These results hint at how a rural pre-industrial socio-environmental system could have adapted to changing socio-cultural and environmental factors. The combination of climatic amelioration and cultural norms that shifted towards olive consumption led contemporaries to cultivate more olives in micro-localities with marginal climates.

It appears that the local inhabitants of the area chose to expand their agricultural practices to these marginal areas in a sustainable manner. The distribution pattern of press weights suggests that pre-industrial societies learned to exploit their local micro-regions to maximize their economic output over the long-term without over-stretching their capabilities. They refrained from investing the substantial amount of resources required to construct new presses in unsustainable locations. Instead, they probably carefully observed micro-localities in their region over time, either passively or actively through experimentation. This method was slow but ensured they minimized risks. As the climate improved and local inhabitants established the best placement for press weights, they would slowly commit the necessary resources to expand their cultivation range. When the climate deteriorated and demand for olives decreased, the limited olive industry in marginal areas could be changed or abandoned without a substantial transformation in local society.

Conclusions

The high-altitude cultivation of olives in Anatolia is a complex problem. Pollen analysis shows the presence of ancient olive pollen in areas where there is no present-day cultivation of olives. This may result from windblown pollen, climate change, or changing cultural choices. Analysis of ancient pollen data alongside present-day pollen-vegetation studies from various altitudinal locations in southwestern Asia Minor (Eastwood et al. 2011, 2012) appears to eliminate the windblown hypothesis. Modeling a +2°C climate change for an area around the ancient city of Hierapolis shows that all the ancient presses there lay within a zone that would have had a low risk of crop damage. This highlights trends in the pre-industrial cultivation of crops for local consumption on marginal lands and suggests that the inhabitants of Roman Anatolia expanded their high-altitude operations only when climate improved and the demand for olives increased for cultural reasons. Rather than overextending themselves for short-term profit, however, local inhabitants seem to have minimized their risk over time, with much of the high-altitude production probably being for domestic consumption rather than for export. Premodern sustainable practices were inherently flexible, with built-in resilience through wide error margins. The systems thus buffered themselves from unexpected risk at the cost of reducing economic efficiency relative to present-day production. Further research is needed, however, both on the spatial extent of micro-climates and their impacts on ancient agriculture, and on the effects of higher-altitude olive cultivation on productivity.

Works Cited

Aydınoğlu, Ümit. 2009. *Dağlık Kilikia Bölgesinde Antik Çağda Zeytinyağı ve Şarap Üretimi.* Istanbul: Ege Yayınları.

Aydınoğlu, Ümit, and Kaan Şenol, eds. 2010. *Olive Oil and Wine Production in Anatolia during Antiquity.* Istanbul: Ege Yayınları.

Bottema, Sytze, and Hendrik Woldring. 1984. "Late Quaternary Vegetation and Climate of Southwest Turkey II." *Palaeohistoria* 26: 123–49.

———. 1990. "Anthropogenic Indicators in the Pollen Record of the Eastern Mediterranean." In *Man's Role in the Shaping of the Eastern Mediterranean Landscape*, edited by Sytze Bottema, Gertie Entjes-Nieborg, and Wim van Zeist, 231–64. Rotterdam: Balkema.

Bottema, Sytze, Hendrik Woldring, and Burhan Aytuğ. 1986. "Palynological Investigations on the Relations between Prehistoric Man and Vegetation in Turkey: The Beyşehir Occupation Phase." In *Proceedings of the 5th Optima Congress*, 315–238. Istanbul.

Bottema, Sytze, Hendrik Woldring, and Burhan Aytuğ. 1993. "Late Quaternary Vegetation History and Climate in Northern Turkey." *Palaeohistoria* 35/36: 13–72.

Butzer, Karl W. 2005. "Environmental History in the Mediterranean World: Cross-Disciplinary Investigation of Cause-and-Effect for Degradation and Soil Erosion." *Journal of Archaeological Science* 32: 1773–1800.

Diler, Adnan, Kaan Şenol, and Ümit Aydınoğlu, eds. 2015. *Olive Oil and Wine Production in the Eastern Mediterranean during Antiquity.* Izmir: Ege University Press.

Eastwood, Warren J. 1997. "The Palaeoecological Record of Holocene Environmental Change in Southwest Turkey." PhD diss, University of Wales.

Eastwood, Warren J., Neil Roberts, and Henry F. Lamb. 1998. "Palaeoecological and Archaeological Evidence for Human Occupance in Southwest Turkey: The Beyşehir Occupation Phase." *Anatolian Studies* 48: 69–86.

Eastwood, Warren J., Neil Roberts, Henry F. Lamb, and John C. Tibby. 1999. "Holocene Environmental Change in Southwest Turkey: A Palaeoecological Record of Lake and Catchment-Related Changes." *Quaternary Science Reviews* 18, no. 4-5: 671–695.

Eastwood, Warren J., Çetin Şenkul, Ralph Fyfe, Anneley McMillan, and Andrew Moss. 2012. "Quantitative Vegetation Modelling in Southwest Turkey." *Heritage Turkey* 2: 5–6.

Eastwood, Warren J., Ralph Fyfe, Çetin Şenkul, and Anneley McMillan. 2011. "Quantitative Vegetation Modelling in Southwest Turkey." *Heritage Turkey* 1: 6–7.

England, Ann, Warren J. Eastwood, C. Neil Roberts, Rebecca Turner, and John F. Haldon. 2008. "Historical Landscape Change in Cappadocia (Central Turkey): A Palaeoecological Investigation of Annually Laminated Sediments from Nar Lake." *The Holocene* 18, no. 8: 1229–45. https://doi.org/10.1177/0959683608096598.

England, Josh. 2021. "Olives in the Mountains: A Case Study of the Roman city of Sagalassos." MA thesis, Trent University.

Florenzano, Assunta, Anna Maria Mercuri, Rossella Rinaldi, Eleonora Rattighieri, Rita Fornaciari, Rita Messora, and Laura Arru. 2017. "The Representativeness of Olea Pollen from Olive Groves and the Late Holocene Landscape Reconstruction in Central Mediterranean." *Frontiers in Earth Science* 5: 1–11. https://doi.org/10.3389/feart.2017.00085.

Huntley, Brian J., and Harry J. B. Birks. 1983. *An Atlas of Past and Present Pollen Maps for Europe, 0–13,000 Years Ago*. Cambridge: Cambridge University Press.

Kaniewski, David, Etienne Paulissen, Véronique De Laet, Koen Dossche, and Marc Waelkens. 2007. "A High-Resolution Late Holocene Landscape Ecological History Inferred from an Intramontane Basin in the Western Taurus Mountains, Turkey." *Quaternary Science Reviews* 26, no. 17: 2201–18. https://doi.org/10.1016/j.quascirev.2007.04.015.

Knipping, Maria, Marc Müllenhoff, and Helmut Brückner. 2008. "Human Induced Landscape Changes around Bafa Gölü (Western Turkey)." *Vegetation History and Archaeobotany* 17, no. 4: 365–80. https://doi.org/10.1007/s00334-007-0132-8.

McMillan, Anneley. 2013. "A GIS Approach to Palaeovegetation Modelling in the Mediterranean: The Case Study of Southwest Turkey." PhD diss, University of Birmingham.

Miebach, Andrea, Phoebe Niestrath, Patricia Roeser, and Thomas Litt. 2016. "Impacts of Climate and Humans on the Vegetation in Northwestern Turkey: Palynological Insights from Lake Iznik since the Last Glacial." *Climate of the Past* 12, no. 2: 575–93. https://doi.org/10.5194/cp-12-575-2016.

Mitchell, Stephen. 2005. "Olive Cultivation in the Economy of Roman Asia Minor." In *Patterns in the Economy of Roman Asia Minor*, edited by Stephen Mitchell and Constantina Katsari, 83–113. Swansea: Classical Press of Wales.

Roberts, Neil. 2018. "Re-Visiting the Beyşehir Occupation Phase: Land-Cover Change and the Rural Economy in the Eastern Mediterranean during the First Millennium AD." In *Environment and Society during the Long Late Antiquity*, edited by Adam Izdebski and Michael Mulryan, 53–68. Late Antique Archaeology 13. Leiden: Brill.

———. 1990. "Human-Induced Landscape Change in South and South-West Turkey during the Later Holocene." In *Man's Role in the Shaping of the Eastern Mediterranean Landscape*, edited by Sytze Bottema, Gertie Entjes-Nieborg, and Wim van Zeist, 53–67. Rotterdam: Balkema.

Scardozzi, Giuseppe. 2010. "Oil and Wine Production in Hieropolis of Phrygia and its Territory during the Roman and Byzantine Age: Documentation from Archaeological Excavations and Surveys." In *Antik Çağda Anadolu'da Zeytinyağı ve Şarap Üretimi: Sempozyum Bildirileri, 6–8 Kasım 2008, Mersin, Türkiye / Olive Oil and Wine Production in Anatolia during Antiquity: Symposium Proceedings, 6–8 November 2008, Mersin, Turkey*, edited by Ümit Aydınoğlu, Ahmet Kaan Şenol, 277–302. Mersin: KAAM Kilikia Arkeolojisini Araştırma Merkezi.

Sullivan, Donald G. 1989. "Human-Induced Vegetation Change in Western Turkey: Pollen Evidence from Central Lydia." Ph.D. diss, University of California, Berkeley.

Thonemann, Peter. 2012. "Abercius of Hierapolis." In *Historical and Religious Memory in the Ancient World*, edited by Beate Dignas and Roland. R. R. Smith, 257–82. Oxford: Oxford University Press.

van Zeist, Wim, Hendrik Woldring, and Dick Stapert. 1975. "Late Quaternary Vegetation and Climate of Southwestern Turkey." *Palaeohistoria* 17: 55–143.

Vermoere, Marleen, Erik Smets, Marc Waelkens, Hannelore Vanhaverbeke, Ireen Librecht, Etienne Paulissen, and Leo Vanhecke. 2000. "Late Holocene Environmental Change and the Record of Human Impact at Gravgaz near Sagalassos, Southwest Turkey." *Journal of Archaeological Science* 27, no. 7: 571–95. https://doi.org/10.1006/jasc.1999.0478.

Vermoere, Marleen, Thijs Van Thuyne, Simon Six, Leo Vanhecke, Marc Waelkens, Etienne Paulissen, and Erik Smets. 2002. "Late Holocene Local Vegetation Dynamics in the Marsh of Gravgaz (Southwest Turkey)." *Journal of Paleolimnology* 27, no. 4: 429–451.

The Environmental Consequences of the Coming of the Turks to Anatolia

Adam Izdebski

Introduction

This chapter attempts to assess the effects of the Turcoman occupation of Anatolia with the use of paleoenvironmental data, and hence it needs a theoretical framework that can connect ecological and societal processes in a meaningful way. It thus builds on the concept of "second nature," which originated in North American environmental history. It has been developed most extensively by Martin Melosi and Joel Tarr in research on the industrial city but in fact underpins research by other pioneers in environmental history, including William Cronon and Carolyn Merchant (Melosi 1981; Cronon 1983; Merchant 1989; Tarr 1996). This concept indicates that each community (at the local scale) and each society (at the scale of entire regions, countries, and sometimes even continents), by their very existence in the natural world, creates a new quality. This quality is a new natural environment that becomes both home to, and a reflection of, the society. This "second nature"—the new environment—results on the one hand from processes that are completely beyond humans (or that are at least beyond their will and control) such as ecological succession, climate fluctuations, and the development of pathogenic microorganisms, and on the other from conscious human efforts to transform and exploit the surrounding nature.

Thinking in terms of "second nature" is extremely helpful in that it allows reconstructions of past landscapes based on environmental data to be combined with socio-economic history: its central premise is the unity of the social, which is studied via textual sources and the remnants of past material culture, and the natural (environmental and biological—also with regard to humans themselves), which in turn is reconstructed primarily with natural scientific methods. It also emphasizes that the environment in which a society lives (the "niche" or "environmental infrastructure" it has constructed, depending on whether we want to take an evolutionary biologist's or environmental historian's term for it [Kreike 2013; Laland, Matthews, and Feldman 2016]) can be passed on to subsequent generations, and hence also to successor communities, political and economic systems, and such. These past "niches," or "second natures," pose constraints and offer opportunities for those who happen to live their lives in them, from the individual to the social level. In this case study, the "second nature" approach allows us to ask to what extent the medieval Romans, known today as the "Byzantines," preserved the Roman "second nature" as we know it from Late Antiquity and what happened to it when the Turks arrived in Anatolia. In answering that question, we can use both historical sources and natural data. To paraphrase Cronon (1983): did the land change? Did the country, the socio-ecological whole, the landscape in the broader sense, change? Any attempt to observe a change (or to show that one did not take place in a given period) obviously requires that we determine the initial state. Addressing this question thus forces us to consider what the Roman or, more broadly, the ancient landscape was in order to then be able to find out whether it still existed in the Roman-Byzantine Middle Ages. Only through this long-term approach can we interpret properly the environmental changes that took place in Anatolia in the eleventh through fifteenth centuries, that is, following the coming of the Turks to this area.

The Persistence of "Roman" Landscapes from Late Antique to Medieval Times

Indeed, we can talk about a certain type of vegetation structure—or landscape structure—that was characteristic of Greco-Roman Antiquity according to research being carried out by paleoecologists for more than half a century. Here we come to the so-called Beyşehir Occupation Phase (BOP), named for the lake in southwestern Turkey, with a vegetation structure that palynologists back in the 1980s identified in pollen profiles. The BOP landscape type was dominated by mixed agriculture, which included the cultivation of cereals, the grazing of animals, and the growing of fruit, including olives and vines. Here and there, the BOP can be seen as early as the end of the Bronze Age (around 1000 BCE), while it becomes widespread in the Hellenistic period (Bottema, Woldring, and Aytuğ 1986; Eastwood, Roberts, and Lamb 1998). As pollen profiles from successive parts of the Mediterranean world were published, it became clear that the type of landscape defined by naturalists as BOP was present throughout the eastern Mediterranean and should be associated with the Hellenistic-Roman agricultural model (Izdebski 2011, 2012, 2013a, 2013b; Haldon et al. 2014). Moreover, the BOP landscape continued to dominate overwhelmingly in the eastern Mediterranean in the sixth century CE and even intensified in Late Antiquity: the BOP landscape can therefore be considered a key element, a marker, of Roman "second nature" in this part of the world. Accordingly, the question of the Roman character of medieval landscapes can therefore be framed as the question of the persistence of the profoundly anthropogenic landscape that dominated the eastern Roman Empire in Late Antiquity.

It should be emphasized that this is not a purely environmental question. Specific ecosystems and landscapes exist in particular places because they are inhabited by certain people and not others (put bluntly, "because people want it so," although this phrase overemphasizes the human ability to influence nature at their will). Did the people we call the Byzantines maintain or create social and economic structures and cultural patterns and institutions that sustained a landscape "inherited" from Roman (Late) Antiquity? Did it matter to the Byzantines that the surrounding ecosystems should provide the same goods that they had provided to their Late Roman predecessors? The question of the Roman landscape's persistence in Byzantium therefore concerns both the outcome of the interaction of man and nature, that is, the actual conglomerate of ecosystems (the landscape that we can observe through natural sources), and the very socio-ecological processes that had produced and then maintained that landscape.

In this context, it is worth asking a question inspired by the works of Merchant (1989): did the world of the Roman Middle Ages that we call "Byzantium" experience ecological revolutions, that is, profound changes in the landscape and society that led to at least a partial break with the previous state? This question leads us to look anew at the turning points in the traditionally understood history of Byzantium. We may ask whether a natural dimension to these turning points can be found or whether, despite political crises or transformations in social structure, little in man's relationship with nature changed; in other words, we can ask whether Fernand Braudel was right in thinking that environmental processes occur only at the level of "deep and slow time" (Braudel 1949), and thus the profound political and economic transformations of the late first and early second millennium CE would not have significantly changed eastern Mediterranean landscapes.

In the case of Byzantium, this question mainly concerns the seventh century, "the real end of Antiquity," and the eleventh century, that is, the "end of the world" in Anatolia, a region that had until then been the heart of the medieval Empire (Whittow 2017). One may wonder if anything Roman remained in the Empire's natural environment when the Arab conquests completely changed its geographic, political, and social shape—whether great migrations and invasions, such as the emergence of Turcoman groups in Anatolia, actually had an environmental dimension.

Methods: Palynology and History

In order to answer these questions, we will begin by looking at the landscape of Anatolia using pollen data and then discuss them in the context of the medieval Roman (Byzantine) political, cultural, and economic systems. As we have already learned, the pollen data allow us to reconstruct approximately the vegetation of areas from which the lake and peat bog cores analyzed by palynologists are taken. In order to handle these data on a larger scale, appropriately selected chronological categories need to be used in historical analysis. Since the overwhelming majority of the chronology of palynological data is based on radiocarbon dating and mathematical modeling of the age-depth relationship, a margin of error of as much as 100 years should be assumed in dating samples. For this reason also, while the analysis in this chapter is based on information provided by individual palynological samples from several different sites in Anatolia, we will look at changes in the plant landscape over periods of 150-200 years.

This time interval is now accepted as standard in the palynological literature as the unit of analysis in research aiming to compare the dynamics of changes in vegetation cover using data from multiple different sites (Fyfe, Woodbridge, and Roberts 2015; Roberts et al. 2018). Although it is a very wide window relative to traditional political history, such a frame allows us to follow changes taking place over several centuries reliably, making it possible to discern the long-term effects of political, cultural, or economic changes.

For the age of the medieval Turcoman migrations, the time intervals we will adopt are of 150 years, that is, the tenth to eleventh centuries (925–1075), which is the initial period of economic boom in the medieval Empire preceding the most important waves of Turcoman migrations and the Seljuk invasion; the "long" twelfth century (1075–1215), which shows the direct environmental effects of the invasion; and the thirteenth to fourteenth centuries (1215–1375), which will allow our analysis to assess the persistence of natural changes associated with the Seljuk invasion.

The starting point for the interpretation of pollen data in this chapter is the generic concept of human pressure, that is, human impact on ecosystems (and hence its ability to create "second natures"). This expresses itself both as existing ecosystems being directly transformed into agro-ecosystems that require constant work and maintenance (in the form of direct cultivation) and as existing ecosystems being altered through the exploitation of local natural resources in uncultivated, semi-wild environments (such as woodlands). Thus, the pollen data reveals human pressure in several ways: 1) the clear presence of the pollen of various crops (from cereals to fruit trees); 2) the decreasing share of forest plant pollen in the total pollen in successive samples; and 3) an increase in the share of pollen from plants typical of open landscapes, in tandem with increasing shares of pollen from plants consistent with the presence of grazing animals; in short, in high values of pollen types indicative of pastoral farming and of synanthropic plants.

In the present analysis we will distinguish three degrees of human pressure, of which the two extremes can be most precisely defined. "Weak," which I describe as "negligible" anthropic pressure, is the absence (or near absence) of signs of human activity, and, thus, only trace levels of or even no crop and fodder plant pollen, with a simultaneous increase in or high values of tree and shrub pollens. By contrast, "strong" anthropic pressure is the clear presence of a wide range of crops and high values of fodder indicators, with a significant decrease or persistently low values of forest vegetation pollen types. The intermediate degree, that is, "moderate" human pressure, is of course relative in such a categorization, with the pollen types that indicate human activity remaining high enough that it is hard to talk about human activity disappearing. For example, the disappearance of some crop pollens indicates that the range of human activity in the landscape diminishes, but at the same time the pollen of other plants remains at relatively stable levels. Moreover, pollen indicative of human activity should decrease sufficiently for talk of "strong" anthropic pressure to be unwarranted. In order to make this category more clear-cut, in the samples classified as representing moderate human pressure, the most common indicators of human activity must be decidedly lower than the peak values these indicators attain for the entire interval of analysis. In summary, then, using these three categories of strength of anthropic pressure, I look at the dynamics of changes at each of the studied sites between three consecutive sub-periods: the tenth–eleventh centuries; the long twelfth century; and the thirteenth–fourteenth centuries.

Based on the analysis, I have drawn up a set of maps. I have created three basic maps showing the spatial distribution of sites with different intensities of human pressure, one for each of the three consecutive sub-intervals (Fig. 1). They present the results of the analysis of palynological data, allowing preliminary conclusions to be drawn regarding the direction of landscape changes in successive sub-intervals (for the detailed discussion of each pollen site, see Izdebski 2013a, as well as Niewöhner et al. 2016 and Şenkul et al. 2018). I have also prepared one "synthetic" map categorizing the sites according to the dynamics of landscape changes across the three consecutive sub-intervals (Fig. 2, top panel). Therefore, the maps distinguish various trajectories of landscape changes, showing the direction of the environment's transformation at individual sites and in particular regions over several centuries.

For medieval Anatolia, we can define three main transformation models relative to changes in the landscape as compared to the situation preceding the Seljuk invasion (i.e., the first interval, the years 925–1075). We see either no change at all (black on the map), an increase in intensity of agriculture (its scale or diversity, shown as different shades of green), or a decrease (yellow-red shades). Increases and decreases can be further divided into three sub-categories depending on whether the changes are visible in only one of the two intervals (in the "long twelfth century" or in the thirteenth–fourteenth centuries) or whether they were long-term (they persisted from the twelfth–fourteenth centuries).

The analysis of medieval Anatolia is further enriched with one more specific pollen index—the share of pine pollen in the total pollen for samples dated to the "long twelfth century" (Fig. 2, bottom panel). For several years, the literature on the history of Anatolia has indicated that the collapse of the Roman order in the seventh century was accompanied by a significant increase in the share of pine pollen in data from sites in various parts of the region (Bakker et al. 2012; Izdebski 2013a, 2019). Pine is a pioneer species in secondary ecological succession, meaning it requires little time (a quarter of a century or less) to colonize abandoned farmland and pastures (Latałowa, Tobolski, and Nalepka 2004).[1] It would therefore seem that changes in the proportion of pine pollen in the total pollen sum are a good indicator that agriculture has experienced a crisis

1 This phenomenon is further reinforced by the exceptional pollen productivity of pine and easy recognition of its pollen; the disappearance of crop plants, especially heavy pollen-producing plants such as olive (and, locally, cereals), can in itself also increase the proportion of pine pollen in the total pollen of a given sample.

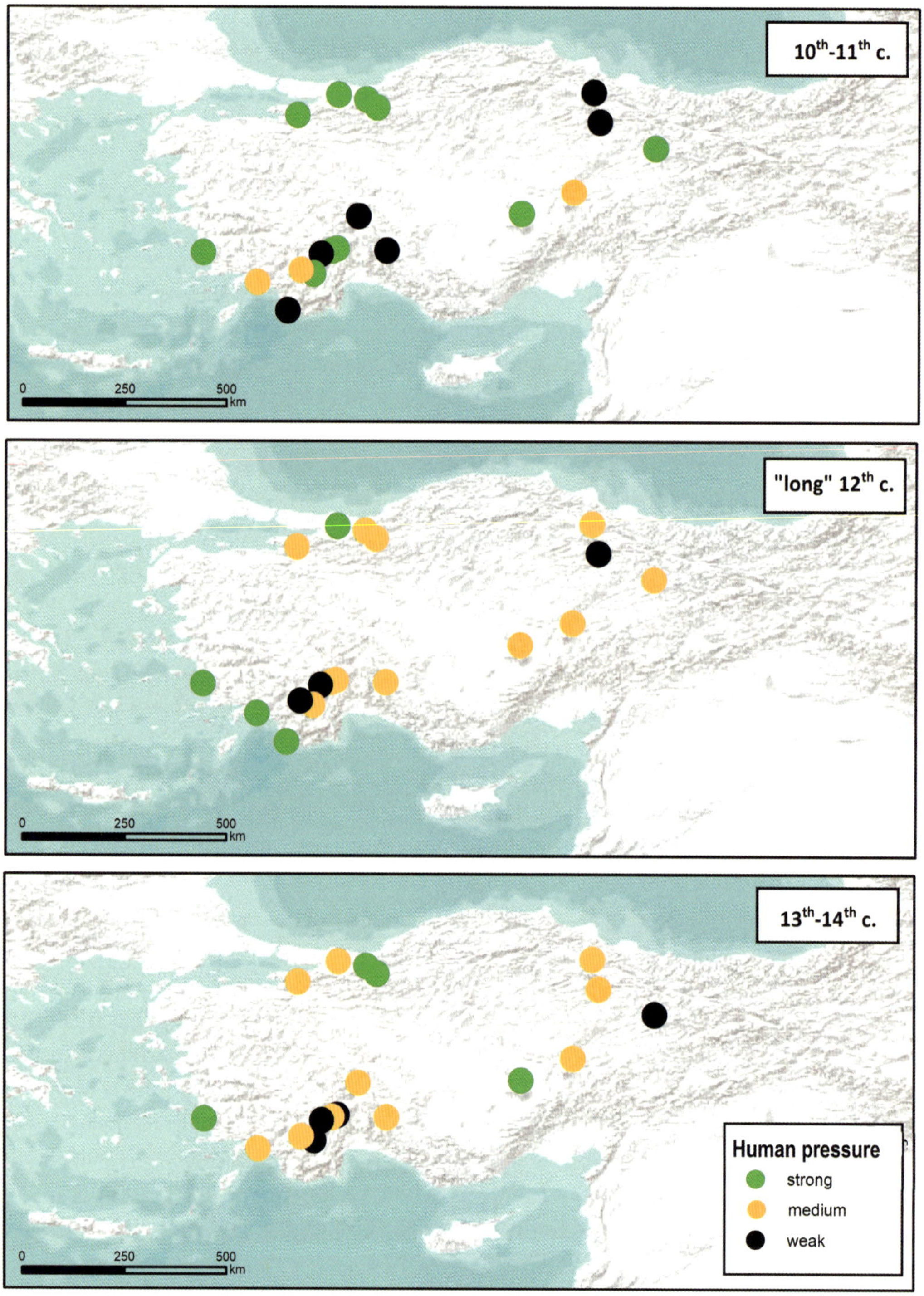

Fig. 1. Human pressure on Anatolian landscapes before and after the Seljuk invasion.

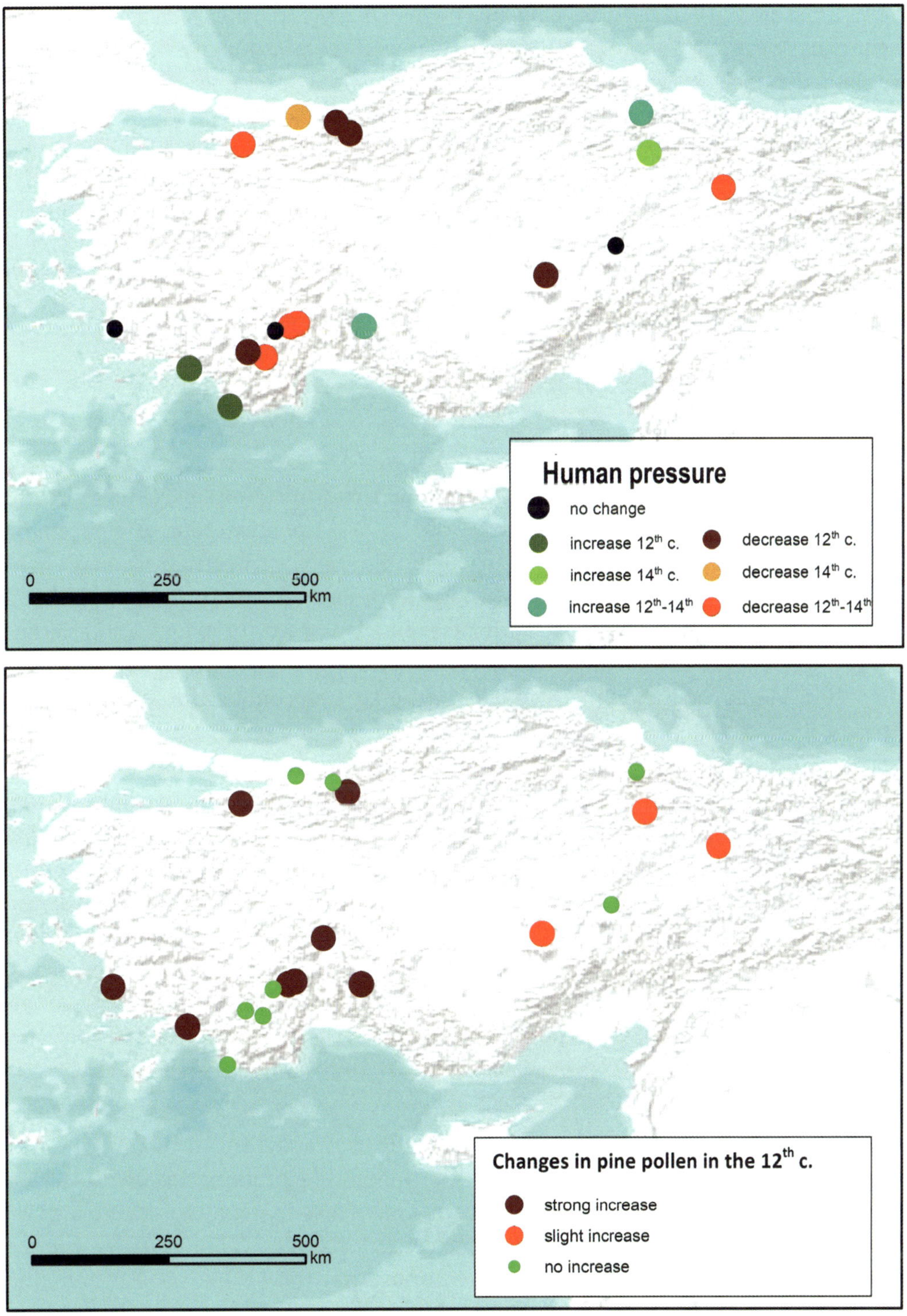

Fig. 2. Trajectories of change in human pressure and secondary ecological succession on Anatolian landscapes before and after the Seljuk invasion.

and that the stability of the landscape has been disturbed in a given area, even if these phenomena were short-lived. (We should remember that pine pollen can be transported over long distances by wind, and thus that changes in its presence at a particular site speak more about a region as a whole than about a given site's immediate vicinity.) By looking at the long-term trajectory of changes in the strength of human pressure alongside the landscape's "initial reaction" to the Seljuk invasion and Turcoman migrations in individual parts of Anatolia, we can bring another socio-ecological dimension into our narrative. We can see where the landscape structure went into crisis as a consequence of the arrival of Turcoman groups and then how this "immediate response" translated into lasting changes in agriculture and the vegetation structure.

Results: "The Fall of Roman Anatolia"

In many areas of the eastern Mediterranean, the Roman "second nature" survived "the Fall of Rome," that is, the collapse of the global world of Late Antiquity in the seventh century CE (Izdebski 2013a). Another turning point that threatened its existence in areas of key importance to the medieval Empire was the loss of political control over Anatolia to Turcoman tribes in the eleventh century. The migration of Turcoman groups in many ways resembled the Germanic invasions and migrations of Late Antiquity and the fall of the Roman Empire in the West. Both cases involve powerful external interference that was beyond the Roman state's capacity to handle, along with displacement of the population. Moreover, historians who refer to the Turkish conquest of Anatolia as "the Second Fall of Rome," such as Mark Whittow, point to many other similarities between the two political watersheds, earlier in the West and later in the East; they pay particular attention to the local Roman elites and their reconciliation with the new political reality: a consenting of sorts to the ending of the old Roman order (Whittow 2017). Moreover, Whittow is not alone in having attempted to reinterpret eleventh-century Anatolia in recent years. While the most extreme archaeological interpretation suggests that Anatolia had already been in decline (as evidenced by the surprising simplicity of Anatolian architectural culture in the tenth–eleventh centuries), other researchers explain the pace and scale of changes in terms of the dynamics of social processes and the poor suitability of the Roman political structure to the new ethno-military situation (Niewöhner 2020; see also Peacock 2010; Peacock and Yıldız 2013; Beihammer 2017; Lauxtermann and Whittow 2017).

It is beyond doubt that the loss of Anatolia was a significant turning point in the history of the Empire. It resulted in the increasing prominence of the Aegean regions of the southern Balkans and western Anatolia taking on an even greater role as the Roman state's political and economic center of gravity (Lilie 1991), the more so as western Anatolia, which for the two centuries after the Seljuk invasion remained in Roman hands, had already been one of the most important parts of the medieval Empire. In contrast to western Anatolia, the eastern territories that were lost in the eleventh century had since the seventh century been a Byzantine-Arab borderland that was only re-colonized in the ninth-tenth centuries (Lilie 1976). Whereas earlier works, including research by Michel Kaplan, indicated that although there were no sources for Cappadocian agriculture and aristocracy in the seventh-ninth centuries, the silence of the sources could be interpreted as an unattested continuity (Kaplan 1981), palynological studies in Cappadocia show that there was no such continuity (England et al. 2008; Şenkul et al. 2018; see also Eger 2015).

The scale of the migration of Turcoman groups to Byzantine Anatolia remains an open question. Neither the fact that almost all the inhabitants of Anatolia today speak Turkish, nor even the fact that they constituted a majority in the nineteenth century, means that they are all descendants of the Turcomans who came to Anatolia in the eleventh-twelfth centuries (for recent synthesis of the ancient DNA [aDNA] evidence, see Wawruschka 2016; for historical sources, see Vryonis 1971, 1975; Peacock, De Nicola, and Yıldız 2015). Nevertheless, while this migration was probably not numerically significant for the entire population of Anatolia, and we should instead assume substantial demographic continuity in most areas, the only (but imperfect) studies of aDNA from medieval Anatolia suggest that the migration of groups bringing with themselves a new and distinct lifestyle was a real phenomenon. Tests on mitochondrial DNA from human remains representing the Sagalassos area in two historical periods—Roman and Byzantine (twelfth-thirteenth centuries)—and from contemporary inhabitants of the nearby Turkish town of Ağlasun indicate that the fall of the Byzantine city and the spread of a new, pastoral, and nomadic way of life around the former Byzantine settlement disturbed the demographic continuity of this territory (Ottoni et al. 2011, 2016; note, however, that the use of mitochondrial DNA rather than the more expensive full genome analysis has serious limitations [Geary and Veeramah 2016; Amorim et al. 2018]).

In light of this, the contribution that pollen analysis makes to the debate on the Seljuk invasion's impact on the history of Anatolia is interesting, as the signals of change and disturbance in the landscape may be able to indicate the extent to which local cultures and demographics experienced continuity and the extent to which they underwent migration and discontinuity. Let us again start with the base maps showing the scale of anthropic pressure in the three successive sub-intervals of the medieval period (Fig. 1). There is no doubt that the landscape was disturbed extensively for one or two generations as a result of the Turkish conquest. The changes between the tenth-eleventh centuries and the "long" twelfth century are significant, and the structural change is noticeable: the green (strong human pressure) disappears in many parts of Anatolia, including in places where it had earlier reappeared in the tenth century, after a period of abandonment during the time of the Arab invasions. Thus, in the twelfth century, strong anthropic pressure remains only at sites near the west coast of Anatolia. At the same time, it is the western, and not the eastern, part of Anatolia that is most heavily disturbed as soon as we move away from the Aegean coast, and this is visible in both the northern and southern sites. The "pine signal" confirms this scenario (Fig. 2, bottom panel). While a weak increase in the proportion of pine in the pollen total is visible at approximately half the sites in eastern Anatolia, the signal is more common and, crucially, stronger in western sites. Of course, this does not necessarily indicate that pine forests and shrubs were present in the immediate vicinities of these sites (since wind can carry very light pine pollen tens of kilometers from the trees that produced it). However, the percentage increase in pine pollen suggests an intensification of secondary ecological succession processes in the region in general and a crisis in agriculture around a given site, as the declining presence of cultivated pollen would also have rendered pine pollen coming from further afar more visible in the entire pollen sample. This east-west contrast, with the west showing stronger signs of landscape disturbance, is probably related to the fact that the first Turkish-Byzantine military confrontation lasted longest in western Anatolia (about 40 years, 1080–1120), until the new border stabilized in these territories, while the east was spared the heaviest warfare.

The landscape in the thirteenth and fourteenth centuries reveals that the emergence of new Turcoman political centers and the arrival of new inhabitants into Byzantine Anatolia brought about socio-ecological changes that were to last. The geography of human activity in the landscape two to three centuries after the fall of Byzantine power in Anatolia had changed: the former centers of agricultural activity and anthropic pressure had mostly lost their importance, and new ones had appeared. However, we should remember that, compared to Late Antiquity (more precisely, the fifth–sixth centuries), anthropic pressure is generally weaker in the Middle Ages, which means the general cultivation and production levels were in all probability lower, as were population numbers (Izdebski, Koloch, and Słoczyński 2015). This means that we must pay attention to the appearance or disappearance of the green color (the strongest anthropic pressure), but also, for example, to the increase in human impact on the landscape in central and eastern Anatolia that is evident from the twelfth century onwards (which is shown in yellow in the upland sites surrounding the Anatolian plateau to the southwest and northeast).

Discussion: The Social Factors Behind Landscape Changes

Having discussed the pollen, let us return to the two explanations for the scale of the political catastrophe in Byzantine Anatolia that features most prominently in the scholarly literature, namely, the claim that this part of the Empire had already been in long-term decline, and the hypothesis that a key factor in the Turcoman success had been the willingness of local elites, or at least local communities, to accept new overlords. As Figure 1 makes clear, it cannot be said that the pre-conquest landscape (the upper panel on this map) bore any omens of an ongoing cataclysm. Comparing the distribution of anthropic pressure in Anatolia in the tenth–eleventh centuries with the situation in the seventh–eighth centuries (Izdebski 2013a), the situation can even be said to have improved. Where environmental continuity had held in the seventh–eighth centuries, it persisted over the following centuries, well into the medieval period; and, in addition, there is a clear revival in southeastern Anatolia, which in the seventh–eighth centuries was the main theater of the Byzantine-Arab war (Haldon and Kennedy 1980; Eger 2015). In the tenth century, as the imperial frontier moved southeast to Syria, eastern and central Anatolia were heavily colonized under the control of Eastern Roman aristocratic-military families, resulting in intensive agricultural development (Kaplan 1981; Cheynet 1990, 213–29; Ousterhout 1998; Whittow 2008). The relatively good social and environmental situation that the pollen data reveal in Anatolia prior to the Seljuk invasion forces us to re-interpret the waning importance, and sometimes even disappearance, of many ancient cities in Anatolia in the early Middle Ages. It also obliges us to reconsider the related lack of monumental construction activity that would have been comparable to that apparent in Greece or Constantinople in the tenth–eleventh centuries (Niewöhner 2017, 2020). Thus, it needs emphasizing that not all cities in Anatolia experienced the fate of Miletus, which is a paradigmatic case of urban decline: the city practically ceased to exist in the eighth–ninth centuries and was only reborn (as a new city-fortress settlement) after, and as if in response to, the Turcoman invasion (Niewöhner 2013, 2016; Niewöhner et al. 2016). For instance, in the case of Ephesus, recent archaeological research suggests that settlement continued uninterrupted throughout the early Middle Ages and then on after the Seljuk invasion (Ladstätter 2018). Moreover, it is not the case that Anatolia was unique in the changing role and partial disappearance of urban centers in the Eastern Roman settlement network: a similar phenomenon can be seen on the other side of the Aegean Sea (Veikou 2012, 2013).

The early medieval ruralization of Anatolia must therefore be viewed differently: not as having caused the region's decline, but as symptomatic of the socio-economic restructuring of the Empire that led over several centuries to urban centers losing the key role they had played in the Greek and Roman periods (Brandes 1999; Brandes and Haldon 2000), with some exceptions, for instance Thessaloniki (Brown 2010), Amorium (Ivison 2007; Lightfoot 2009), Euchaita (Haldon, Elton, and Newhard 2017). When in the eleventh and twelfth centuries the cities began to develop again, they performed different functions more akin to those of the new urban settlements of Latin Europe (Dagron 2002). At the same time, an additional factor counteracting the material rebirth of cities before the Seljuk invasion was the growing political and economic importance of aristocratic families in Eastern Roman society (Cheynet 1990; Haldon 2009, 2016). In the tenth century they began to control large areas of arable land, gaining enormous influence as great landowners in Byzantine agriculture. Aristocratic families born of military pasts were extremely important in the Byzantine "interior" colonization of eastern Anatolia, and their importance in the local management of agriculture is also evident in other parts of Anatolia. For example, the involvement of aristocracy in farming (Evert-Kappesowa 1963; Oikonomides 1996) and the role of land ownership and agricultural income for the social position of aristocratic families (Frankopan 2009; Thonemann 2011, 204–59) are well established. The *topos* (literary theme) of the Byzantine aristocratic landowner engaged in the life and economy of the countryside appears already in a ninth-century hagiography composed in Constantinople (Niketas 2002).

It seems that the way the Eastern Roman aristocracy of central and eastern Anatolia reacted to the Seljuk invasion is the very key to understanding the reasons for both the political "fall" of Byzantine Anatolia and the changes in nature themselves. Jean-Claude Cheynet's research into the family names of Byzantine aristocrats in Constantinople and the Balkans in the decades following the Turcoman invasion of Anatolia revealed the appearance of a number of individuals and entire families with new names that had earlier been attested on the other side of the Aegean. This evidence from anthroponymy thus shows that the Byzantine aristocracy—both the middle and lower ranks—simply fled most of Anatolia (Cheynet 2016). This conclusion is confirmed in the archaeological material. So far, only two long-term excavations have been carried out on fortified Byzantine sites not of urban character in Anatolia (i.e., probably aristocratic residences), and these are located in two different regions. They have provided numerous finds clearly proving that their inhabitants fled at the end of the eleventh century. In both cases we have traces of flight and destruction but no signs of plunder. Presumably, the townspeople abandoned them suddenly, before the Seljuk invaders appeared. This damage is more indicative of panic than of the scale of the Turcoman migration. One of the sites (Çadır Höyük, on the northern periphery of Cappadocia) was re-inhabited shortly thereafter, evidently by new inhabitants who used the same methods of land management but had no need to continue or even make use of the material culture of the wealthy Byzantine former inhabitants of the settlement (Böhlendorf-Arslan 2004; Cassis 2017).

In conclusion, although various Byzantine administrative structures in Anatolia continued for a decade after the 1071 defeat at Manzikert (Cheynet 1998), the exodus of the aristocracy in combination with the very poor development of the cities effectively cut the head off Byzantine Anatolia. On the one hand, this meant that the social group potentially capable of defending the Empire in Anatolia ceased to exist and, on the other, that the social class managing the economy suddenly disappeared. The Byzantine state structures decentralized and then collapsed across most of Anatolia (outside the Aegean area and part of the interior in the west) (Beihammer 2017). In this way, another factor that would have favored the maintenance of the region's Eastern Roman socio-ecological model—that is, the political system that produced and sustained the Roman "second nature"—disappeared. This collapse is well exemplified by the garrison town of Euchaita in the northeastern interior of Anatolia. Its importance waned, and it ceased to exist as a city in the first decades of the twelfth century, when the Roman state withdrew from the area (Haldon, Elton, and Newhard 2017).

Now let us look at the "dynamic" map showing the long-term direction of changes in the landscape of Anatolia between the tenth and fourteenth centuries (Fig. 2, upper panel). First, we still see signs of environmental continuity on the western and southern coasts of Anatolia. This is not surprising, considering that, for example, at the mouth of the Meander, Turks do not appear until the thirteenth century, although local communities are preparing for their arrival. A new, fortified small town is built on the site of previously abandoned Miletus (renamed Palatia), as well as numerous smaller fortifications and towers in rural areas (Lohmann 1999; Niewöhner 2016). Similar processes can also be seen elsewhere in western Anatolia, which undergoes a phase of intense urbanization in the eleventh–thirteenth centuries (Foss 1977, 1990, 1994). Despite these defenses, a few decades after the fall of Constantinople in 1204, control of this Byzantine landscape that preserved the Roman "second nature" longest is taken over by Turkish *beyliks* (regional, dynastic, political, and military structures centered around Turcoman families), ensuring essential stability and security (Arel 1995, 2004; Korobeĭnikov 2014). Paradoxically, they extend the duration of Roman social and natural systems in these areas for another 100–150 years.

Secondly, the palynological data divide the interior of Anatolia clearly into an eastern-central part (where Turkish rule stabilizes early) and a western part. In the east and center, the landscape crisis is fairly short-lived, as evidenced by changes in anthropic pressure and the weakness of the pine signal. At many sites a heavy increase in anthropic pressure is seen in the twelfth–fourteenth centuries, especially in the Amasya region in the northeast and in Beyşehir in the southwest. These were key areas of consolidation of the Turkish government and of intense Islamic urbanization during this period (Cahen 1988; Redford, Beach, and Luzzadder-Beach 2000). It can therefore be assumed that, in these territories, after the exodus of the Byzantine aristocracy new elites who were able to maintain or even develop the earlier socio-ecological model emerged relatively quickly.

A strong pine signal and a long-term decline in agriculture are, however, to be seen in Bithynia and Pisidia in the northwest and the southwest interior. In earlier centuries, from the seventh-century transformation to the Seljuk invasion, these were environmental "islands of continuity," while at the same time being relatively untouched by war and invasions during the Arab threat, except for some short-lived episodes of warfare, as opposed to recurrent raids over several decades in eastern Anatolia. At the end of the eleventh century, these western regions unexpectedly found themselves in a state of war and political turmoil, which lasted from around 1080–1120. After the political situation had stabilized, these areas became launch points for successive Byzantine military expeditions against the Turks or even key theaters of military confrontation (Cheynet 2002; Beihammer 2017). Therefore, it is no surprise that the scale of destruction and disturbances in economic and social life inevitably brought permanent environmental and social changes, that is, the permanent degradation of the Roman environmental infrastructure. It can be assumed that in many places the existing inhabitants had to leave their villages for some time at least, leading to a much deeper and long-lasting discontinuity in agriculture and landscape-shaping activities than in those areas where the only "refugees" who left were the Byzantine aristocrats. In these western interior areas, the Seljuk invasion ends the long afterlife of the Roman "second nature."

Conclusions

The coming of the Turks to Anatolia set in motion a socio-ecological process that led to a profound environmental transformation. It was not necessarily the type of change that one would hypothesize, however, based on the image of Turks as wandering groups of predominantly pastoralist nomads: there was no widespread shift to pastoralism across Anatolia "immediately" (within decades) after their arrival. The Turks do not seem to have brought with them a new "landscape" that they would have "planted" or "promoted" on a noticeable scale. Rather, the environment changed either as a result of disruption to the human practices that sustained specific productive landscapes (agro-ecosystems and managed wildlands) or as a result of the abandonment of the landscape by most local Byzantine elites. In other words, it was social change—the empire's political collapse in Anatolia—that caused the environment to change: the interruption was too much for the "second nature" to maintain its structure without continued human action. Social and ecological worlds were connected in such a complex way that any shift in the balance and structure on the social side must have led to change in ecological structures, too.

As a result, the coming of the Turks can indeed be described as the final "Fall of Rome" in Anatolia, also from the environmental point of view. Whereas the seventh century CE saw the collapse or transformation of (Late) Roman ecological structures and landscapes predominantly in the eastern parts of Anatolia, as a result of continued Arab raiding, the eleventh and twelfth centuries CE were a time when Anatolia's western regions experienced military, political, and social disruption that was both long and severe enough to make impossible or at least discourage the maintenance of Late Roman ecological structures. The "Fall of Rome" probably took longer in Anatolia than anywhere else in the Mediterranean; it started with the Arab raiding of the seventh century and ended with the coming of the Turks. At the same time, thanks to the paleoecological evidence produced by generations of palynologists and geologists, we are in a unique position to trace the ecological dimension of Rome's fall, in which the social and the ecological were inseparably entangled.

Works Cited

Amorim, Carlos Eduardo G., Stefania Vai, Cosimo Posth, Alessandra Modi, István Koncz, Susanne Hakenbeck, Maria Cristina La Rocca, et al. 2018. "Understanding 6th-Century Barbarian Social Organization and Migration through Paleogenomics." *Nature Communications* 9, no. 1: 3547. https://doi.org/10.1038/s41467-018-06024-4.

Arel, Ayda. 1995. "Architecture and the Turkmene Emirates of the Aegean Area: A General Appraisal." In *9. Milletlerarası Türk Sanatları Kongresi (23-27 Eylül 1991, Istanbul), Bildiriler* 1, 163-72. Ankara.

———. 2004. "Land Appropriation, Acculturation, Innovation: The Turkish Conquest of Western Asia Minor and Its Aftermath." In *İstanbul Üniversitesi 550. Yıl, Uluslararası Bizans ve Osmanlı Sempozyumu (XV. Yüzyıl): 30-31 Mayıs 2003*, edited by Sümer Atasoy, 75-97. Istanbul: İstanbul Üniversitesi.

Bakker, Johan, Etienne Paulissen, David Kaniewski, Véronique De Laet, Gert Verstraeten, and Marc Waelkens. 2012. "Man, Vegetation and Climate during the Holocene in the Territory of Sagalassos, Western Taurus Mountains, SW Turkey." *Vegetation History and Archaeobotany* 21: 249-66.

Beihammer, Alexander Daniel. 2017. *Byzantium and the Emergence of Muslim-Turkish Anatolia, ca. 1040-1130*. Birmingham Byzantine and Ottoman Studies. Abingdon: Routledge.

Böhlendorf-Arslan, Beate. 2004. *Glasierte Byzantinische Keramik Aus Der Türkei*. 3 vols. Istanbul: Ege Yayınları.

Bottema, Sytze, Henk Woldring, and Burhan Aytuğ. 1986. "Palynological Investigations on the Relations between Prehistoric Man and Vegetation in Turkey: The Beyşehir Occupation Phase." In *Proceedings of the 5th Optima Congress*, 315-238. Istanbul.

Brandes, Wolfram. 1999. "Byzantine Cities in the Seventh and Eighth Centuries - Different Sources, Different Stories?" In *The Idea and Ideal of the Town between Late Antiquity and the Middle Ages*, edited by Gian Pietro Brogiolo and Bryan Ward-Perkins, 25-57. Transformation of the Roman World 4. Leiden: Brill.

Brandes, Wolfram, and John Haldon. 2000. "Towns, Tax and Transformation: State, Cities and Their Hinterlands in the East Roman World, C. 500-800." In *Towns and Their Territories between Late Antiquity and the Early Middle Ages*, edited by Gian Pietro Brogiolo, Nancy Gauthier, and Neil Christie, 141-72. Transformation of the Roman World 9. Leiden: Brill.

Braudel, Fernand. 1949. *La Méditerranée et le monde méditerranéen à l'époque de Philippe II*. Paris: Colin.

Brown, Amelia Robertson. 2010. "Islands in a Sea of Change? Continuity and Abandonment in Dark Age Corinth and Thessaloniki." *International Journal of Historical Archaeology* 14, no. 2: 230-40. https://doi.org/10.1007/s10761-010-0103-0.

Cahen, Claude. 1988. *La Turquie Pre-Ottoman*. Istanbul-Paris: Institut Français d'Etudes Anatoliennes.

Cassis, Marica. 2017. "Çadır Höyük." In *The Archaeology of Byzantine Anatolia: From the End of Late Antiquity until the Coming of the Turks*, edited by Philipp Niewohner, 368–74. Oxford: Oxford University Press. https://doi.org/10.1093/acprof:oso/9780190610463.001.0001.

Cheynet, Jean-Claude. 1990. *Pouvoir et contestations à Byzance (963–1210)*. Paris: Publications de la Sorbonne.

———. 1998. "La résistance aux Turcs en Asie Mineur entre Manzikert et la Première Croisade." In *Eupsychia: mélanges offerts à Hélène Ahrweiler* 1, 131–47. Paris: Publications de la Sorbonne.

———. 2002. "Le rôle de la Pisidie entre Byzance et les Turcs aux XIe-XIIe siècles." In *Actes du Ier congres international sur Antioche de Pisidie*, edited by Thomas Drew-Bear, Mehmet Taşhalan, and Christine M. Thomas, 447–55. Lyon: Université Lumière-Lyon 2.

———. 2016. "La Perte de l'Asie Mineure Au Xie Siècle a-t-Elle Laissé Des Traces Dans l'anthroponymie Familiale ?" In *Studies in Byzantine Sigillography*, edited by Jean-Claude Cheynet and Claudia Sode, vol. 12, 1–12. Berlin, Boston: De Gruyter. https://doi.org/10.1515/9783110474589-003.

Cronon, William. 1983. *Changes in the Land: Indians, Colonists, and the Ecology of New England*. New York: Hill and Wang.

Dagron, Gilbert. 2002. "The Urban Economy, Seventh–Twelfth Centuries." In *The Economic History of Byzantium: From the Seventh through the Fifteenth Century*, edited by Angeliki E. Laiou, vol. 2, 393–461. Dumbarton Oaks Studies 39. Dumbarton Oaks: Harvard University Press.

Eastwood, Warren J., Neil Roberts, and Hubert F. Lamb. 1998. "Palaeoecological and Archaeological Evidence for Human Occupance in Southwest Turkey: The Beyşehir Occupation Phase." *Anatolian Studies* 48: 69–86.

Eger, A. Asa. 2015. *The Islamic-Byzantine Frontier: Interaction and Exchange among Muslim and Christian Communities*. Library of Middle East History 34. London: IB Tauris.

England, Ann, Warren J. Eastwood, N. Roberts, R. Turner, and John F Haldon. 2008. "Historical Landscape Change in Cappadocia (Central Turkey): A Palaeoecological Investigation of Annually-Laminated Sediments from Nar Lake." *The Holocene* 18, no. 8: 1229–45.

Evert-Kappesowa, Halina. 1963. "Une Grande Propriété Foncière Du VIIIe s. à Byzance." *Byzantinoslavica* 24: 33–40.

Foss, Clive. 1977. "Archaeology and the "Twenty Cities" of Byzantine Asia." *American Journal of Archaeology* 81, no. 4: 469–86. https://doi.org/10.2307/503279.

———. 1990. "Byzantine Malagina and the Lower Sangarius." *Anatolian Studies* 40: 161–83.

———. 1994. "The Lycian Coast in the Byzantine Age." *Dumbarton Oaks Papers* 48: 1–52.

Frankopan, Peter. 2009. "Land and Power in the Middle and Later Period." In *A Social History of Byzantium*, edited by John F. Haldon, 112–42. Oxford: Wiley-Blackwell.

Fyfe, Ralph M., Jessie Woodbridge, and Neil Roberts. 2015. "From Forest to Farmland: Pollen-Inferred Land Cover Change across Europe Using the Pseudobiomization Approach." *Global Change Biology* 21, no. 3: 1197–1212. https://doi.org/10.1111/gcb.12776.

Geary, Patrick, and Krishna Veeramah. 2016. "Mapping European Population Movement through Genomic Research." *Medieval Worlds* 4: 65–78. https://doi.org/10.1553/medievalworlds_no4_2016s65.

Haldon, John. 2009. "Social Elites, Wealth, and Power." In *A Social History of Byzantium*, edited by John Haldon, 168–211. Oxford: Wiley-Blackwell.

———. 2016. "Bureaucracies, Elites and Clans: The Case of Byzantium, c.600–1100." In *Empires and Bureaucracy in World History: From Late Antiquity to the Twentieth Century*, edited by Peter Crooks and Timothy Parsons, 147–69. Cambridge: Cambridge University Press.

Haldon, John, Hugh Elton, and James Newhard. 2017. "Euchaïta." In *The Archaeology of Byzantine Anatolia: From the End of Late Antiquity until the Coming of the Turks*, edited by Philipp Niewohner, 375–88. Oxford: Oxford University Press. https://doi.org/10.1093/acprof:oso/9780190610463.001.0001.

Haldon, John, and Hugh Kennedy. 1980. "The Arab-Byzantine Frontier in the Eighth and Ninth Centuries: Military Organisation and Society in the Borderlands." *Recueil Des Travaux de l'Institut d'Etudes Byzantines* 19: 79–116.

Haldon, John, Neil Roberts, Adam Izdebski, Dominik Fleitmann, Michael McCormick, Marica Cassis, Owen P. Doonan, et al. 2014. "The Climate and Environment of Byzantine Anatolia: Integrating Science, History and Archaeology." *Journal of Interdisciplinary History* 45: 113–61.

Ivison, Eric A. 2007. "Amorium in the Byzantine Dark Ages (Seventh to Ninth Centuries)." In *Post-Roman Towns, Trade and Settlement in Europe and Byzantium*, edited by Joachim Henning, Byzantium, Pliska, and the Balkans 2, 25–60. Berlin: Walter de Gruyter.

Izdebski, Adam. 2011. "Why Did Agriculture Flourish in the Late Antique East? The Role of Climate Fluctuations in the Development and Contraction of Agriculture in Asia Minor and the Middle East from the 4th till the 7th c. AD." *Millenium. Jahrbuch Zu Kultur Und Geschichte Des Ersten Jahrtausends n. Chr.* 8: 291–312.

———. 2012. "The Changing Landscapes of Byzantine Northern Anatolia." *Archaeologia Bulgarica* 16: 47–66.

———. 2013a. *A Rural Economy in Transition: Asia Minor from Late Antiquity into the Early Middle Ages*. Journal of Juristic Papyrology Supplement Series 18. Warsaw: Taubenschlag Foundation.

———. 2013b. "The Economic Expansion of the Anatolian Countryside in Late Antiquity: The Coast versus Inland Regions." *Late Antique Archaeology* 10: 343–76. https://doi.org/10.1163/22134522-12340036.

———. 2019. "The "Pine Signal" as a Symptom of Crisis in Byzantine History." In *Byzantine et Slavica. Studies in Honour of Professor Maciej Salamon*, edited by Stanisław Turlej, Michał Stachura, Bartosz Kołoczek, and Adam Izdebski, 155–56. Krakow: Historia Iagiellonica.

Izdebski, Adam, Grzegorz Koloch, and Tymon Słoczyński. 2015. "Exploring Byzantine and Ottoman Economic History with the Use of Palynological Data: A Quantitative Approach." *Jahrbuch Der Österreichischen Byzantinistik* 65: 67–110.

Kaplan, Michel. 1981. "Les Grands Propriétaires de Cappadoce (VIe-XIe Siècles)." In *Le Aree Omogenee Della Civiltà Rupestre Nell'ambito Dell'impero Bizantino: La Cappadocia: Atti Del Quinto Convegno Internazionale Di Studio Sulla Civiltà Rupestre Medioevale Nel Mezzogiorno d'Italia (Lecce-Nardò, 12–16 Ottobre 1979)*, edited by Cosimo Damiano Fonseca, 125–58. Galatina: Congendo.

Korobeĭnikov, Dimitri. 2014. *Byzantium and the Turks in the Thirteenth Century*. First edition. Oxford Studies in Byzantium. Oxford: Oxford University Press.

Kreike, Emmanuel. 2013. *Environmental Infrastructure in African History: Examining the Myth of Natural Resource Management in Namibia*. Studies in Environment and History. Cambridge: Cambridge University Press.

Ladstätter, Sabine. 2018. "Ephesos im 11. Jahrhundert. Eine archäologische Bestandsaufnahme." In *Space, Landscapes and Settlements in Byzantium. Studies in Historical Geography of the Eastern Mediterranean*, edited by Andreas Külzer and Mihailo Popović, 215–36. Novi Sad: Akademska knjiga.

Laland, Kevin, Blake Matthews, and Marcus W. Feldman. 2016. "An Introduction to Niche Construction Theory." *Evolutionary Ecology* 30: 191–202. https://doi.org/10.1007/s10682-016-9821-z.

Latałowa, Małgorzata, Kazimierz Tobolski, and Dorota Nalepka. 2004. "Pinus L. Subgenus Pinus (Subgen. Diploxylon (Koehne) Pilger) - Pine." In *Late Glacial and Holocene History of Vegetation in Poland Based on Isopollen Maps*, edited by Magdelena Ralska-Jasiewiczowa, 165–77. Krakow: W. Szafer Institute of Botany, Polish Academy of Sciences.

Lauxtermann, Marc Diederik, and Mark Whittow, eds. 2017. *Byzantium in the Eleventh Century: Being in Between*. Publications of the Society for the Promotion of Byzantine Studies 19. London: Routledge.

Lightfoot, Christopher S. 2009. Excavations at Amorium: Results from the Last Ten Years (1998-2008)." In *Archaeology of the Countryside in Medieval Anatolia*, edited by Tasha Vorderstrasse and J. J Roodenberg, 139–53. Leiden: Nederlands Instituut voor het Nabije Oosten.

Lilie, Ralph-Johannes. 1976. *Die Byzantinische Reaktion Auf Die Ausbreitung Der Araber: Studien Zur Strukturwandlung Des Byzantinischen Staates Im 7. Und 8. Jahrhundert*. Miscellanea Byzantina Monacensia, Heft 22. München: Institut für Byzantinistik und Neugriechische Philologie der Universität.

———. 1991. "Twelfth-Century Byzantine and Turkish States." In *Manzikert to Lepanto: The Byzantine World and the Turks, 1071 - 1571*, edited by Anthony Bryer, 35–51. Byzantinische Forschungen 16. Amsterdam: Hakkert.

Lohmann, Hans. 1999. "Survey in Der Chora von Milet: Vorbericht Über Die Kampagnen Der Jahre 1996 Und 1997." *Archäologischer Anzeiger*, 439–73.

Melosi, Martin V. 1981. *Garbage in the Cities: Refuse, Reform, and the Environment*. Revised edition. History of the Urban Environment. Pittsburgh: University of Pittsburgh Press.

Merchant, Carolyn. 1989. *Ecological Revolutions: Nature, Gender, and Science in New England*. Chapel Hill: University of North Carolina Press.

Niewöhner, Philipp. 2013. "Neue Spät- Und Nachantike Monumente von Milet Und Der Mittelbyzantinische Zerfall Des Anatolischen Städtewesens." *Archäologischer Anzeiger* 2: 165–233.

———. 2016. "The End of the Byzantine City in Anatolia. The Case of Miletus." In *Städte Im Lateinischen Westen Und Im Griechischen Osten Zwischen Spätantike Und Früher Neuzeit : Topographie, Recht, Religion*, edited by Elisabeth Gruber, Mihailo Popović, Martin Scheutz, and Herwig Weigl, 63–77. Wien: Böhlau.

———. 2017. "Urbanism." In *The Archaeology of Byzantine Anatolia. From the End of Late Antiquity until the Coming of the Turks*, edited by Philipp Niewöhner, 39–59. Oxford: Oxford University Press.

———. 2020. "What Went Wrong? Decline and Ruralisation in Eleventh Century Anatolia. The Archaeological Record." In *Eleventh-Century Byzantium. Social Change in Town and Country*, edited by James Howard-Johnston, 98–132. Oxford: Oxford University Press.

Niewöhner, Philipp, Arzu Demirel, Adam Izdebski, Hacer Sancaktar, Nico Schwerdt, and Harald Stümpel. 2016. "The Byzantine Settlement History of Miletus and Its Hinterland - Quantitative Aspects. Stratigraphy, Pottery, Anthropology, Coins, and Palynology." *Archäologischer Anzeiger* 2: 225–90.

Niketas. 2002. *The Life of St Philaretos the Merciful Written by His Grandson*. Edited by Lennart Rydén. Acta Universitatis Upsaliensis. Uppsala: Uppsala University Library.

Oikonomides, Nikos. 1996. "The Social Structure of the Byzantine Countryside in the First Half of the Xth Century." *Byzantina Symmeikta* 10: 105–25. https://doi.org/10.12681/byzsym.808.

Ottoni, Claudio, Rita Rasteiro, Rinse Willet, Johan Claeys, Peter Talloen, Katrien Van de Vijver, Lounès Chikhi, Jeroen Poblome, and Ronny Decorte. 2016. "Comparing Maternal Genetic Variation across Two Millennia Reveals the Demographic History of an Ancient Human Population in Southwest Turkey." *Royal Society Open Science* 3, no. 2: 150250. https://doi.org/10.1098/rsos.150250.

Ottoni, Claudio, Francois-X Ricaut, Nancy Vanderheyden, Nicolas Brucato, Marc Waelkens, and Ronny Decorte. 2011. "Mitochondrial Analysis of a Byzantine Population Reveals the Differential Impact of Multiple Historical Events in South Anatolia." *European Journal of Human Genetics* 19, no. 5: 571–76.

Ousterhout, Robert G. 1998. "Questioning the Archaeological Evidence: Cappadocian Monasteries." In *Work and Worship at the Theotokos Evergetis, 1050–1200*, edited by M. Mullet and A. Kirby, 420–31. Belfast: The Queen's University of Belfast.

Peacock, Andrew C. S. 2010. *Early Seljūq History: A New Interpretation*. Routledge Studies in the History of Iran and Turkey 7. London, New York: Routledge.

Peacock, A. C. S., Bruno De Nicola, and Sara Nur Yıldız, eds. 2015. *Islam and Christianity in Medieval Anatolia*. Burlington, VT: Ashgate.

Peacock, A. C. S., and Sara Nur Yıldız, eds. 2013. *The Seljuks of Anatolia: Court and Society in the Medieval Middle East*. Library of Middle East History 38. London, New York: IB Tauris.

Redford, Scott, Timothy Paul Beach, and Sheryl Luzzadder-Beach. 2000. *Landscape and the State in Medieval Anatolia: Seljuk Gardens and Pavilions of Alanya, Turkey*. BAR International Series 893. Oxford: Archaeopress.

Roberts, Neil, R. M. Fyfe, J. Woodbridge, M.-J. Gaillard, B. a. S. Davis, J. O. Kaplan, L. Marquer, et al. 2018. "Europe's Lost Forests: A Pollen-Based Synthesis for the Last 11,000 Years." *Scientific Reports* 8, no. 1: 716. https://doi.org/10.1038/s41598-017-18646-7.

Şenkul, Çetin, Aziz Ören, Uğur Doğan, and Warren John Eastwood. 2018. "Late Holocene Environmental Changes in the Vicinity of Kültepe (Kayseri), Central Anatolia, Turkey." *Quaternary International* 486: 107–15. https://doi.org/10.1016/j.quaint.2017.12.044.

Tarr, Joel A. 1996. *The Search for the Ultimate Sink: Urban Pollution in Historical Perspective.* Technology and the Environment (Akron, Ohio). Akron, OH: University of Akron Press.

Thonemann, Peter. 2011. *The Maeander Valley: A Historical Geography from Antiquity to Byzantium.* Cambridge: Cambridge University Press.

Veikou, Myrto. 2012. *Byzantine Epirus: A Topography of Transformation. Settlements of the Seventh–Twelfth Centuries in Southern Epirus and Aetoloacarnania, Greece.* Leiden: Brill.

———. 2013. "Settlements in the Greek Countryside from the 4th to the 9th Centuries: Forms and Patterns." *Antiquité Tardive* 21: 125–33.

Vryonis, Speros. 1971. *The Decline of Medieval Hellenism in Asia Minor and the Process of Islamization from the Eleventh through the Fifteenth Century.* Berkeley, London: University of California Press.

———. 1975. "Nomadization and Islamization in Asia Minor." *Dumbarton Oaks Papers* 29: 41–71.

Wawruschka, Celine. 2016. "Genetic History and Identity: The Case of Turkey." *Medieval Worlds* 4: 123–61. https://doi.org/10.1553/medievalworlds_no4_2016s123.

Whittow, Mark. 2008. "The Middle Byzantine Economy (600–1204)." In *The Cambridge History of the Byzantine Empire c. 500–1492*, edited by Jonathan Shepard, 465–92. Cambridge: Cambridge University Press.

Whittow, Mark. 2017. "The Second Fall: The Place of the Eleventh Century in Roman History." In *Byzantium in the Eleventh Century: Being in Between*, edited by Marc Diederik Lauxtermann and Mark Whittow, 109–26. Publications of the Society for the Promotion of Byzantine Studies 19. London: Routledge.

Of Wetlands and Reclamation Regimes: Climate Change, Social Upheaval, and Political Practice in Western Anatolia in the Long Nineteenth Century

Semih Çelik
Christina Luke

Introduction

The extreme weather events of the long eighteenth century in the Ottoman world have long been studied as a catalyst for widespread social and political changes. According to Faruk Tabak, the inundation of early-modern Mediterranean waterscapes during this period dramatically transformed settlement and production patterns. Those who moved from valley floors to foothills escaped catastrophic flooding and secured greater protection from rogue bandits, who took advantage of the corresponding social and political upheavals to wreak havoc (Tabak 2008). By the middle of the nineteenth century, climate changes resulted in drier periods, altering yet again local and regional responses. Our focus here is to situate lakes and wetlands into the environmental and political historiographies of western Anatolia from the mid-nineteenth century through the early twentieth century.

While wetlands are often at the heart of microregions that define Mediterranean river basins and deltas (Horden and Purcell 2000, 187), only recently have scholars begun to turn their attention to inland lakes, rivers, and abutting marshes. This new work complements environmental scholarship on early-modern Ottoman histories of potable water, navigable rivers, and infrastructure for ports and railways (Christensen 2017; Cole 2016; McPhillips, Meier, and Rabo 2019; Shopov 2019; Husain 2014, 2016). Seeing limited (if any) potential for large-scale cultivation that would generate sizeable taxable activities, Ottoman officials in Constantinople had for centuries transferred the control and management of the micro-ecologies of wetlands—marshes, bogs, riverbanks, and shorelines—to local powerbrokers. The result most often was the invisibility of wetlands in the correspondence of the central bureaucracy. Given this peripheral interest from authorities, researchers have tended to consider early-modern wetlands as empty spaces. Defining these zones as "dead" (*mevt*) lands even finds a legacy in Roman law (İnalcık 1994, 120–21). The nineteenth century wetlands, on the other hand, have been conceptualized as spaces of putrefaction, decay, and breeding grounds for illness. These "swamps" (*bataklık*) have been associated with invasive reeds and "unproductive" mud.

Contrary to these perspectives, Ottoman-era wetlands were hubs of autonomous social networks and political structures that responded to specific local conditions that realized significant and diverse economic opportunities (Horden and Purcell 2000, 182). In the eastern provinces (far away from the encroachment of the Ottoman center) place-making in swamps, wetlands, and reedbeds embraced seasonal patterns of mobility. Even so, in the capital many viewed such lifeways as rogue and potentially inciting conflict and rebellion. Mid-nineteenth century accounts suggest that liminal zones of wetlands became more and more strategic for transhumant groups who bucked policies aimed at permanent settlement. In turn, Ottoman bureaucrats who wished to settle (i.e., permanently) transhumant tribes often gave these groups disparaging labels, which further discredited their agency. In fact, these "unruly" and "roaming" tribes were undercut by reference to their annual patterns of migration, often dictated by seasonal inundation. In the words of Derviş Paşa, a contemporary bureaucrat, they were tribes who "rise up with the rising up of water" (Derviş Paşa and Mehmed Hurşid Paşa 1997 [1848], 42–43) and fostered a "natural republic of disorder" (Derviş Paşa and Mehmed Hurşid Paşa 1997 [1848], 25).

In western Anatolia Ottoman officials saw even greater threats to their political power, coming not only from transhumant communities, but more specifically from actors with economic power or socio-ecological knowledge over wetland ecologies. The ad-hoc and predominantly decentralized land-tenure systems further shifted in the mid-nineteenth century. Investments in wetlands came to include the interests of provincial leadership, Levantine and other European groups, and refugees from Cossack and Balkan regions. What emerges in these histories is that in western Anatolia, wetlands were not areas of conflict between nomadic tribes and the central government, but spaces of political and economic competition between a growing landowning class, Ottoman bureaucrats, and the sultan. Reports of subversive political networks, rogue communities of foreigners, and zones of malaria and other diseases were pervasive. Ottoman officials leveraged these threats to security—political, social, and health—to justify state seizures and to implement modernization policies of reclamation and agriculture.

The confluence of mid- to late nineteenth century ruptures in political structures, growing social discontent, and new technological ideals of landscape management confronted major climate changes. In turn, ecosystems were targeted for greater "productivity." Here, we see undercurrents of how a range of actors grappled with the emerging concepts of the environment and its management in the Ottoman Empire. Blended with economic and administrative concerns, a self-conscious discourse on the environment emerged as a result of conflict and competition, as well as cooperation between actors from various scales—local, regional, imperial, and inter-imperial. We argue that political and economic competition in western Anatolia triggered unprecedented harm to the integrity of not only "natural" but also managed wetland ecosystems. Significantly and ironically, this competition fostered a body of knowledge and study of the environment, which in turn became catalysts for conversations about conserving such ecosystems. Here we focus on the *longue durée* of the Ottoman era to discuss the natural and infrastructural histories of interior lakes and wetlands in western Anatolia and the varied success of reclamation efforts. By the late nineteenth century these tensions are clearly articulated in the careful manipulation of the 1858 Land Code, whether at the cost of environmental integrity or in efforts to preserve it.

Spheres of the Sultan in Western Anatolia

From the mid-nineteenth century onwards, wetlands in Aydın province became flashpoints of political competition for the sultan. Seeking a "natural" reason to exert greater control, outbreaks of malaria in the late nineteenth century became convenient guises for state intervention (Gamal-Eldin 2019; Gratien 2017; Mitchell 2002, 19–54). Modernization programs aimed at reclamation of wetlands served multiple purposes: they allowed the state to meddle in foreign interests as well as pursue social engineering agendas that fractured regional social and economic structures (Kaya 2021; Yücesoy 2018). These paternalistic policies legitimized state actions in the name of "civilizing" nomadic tribes and "humanizing" refugee communities (Gratien 2017). In pursuing these agendas, however, the state confronted policies that had for centuries sustained the resilient micro-ecologies of wetlands, policies that had been upheld and followed by *waqfs* (pious foundations), elite Ottoman and Levantine families, and provincial authorities (Çelik, Luke, and Roosevelt, in review; Tabak 2008, 193).

The situation in the hinterlands of western Anatolia was also greatly influenced by the expansion of Smyrna/Izmir, a bustling commercial and international hub (see Goffman, Eldem, and Masters 1999, 79–134; Frangakis-Syrett 1988, 2009; Kuru 2017; Küçükkalay 2008). As Ottoman officials took greater interest in the city, they also learned about the economic cycles, social networks, and political structures of lucrative wetlands. Fishing concessions could be taxed, but the expansive economy of seasonal reeds could not. Reeds grew in zones registered as non-taxed "commons" (*hüda-yı nabit*)—those zones that were deemed uninhabitable as well as "unproductive." The annual cycles of healthy (resilient) wetlands and unskilled labor required for harvesting reeds meant that local communities received significant profits without major investment or management (Aşkın 2017, 44). The importance of this economy is further evidenced by the relatively rich vocabulary of reed types that often corresponded to the territories of cultivation and target markets.[1] Reeds also provided ample (and low-cost) building materials, including seasonal housing for agricultural laborers (Sarıbey 2013, 43–44).

To secure greater revenues from the fertile landscape, as well as from fishing and niche markets (such as leeches), and to control reed economies, the state pursued legal loopholes for landgrabs and the establishment of sultanic *çiftliks* (*çiftlik-i hümayun*) in western Anatolia. The economic profits would directly fill the privy purse (*hazine-i hassa*) of Sultan Abdülhamid II. A collection of various villages, these *çiftliks* were located on fertile and well-watered valleys (Kokdaş 2018), and a good deal of their subsistence and local revenue derived from vibrant, productive wetlands. Despite the high returns of specific agricultural cash crops (such as wheat) typically associated with *çiftliks* (see Haykıran 2013, 164–66; Nagata 1995, 124), archival data demonstrates that wetland-based activities made up an equal part, *if not the majority*, of their initial economic returns in western Anatolia. In fact, wetland zones were such a significant catalyst in expanding territorial interests to protect imperial integrity that the state began to pursue confiscation of all lands in close proximity. Yet difficulties arose, pointing further to the power of influential landowners, both foreign and provincial Ottoman citizens as well as groups who held fishing concessions.

The state justified its actions carefully. Peasants, they claimed, were subject to ruthless private (often foreign) landowners; in turn, the state promised to save them from such exploitation.[2] In other cases, the state pledged to cancel debts held by foreigners in exchange for land (Kaya 2019). Even more complex negotiations arose for refugee groups. For decades Cossack communities had been given safe haven in the Ottoman Empire, and they had coveted concessions for lakes and wetlands in Anatolia, owing to their legacy lifeways in the Russian empire (Pianciola 2019, 62; Fındıkoğlu 1962). The influx of Balkan refugees into western Anatolia, however, upset this balance. Once the state understood the major economic profits realized by Cossack communities, in spite of their hardship claims, the state shifted its policies. In some cases, Cossack concessions were canceled, and in other cases, they were outbid by the state. These zones and the corresponding profits were then used for programs to resettle Balkan refugees. In sum, then, by the mid-nineteenth century western Anatolian wetlands became hotspots for economic and political competition among locals, Sultan Abdulhamid II, high-ranking pashas, refugees, and independent landowners. The result of the tensions that emerged had both intended as well as unintended ecological consequences.

1 At least four different names for reeds can be found in the archival material cited in this article.

2 For examples from the Balkans, see Kaya 2021, 42. For an example on the Baltazzi family *çiftliks* in western Anatolia, see Istanbul Municipality Atatürk Library, Pertevniyal Valide Sultan (PVS) Collection, PVS 585; PVS 4779; PVS 4873.

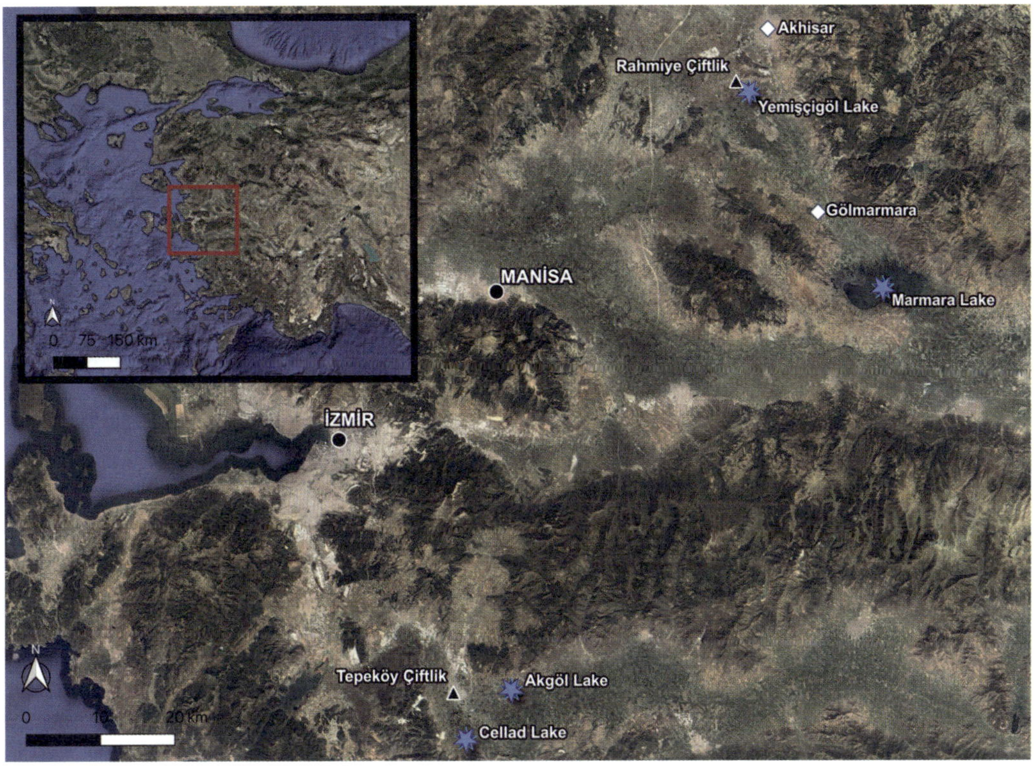

Fig. 1. Map of case study locations.

Adjusting to New Climates

In addition to the political intrigue of these areas, there was also considerable change in the climate. For the wetlands and shallow lakes well watered by the Gediz and Küçük Menderes Rivers, winter rains resulted in expanded lakeshores and/or the merger of nearby ponds with larger lakes. In the summer heat, shorelines significantly contracted and adjacent marshes nearly dried. These fluctuating and extremely fragile micro-ecologies were vibrant areas for fishing and hunting during the late autumn, winter, and early spring. They were also pastoral grounds for transhumance groups migrating between the uplands of the surrounding mountain ranges in both spring and fall, and the shorelines were fertile agricultural zones in late spring through the autumn harvest. Annual inundation and manure cycles replenished soils each year. Profitable agricultural ventures altered seasonal shifts, which compromised the fragile local ecosystems. Overexploitation of reeds, fish, and leeches, the introduction of "new" species, often alien to the region and invasive, and the expansion of herds and their grazing territories further deteriorated wetland ecologies. Irrigation canals and agricultural expansion exacerbated the problem. The result was large-scale desiccation and the metamorphosis and slow death of fragile ecosystems.

The climate shifts of the mid- to late nineteenth century added fuel to the burning fire. Despite local differences, climatic instability in most parts of Anatolia after the 1830s made both rural and urban communities experience environmental shifts. All over Anatolia, numerous discourses claimed that climate events were new, "unseen before" (Çelik 2017; 2019). The microclimates of Aydın province, however, followed unique trajectories at times. At the height of the 1845–50 drought and famine that devastated almost all Anatolia, and in the warmest year of the first half of the century, 1847, the middle Gediz Valley was humid, and extreme rainfall caused local waterscapes[3] to overflow, while large parts of Anatolia and Istanbul lacked water (Çelik 2017). Similar trends followed in the 1860s. The hinterlands of Izmir and the middle Gediz Valley suffered during the drought of 1862, followed by massive flooding in 1863.[4] The microecologies of both the middle Gediz and Küçük Menderes valleys nevertheless continued to support resilient patterns of waterscapes and wetland ecologies.

Growing political and economic competition in the western Anatolian wetlands in the following decades resulted in more human intervention, breaking previous patterns of resilience. The negative impacts from the 1886 drought in Aydın province were magnified by the 1887–89 droughts that swept across western and central Anatolia, even resulting in famine conditions.[5] This two year period of extreme drought was made particularly painful in the latter half of 1889, when the skies were dry from June through December (Haykıran 2013, 140). Yet again, in 1890, 1893, and 1898 large areas of western Anatolia went dry (Pehlivan 2020). The small lakes and adjacent wetlands were no match for the extreme drought of 1893. The scale of the problems even extended to the relatively large lakes in western and northwestern Anatolia, such as Lake Manyas, Lake Sazlı, and Lake Gökçeviran.[6] The nearby Lake Marmara, too, went partially dry, as did the delta of the Gediz River.[7] By the turn of the twentieth century, the fragile lakes and wetlands that had been part of the landscapes of western Anatolia for millennia were on the brink of extinction.

3 Prime Ministry Ottoman Archives (PMO), Istanbul, İ.MVL 110-2528.

4 PMO. ŞD 398-57; ŞD 2987-9.

5 PMO. Y.PRK.KOM 5-8. In 1888, however, a rather wet spring season caused the wetlands to extend temporarily. The water around the Meşhed *çiftlik* rose above three meters as a result of the flood in March. PMO. HH.THR 324-81.

6 For Lake Manyas, see PMO. DH.MKT 199-53. For Sazlıgöl, total drainage and reclamation was attempted for refugees from Lofça, see PMO. DH.MKT 190-67. For Gökçeviran Lake, see PMO. DH.MKT 410-26. For an example from southern Anatolia (Lake Amik), see Efe 2016.

7 See the map series held by the University of Wisconsin: Great Britain. Hydrographic Office. Admiralty charts. These were gifted by the tycoon and explorer Alison Armour in preparation for the Paris Peace Conference, and thus plans for territorial (re)alignment. Note the quality of the topographic details and notes on wetlands and rivers, including periods of drought, for western Anatolia (as well as other areas of the Mediterranean). See the following maps in the collection: Gulf of Izmir, Turkey 1881; Izmir, Turkey 1902; Asia Minor 1911; Turkey 1905; Turkey coast 1877; Gulf of Kuşadası, Turkey 1898. Also for extreme dry weather conditions in Tepeköy, see PMO. ŞD 1386-18.

We now turn to three case studies from Aydın province that exemplify the issues at hand. Our first example draws from the Tepeköy *çiftlik*, associated with the Akgöl, Cellad, and Ibrahim Lakes near modern-day Torbalı, Izmir; the second is from Yemişçigöl, ca. 10 km south of modern-day Akhisar, Manisa; and the third pertains to Lake Marmara, ca. 10 km south of the town of Gölmarmara, Manisa (Fig. 1). Each case demonstrates the significant impacts that imperial programs of infrastructure and reclamation had on the landscape as well as the ensuing ruptures in the body politic of not only wetlands but also rural management writ large in western Anatolia. As much as the overarching themes discussed above defined the policies, the unique micro-ecologies of each zone dictated the relative trajectories of landscape transformation, and we aim to illustrate these in each of the sections below. We demonstrate how specific actors attempted, with mixed success, to leverage the broad trends in late nineteenth-century climate changes to exploit or preserve regional wetlands and lakes. Of the many lakes and wetlands discussed, ultimately only Lake Marmara survived.

Draining Agency in Aydın

Tepeköy çiftlik/*Akgöl, Cellad, and Ibrahim Lakes*

The story of the Tepeköy *çiftlik* provides a window into the administrative center of the sultanic *çiftliks* in Aydın province and specifically the Akgöl and larger Cellad Lakes. Fed by the torrents of the Küçük Menderes River that cut across the *çiftlik* lands, these lakes and their extended wetlands expanded during winter rains, and Cellad Lake often merged with the waters of the smaller, adjacent Ibrahim Lake.[8] The 1861 construction of the Izmir-Aydın railway passed along the northeastern shoreline of Cellad Lake. The associated earthworks served as levees and extended the seasonal wetlands, which spurred investments in water buffalo and water-loving crops.[9] In turn, Tepeköy became an even more lucrative region, garnering the attention of the sultan.[10] In 1883 the sultan acquired the lands of Aristidi Baltazzi in exchange for absolving the Levantine Baltazzi family debts owed to the state (Kaya 2019).[11] In this process, however, the imperial hand reached into the wetland economies of the eleven (plus a half) semi-autonomous small *çiftliks* subsumed under the Tepeköy *çiftlik* administrative structure. For generations the extensive natural wetlands had been sustained by the delicate balance of significant winter rains, cool and wet springs, relatively hot and dry summers, and temperate autumns with light rain. The most extensive component was the Meşhed *çiftlik*, which included 8,353 *dönüms* of wetland and 5,297 *dönüms* of cultivable land (Fig. 2).[12] In fact,

8 PMO. ML.EEM 782-22.
9 PMO. ŞD 1386-18.
10 PMO. ML.EEM 805-55.
11 The Baltazzi family became one of the most powerful economic actors in the empire, controlling nearly the whole tax base of western Anatolia and owning a great deal of land in the region between the 1850s and 1880s. PMO. ML.EEM 90-14 and PMO. ML.EEM 72-7; PMO. HH.THR 9-43.
12 PMO. ML.EEM 72-7.

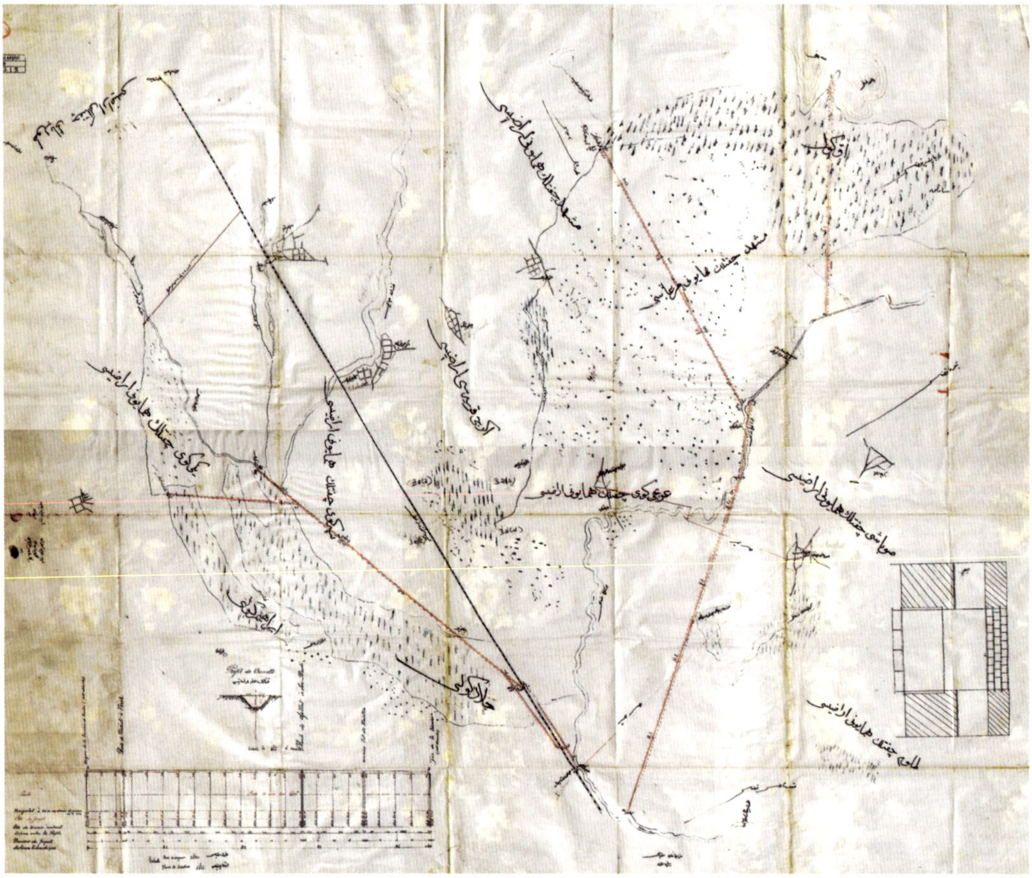

Fig. 2. Map of the Tepeköy *çiftlik* complex dating to 1897. Lakes are indicated by blue lines, swamps by blue dots. North at top. Source: PMO. ŞD 1211-22.

only a quarter of the landscape itself was cultivable year-round: the vast majority of the Tepeköy holdings were extensive winter lakes that each spring gave way to thousands of *dönüms* of wetland.[13]

From these areas, numerous resources supported communities. Informal, small, non-market economies sustained household-level diets and baseline resources, such as meat and fish, as well as hides and implements made from animal bone. Accounts report abundant deer and hare, as well as waterfowl. Partridge lived in the surrounding hills; ducks, woodcock, and quail were found in the marshlands north of Cellad Lake; and woodpigeons flocked to the seasonal grasses in the early spring (Giraud 1940, 32–39). Significant market-driven economies stemmed from the rich and varied natural vegetation (*şemsü'l-evrak pek çok nebatat tenebbüt etmektedir*).[14] In the

13 PMO. HH.THR 524-129.

14 PMO. HH.THR 525-21.

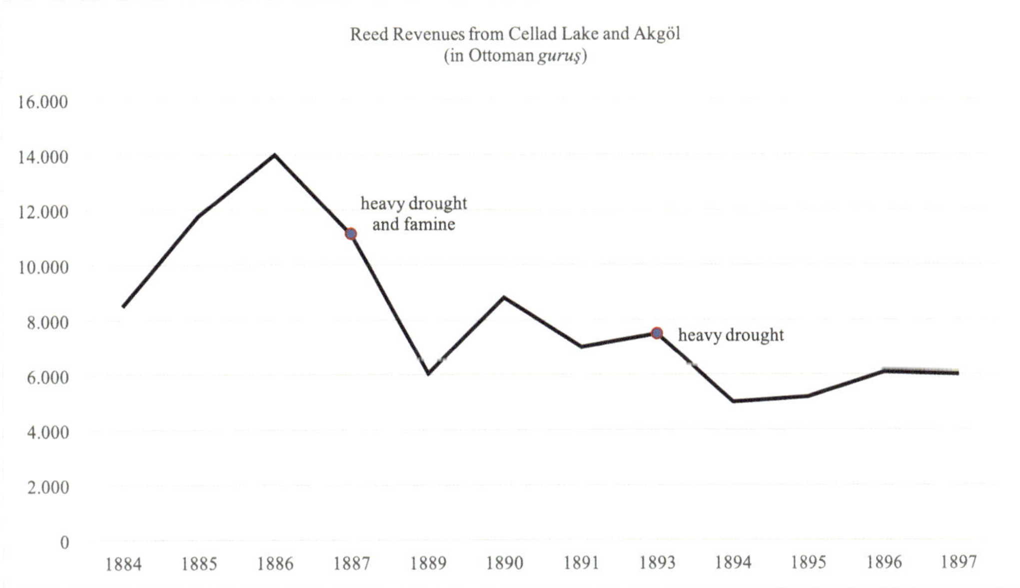

Fig. 3. Graph of reed revenues in Ottoman *guruş* from Cellad Lake and Akgöl, compiled from the 1884–1897 financial registers of the Tepeköy *çiftlik*. Source: PMO. HH.THR files.

western wetlands of the Cellad and Ibrahim Lakes, a specific kind of reed (*havut sazı*) was used for camel saddles. Common reeds grew in the wetlands northwest of Akgöl, and in areas between it and the Meşhed *çiftlık* grew a different rare reed species.[15]

The balance of these revenue-generating wetlands coupled with sustainable management and harvesting practices were implicit: local actors were keenly aware of these fragile micro-ecologies. As economic demand increased in the late nineteenth century, however, reedbeds were over-exploited. A series of revenue registers between 1884 and 1889 reveal the stress of the local reed economy and indicate its decline (see Fig. 3). Reed revenues from the Akgöl and Cellad Lakes fell by more than half between 1886 and 1897. Importantly, overexploitation was only one of several significant changes. Reclamation programs for agricultural production shifted the relative water table, and the expansion of infrastructure for irrigation canals and diversion channels further altered wetland zones. Climate shifts, too, exacerbated the problem. Following the especially dry winter of 1886, wetlands shrank, and it only rained abundantly in May that year.[16] In August, the reeds all "burned before the harvest."[17] Responses to the regional droughts between 1887 and 1889 transformed local economies. Agricultural production recovered in large part due to enhanced reclamation policies, yet such investments further damaged wetlands, as did the 1890 drought. Even with the winter

15 PMO. HH.THR 524-129.
16 PMO. HH.THR 524-17.
17 PMO. HR.THR 524-29.

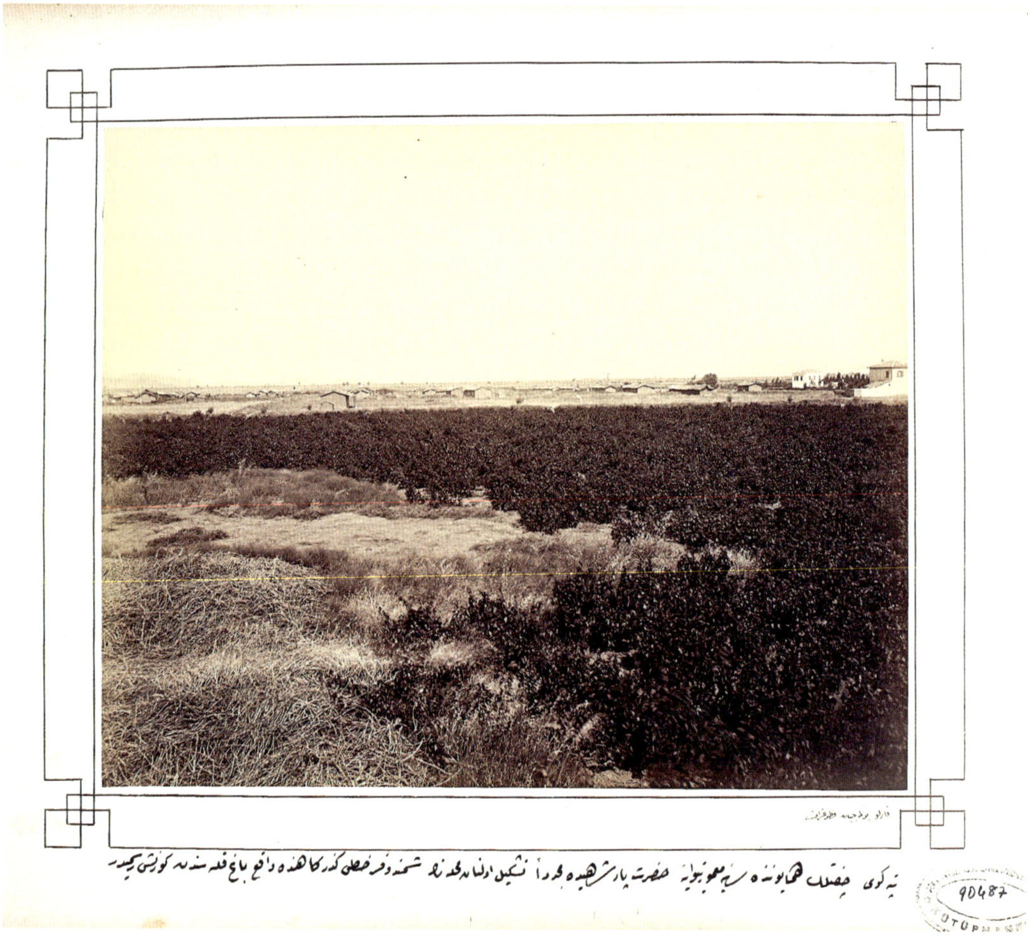

Fig. 4. The changing flora in Tepeköy in 1895, with a possible vineyard in the foreground. Photographer: Carlo Bükmeciyan. Source: Istanbul University Rare Books Online Collection, NEKYA90487/42.

rains of 1891, reedbeds did not mature, as the new infrastructure successfully drained the water away. The droughts in 1893 and 1898 further crippled the reedbeds. By 1903, reed revenues from the Tepeköy *çiftlik* amounted to 4,348 *guruş*, the lowest return in 17 years.[18]

Drought at the end of the 1880s is also blamed for drastic increases in malaria cases and other illnesses among the villagers of Tepeköy *çiftlik*. The area around Tepeköy had always been prone to malarial outbreaks. A decade before the confiscation of the Baltazzi property, the Tepeköy villages were hit by a type of malaria that was more deadly compared to other malarial zones in Aydın province.[19] At the end of the 1880s, in particular, very high summer temperatures exceeding 40°C, the lack of potable

18 PMO. HH. THR 550-16.
19 PMO. A.MKT.MHM 421-96; A.MKT.MHM 423-33.

water, undernutrition, and the lack of health facilities in the *çiftlik* resulted in extensive malarial deaths. According to the October 1889 report of the *çiftlik* physician, Dr. Hüsnü, hygienic conditions were very poor, and when the nearby lakes became desiccated, malaria as well as typhus were inevitable.[20]

As lakes, ponds, reedbeds, and marshes deteriorated, a new cultivated and managed landscape emerged. In the late 1880s an irrigation system of iron and lead pipes watered newly planted trees and crops. A series of photos of the Tepeköy *çiftlik* in 1895 taken by the Armenian photographer Carlo Bükmeciyan illustrates the effects of the drought and the changing of flora (e.g., Fig. 4). Formal policies endorsed planting deep-root, water-loving trees to reclaim lingering wetlands. This included invasive species such as Eucalyptus trees imported from Australia, chosen for the explicit purpose of helping to drain railway construction areas; they thrived in the fertile, moist winter soils and the dry, hot summers. The program was successful in drastically lowering the relative water table (İnal 2019; Özgün 2013; Gratien 2017, 602). In 1895 efforts to plant tree species with similar water requirements moved forward, including groves of willows and figs near the Akgöl wetlands of the Meşhed *çiftlik*.[21]

Over the next two decades the results of these interventions would become clear: the regional lakes and wetlands of Aydın became desiccated. In 1889 the surface area of Akgöl covered approximately 6,000 *dönüms*; in 1900, a particularly wet year, lake depth varied between 1.9 and 3 m.[22] By the summer of 1909 the surface area of Akgöl hovered at 1,800 *dönüms*, and its depth reached only 1.5 m.[23] A similar pattern emerges from the archival record for the Cellad and Ibrahım Lakes. In winter 1889 their surface areas collectively measured approximately 12,000 *dönüms*.[24] By winter 1914 the expanse covered a mere 500 *donums*, the surrounding landscape having been planted with cotton, millet, and sesame.[25] Climate change, agricultural choices, and significant investment to ramp up reclamation efforts near the Tepeköy *çiftliks* in 1909 resulted in a perfect storm that would cripple the natural ecosystems. As wetlands dried and remaining water became more stagnant, outbreaks of malaria increased,[26] in turn fueling top-down policies to rid western Anatolia of such "putrid" wetlands. With diminishing wetlands, so too decreased populations of birds, among the most effective natural forms of pestilence-spreading insect control (Giraud 1940). On recently reclaimed land near the İzmir-Aydın railway, the state sponsored a new school of agriculture; among the intended deliverables was the eradication of malaria (Haykıran 2013, 155–56; Tunalı 2007).[27]

20 PMO. HH.THR 525-21.
21 PMO. HH.THR 57-7.
22 PMO. ŞD 1211-22.
23 PMO. ML.EEM 777-52.
24 PMO. HH.THR 524-129.
25 PMO. ML.EEM 1053-23
26 PMO. HH.THR 524-129.
27 PMO. ML.EEM 776-8.

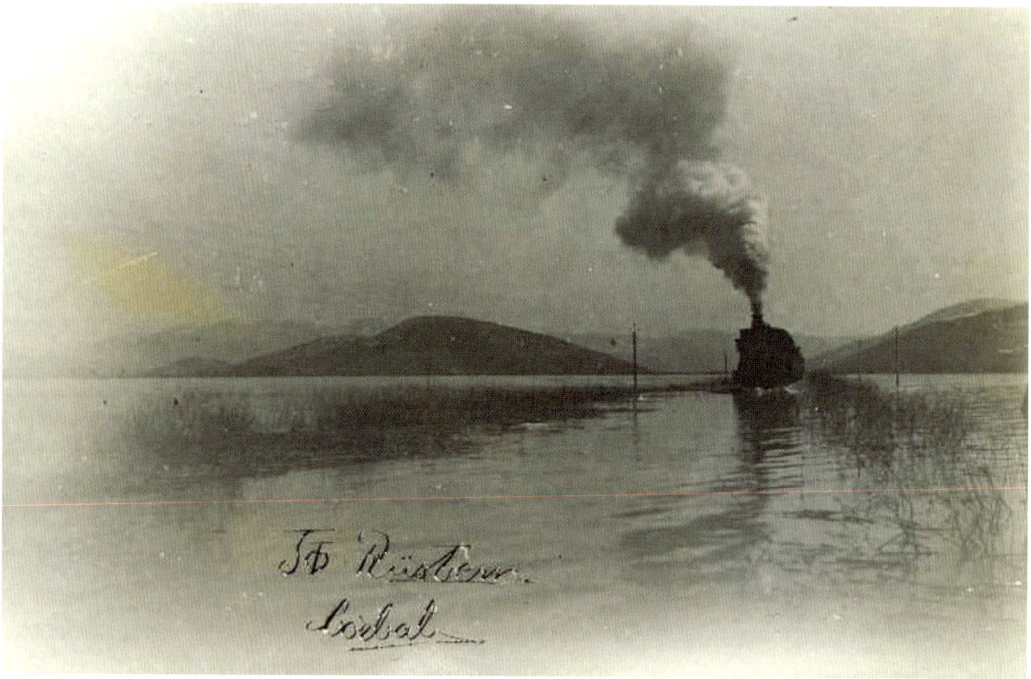

Fig. 5. The Izmir-Aydın passenger train passing through Cellad station by (or within) the Cellad Lake wetlands in the late 1920s. Source: Istanbul Atatürk Library, Postcard Collections, Krt_023690.

The increase of agricultural production substantially increased the value of the Tepeköy *çiftlik*. Small-scale farmers could not compete, especially without the informal and formal wetland economies. They were forced into large-scale, intensive agricultural lifeways. Importantly, these changes did not come without a fight. Local communities defended what they saw as their investments, and in early 1910 villagers of the Tepeköy *çiftlik* protested in response to rumors that the *çiftlik* lands were to be permanently drained and subsequently sold to a German investor. Villagers attempted to block further intervention, fearing that such activities would wreak havoc on the local balance between field agriculture, animal husbandry (especially buffalo), and hopes for future wetland economies. In realizing the extent of environmental transformation, they turned their fight towards protection of the entire wetland ecosystem. Their arguments hinged on their personal *long-term* investments to cultivate and manage this landscape, with "great physical hardship" (*tırnaklarını sökerek*). In fact, as early as 1883, when the Aristidi Baltazzi *çiftlik* was confiscated, villagers had expressed grave concern. They took offence at authorities belittling their efforts and sacrifices in establishing viable wetland economies. Moving forward in this altered landscape and political climate, villagers demanded property rights, refusing to submit to the prior *çiftlik* structures, whether for a foreign private enterprise or the state (cf. Kaya 2021).[28] Even with these interventions, the waters of the Küçük Menderes River continued to inundate the plain, especially during wet winters (Fig. 5). It was only in 1936, after particularly wet and dry cycles, as well as extreme malarial outbreaks, that the draining projects in Tepeköy were fully realized and the region was renamed the "Health Plain" (*Sağlık Ovası*) (Karayaman 2008).

Refugees and Reclamation

Çakıroğlu and Rahmiye çiftliks/*Balkan and Cossack Refugees/Yemişçigöl*

We now turn to the region of Yemişçigöl, 10 km south of Akhisar. During the seventeenth and last quarter of the eighteenth centuries, the wetlands here were along the annual migratory routes of the transhumant Gündeşli tribe, including minor seasonally-used structures.[29] In the first half of the nineteenth century, landowning local notables began to covet this land and built their *çiftlik*s on the high ground along the eastern banks of Yemişçigöl. In addition to their residential complexes, these communities began small-scale infrastructure projects to manage seasonal flooding and to secure the boundaries of the lake itself. From the records, we know that the Çakıroğlu family, local notables from Akhisar, held jurisdiction over the lake and nearby wetlands. The Levantine Baltazzi family had a neighboring *çiftlik*.[30] The Karaosmanloğlu family, whose political and economic power grew in the region in the late eighteenth century, owned the land southeast of the lake. To the southwest, Mehmed Hurşid Paşa, the governor of Aydın province (1875–1876), cultivated rice in the wetlands of the lake as well as in territories extending southwards towards the Gediz River. Construction on the Manisa-Soma railroad, located a few kilometers west, began in the 1880s and was completed in 1890. All these events and people increased the visibility of the region, even attracting the interest of the sultan.

Local actors, from the Gündeşli tribe to the larger *çiftlik*s, had implicitly understood and effectively managed the micro-ecologies of the wetlands for over a century. In 1884 the zones northwest of the lake were put up for auction by the state and sold to the highest bidder, Sultan Abdülhamid II. This area was incorporated into the sultanic *çiftlik*s of Aydın province and became known as the Rahmiye *çiftlik*, named after the sultan's daughter. The area was to be used to resettle refugees from the Balkans. Furthermore, another village was to be established on the northeastern wetlands of the lake. The local branch of the Department of State Property (*emlak-i emiriyye müdüriyeti*) explicitly recognized this dynamic landscape as "a space of pastures, reedbeds, swamps, and water areas, altogether called Yemişçigöl."[31] Since the

28 PMO. ML.EEM 782-22; ML.EEM 805-55.

29 PMO. A.DVNSMHM.d 130-1040.

30 In the late 1850s, the Baltazzi family and the Çakıroğlu would ultimately compete for control of the area. PMO. A.MKT.UM 285-27.

31 PMO. ML.EEM 78-71.

seventeenth century, apparently, this collective definition had been broadly deployed to define lakes with abutting, healthy wetland ecologies. Also in 1884, Çakıroğlu Yanako and his brothers petitioned against the "intrusion" of the newly founded Rahmiye *çiftlik* along the northeastern zones of the lake. Their petition stated that their deed for access and control of Yemişçigöl wetlands included areas seasonally covered with water, pastures, and a reedbed (*saz tabir olunur bir nevi otlar*).[32] Importantly, it is implied that there had been stable wetland zones in the region throughout the seventeenth and eighteenth centuries. Even as the rightful *owner* of the wetland zones, Yanako explained, the Çakıroğlu *çiftlik* did not proactively exploit land from the majority of the liminal wetlands that appeared from under the water seasonally; the exception was one seasonally inundated parcel with extensive reedbeds, from which reeds were collected and sold to the state.

The Rahmiye *çiftlik*, and therefore the sultan's reach, however, began to extend beyond the lands along the northern shore of the lake. The intent may have been to thwart and to eliminate the power of rivals, such as the heirs of Mehmed Hurşid Paşa. The representatives of the Department of State Property argued that the heirs of Mehmed Hurşid Paşa not only unlawfully claimed control over the lake, but they also "modified" the shores of the lake by draining water through a canal (*şarampol*) that they had constructed. As shown in Figure 6, the section for the earthwork and the transect of the canal channeled waters towards rice fields owned by the Mehmed Hurşid Paşa *çiftlik* on the southwestern tip of the lake. Authorities argued that the infrastructure had jeopardized the ecological integrity of the wetlands.[33]

Since the seventeenth century, access to and control of lakes, lakebeds, and seasonal wetlands had apparently been determined by the size of the respective lake and one's proximity to the lakebed (see Çelik, Luke, and Roosevelt, in review). The 1858 Land Code, and specifically Article 123, however, set a new precedent. If full reclamation of "common" wetlands was possible, this new land became the property of the state. During the last quarter of the nineteenth century the Ottoman state used Article 123 to justify state reclamation of seasonal wetlands and in so doing raised a number of controversial issues. In the case of Yemişçigöl, Article 123 was not invoked, much to the frustration of the Çakıroğlu family. Strikingly, the Department of State Property declared that the drainage canal built by the heirs of Mehmed Hurşid Paşa violated the ecological balance of the waterscape. In this case, pseudo-ecological sensitivity became the basis for seizure. Past *inaction* by the state in this region further justified the state's 1884 case. While the state had been aggressive in reclamation of wetlands in many areas of western Anatolia, it had not done so in the nearby vicinity of Yemişçigöl. This precedent implicitly, then, recognized the state's intentions to preserve the ecosystem,

32 This expression hints at the knowledge around the species of reeds; that is, they were not ordinary reeds. PMO. HH.THR 4-66.

33 PMO. ML.EEM 79-61.

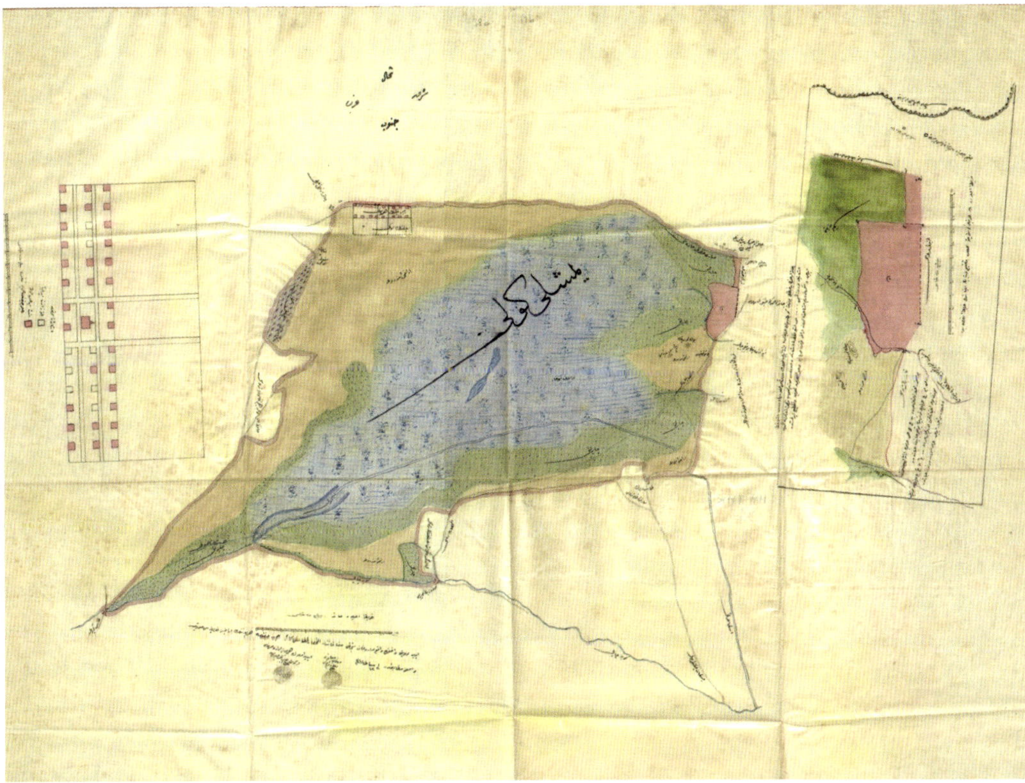

Fig. 6. Map of Yemişçigöl and the surrounding wetlands in 1895. Water and reedbeds are shown in blue, pastures, swamps, and rice fields in green, cultivated land in yellow, and disputed land in red. North at top. Source: PMO. HH.THR 14-48.

notably the lake, reedbeds, and nearby seasonal pastures along the southern and northeastern shorelines (Fig. 6). In fact, in 1887 engineers were hired by the Aydın *çiftlik* administration to enhance the "natural" (ancient) lake basin.[34] Even in 1894, a majority of the Rahmiye *çiftlik* revenues came from reeds that were collected in the fall before the "inundation season" (*tufan mevsimi*).[35] Furthermore, the annual (controlled) flooding of the nearby vineyards protected the vines from phylloxera and other illnesses.[36] In this case, then, we have an example of state seizure on the basis of protecting the larger environment, followed by proactive management of an integrated waterscape.

Following the larger context and scale, however, reveals even more than just the jockeying between provincial and centralized powers and the importance of the environment as a concept; it also provides critical insight about the rights and actions of refugee communities. When the bidding opened for Yemişçigöl and its wetlands

34 PMO. HH.THR 524-94.
35 PMO. HH.THR 14-48.
36 PMO. HH.THR 224-45.

in 1884, a community of Cossacks stepped forward among other anonymous bidders. Cossack communities had a history in western Anatolia, and already by the first half of the nineteenth century one group had settled near Bursa and held concessions for Lake Manyas (Toğrol 1987). Over the decades, Cossacks had become skilled in their dealings with Ottoman authorities, even evading local administrators regarding the profitability of wetland economies. According to a report from the early twentieth century, Cossack communities made fortunes from low-bid concessions awarded for what were in fact extremely profitable areas.[37] Seeing opportunity to continue the same at Yemişçigöl, they entered the bidding in the early months of 1884 with as little as 10 lira; by May bidding for the territory had reached 510 lira. The community representative known in the records as Mustafa Bey requested that the Department of State Property show the "kindness" to grant the concession for the lake and wetlands to this Cossack community for 510 lira on the basis that they were refugees (*muhacirinden bulunduklarına hürmeten*). A competing, as yet anonymous, bid in the archives from Akhisar nearly doubled the bid to 1,000 lira. The sultan's final bid of 1,100 lira secured imperial title.[38]

The intervention of the sultan in obtaining control of this fertile landscape and the revenues from it stirred significant reaction. As the party holding the usufruct of the land and the wetlands for decades, Çakıroğlu Yanako protested that "it was weird" (*garibdir ki*); the sultanic *çiftlik* of Aydın was headquartered in Torbalı, some 30-40 hours in travel time south of Yemişçigöl, and the extensive distance, Yanako argued, exceeded its jurisdictional range. Furthermore, he hinted that he prioritized local control and management of liminal wetlands over their administration from such a distance. In spite of the evidence from newspapers and related correspondence, including the testimony of a dozen local experts (*ehl-i hibre*) that the seasonal wetlands and the fertile reedbed belonged to the Çakıroğlu *çiftlik*, the state did not step back.[39] Allegedly, to stain the image of the Çakıroğlu family further, rumors were spread that Yanako bribed the superintendent of tax affairs in Akhisar by "awarding" him with a hundred sheep and a few *dönüms* of vineyard in exchange for secretly selling the tax-farm of the lands to his men for cheaper prices. A series of court cases found Yanako innocent.[40] Regardless of his efforts and favorable court rulings, he lost the claim and his land.

The sultan's commitment and investment to this wetland environment, however, would not last. The late nineteenth century climatic shifts, as well as the interventions of canals and embankments, slowly eroded the integrity of the wetlands. The resettlement of Balkan refugees included communities much more familiar and comfortable with agriculture (unlike the Cossack communities), and thus the Rahmiye *çiftlik* allowed for

37 PMO. DH.MKT 766-1.

38 PMO. ML.EEM 79-61; PMO. ML.EEM 78-71.

39 PMO. HH.THR 4-66.

40 PMO. HH.THR 14-48.

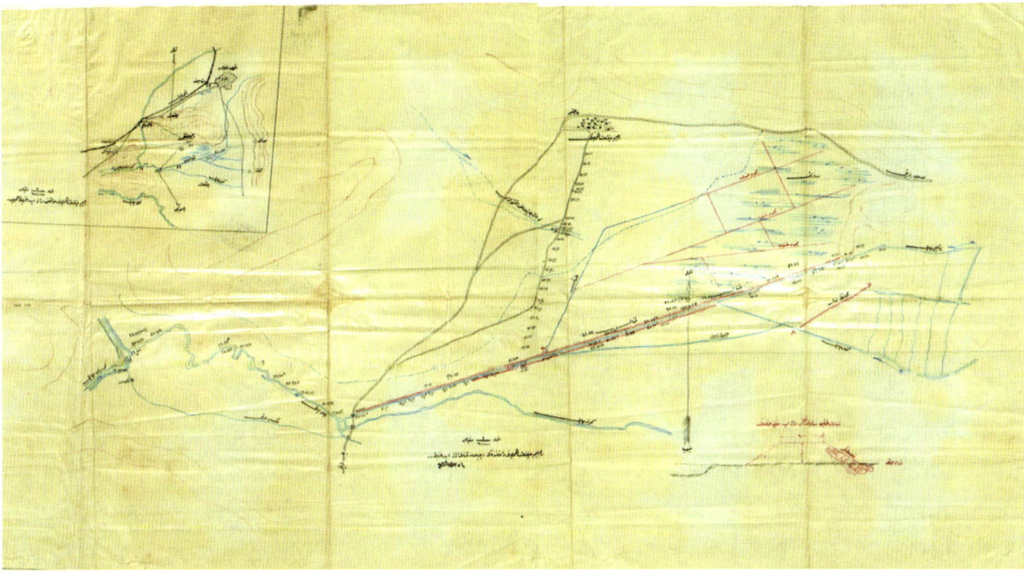

Fig. 7. Yemişçigöl desiccation map from 1906, with the lake now indicated as *sazlık* (reedbed) with clusters of parallel blue lines. The Hurşid Paşa canal is indicated in red hachures alongside a section drawing of the new canal, also in red. North at top. Source: PMO. HH.THR 321-24.

some reclamation and cultivation.[41] When coupled with the climate fluctuations of the 1880s, even these limited interventions shifted the rich flora and fauna around Yemişçigöl. To put this into perspective, archival records indicate that prior to 1884 this region included at least 8,000 *dönüms* of wetlands.[42] This diminished over the next decade. A major turning point was the 1893 drought and another influx of refugees, this time from Plovdiv. Permission was granted to drain the lake, swamps, and pastures to accommodate the new settlement.[43] By the turn of the twentieth century, the face of the landscape had been modified so intensively that the name itself—Yemişçigöl—was no longer used.

On a 1906 map, a small region of the area was designated "reedbed" (*sazlık*) and slated for desiccation under a reclamation project (Fig. 7). In an ironic twist, the Hurşid Paşa canal that had been the justification for state seizure by the sultan was reinvigorated, and plans for expansion were implemented. Modernization projects built after 1903 included new infrastructure with more canals to drain the remaining Yemişçigöl wetlands. These efforts were successful at draining the lake, yet seasonal wetlands remained (*koğalık yeri*). In 1906, the Ottoman state implemented a large-scale program to drain the entire area. As part of the first phase, villagers were asked to cut all existing reeds, and in return, they gained approximately 700–800 *dönüms* of cultivable land. The major climate changes made possible these drastic transformations

41 PMO. HH.THR 321-24.
42 PMO. ML.EEM 78-71.
43 PMO. HH.İ 94-57.

to the landscape. Despite the policy and infrastructure to desiccate Yemişçigöl and its wetlands fully, they survived until 1972, when a final project, which used canals and dikes built by the Ottomans (including the Hurşid Paşa canal), successfully drained the lake and the wetlands.

Establishing the Environment

Halime Hatun waqf/*Gölmarmara/Lake Marmara*

In a final case study, we turn to the area of Lake Marmara, located 10 km south of Gölmarmara. In comparison to Cellad Lake and Yemişçigöl, Lake Marmara was a comparatively large lake that had provided livelihood for the surrounding region for millennia.[44] The Halime Hatun *waqf* oversaw the thriving early modern wetland economies (see Çelik, Luke, and Roosevelt, in review). Halime Hatun had been the wet nurse of Mehmed III, and she established a *waqf* in her hometown, Gölmarmara, in 1602. During the *waqf*'s tenure, the administration oversaw the cycles of agriculture, animal husbandry, reed exploitation, and fishing. By enforcing strategic management policies, the *waqf* sustained the fragile ecological balance of the waterscape throughout the climatic fluctuations of the late seventeenth and eighteenth centuries. By the early nineteenth century, however, competition took a fierce turn, with local actors jockeying for greater access to revenues and land (Çelik, Luke, and Roosevelt, in review).

In 1826 the Halime Hatun *waqf* lost control over the region as the imperial *waqf*s came under the regulation of a new ministry. With the new structure, the ministry of imperial *waqf*s received a percentage from the revenues of each *waqf*. A series of mid- to late nineteenth century events also triggered changes for Lake Marmara; revenues from the wetland economy began to be parsed to private local entrepreneurs, including concessions to Cossack communities and major Ottoman families, such as the Karaosmanoğlu (Nagata 1995; Kaya 2019). Proximity to the new Smyrna/Izmir-Alaşehir railway, as well as the fertile lands of the middle Gediz Valley, also increased interest in the Marmara Lake Basin.

According to early traveler John Durbin, the shallow waters of Lake Marmara hosted "fields of tall, beautiful reeds, of which great quantities of matting are made, and distributed through the country" (Durbin 1847, 153). Robert Walsh and Thomas Allom (1838, 93) also commented on this waterscape,

> "The lake as it now exists, is of considerable extent; the rich mould on its banks, of a muddy consistence [sic], exuberant in reeds, and abundance of such aquatic and pluralistic plants as love such a soil. The water in colour and

44 Unlike the other regions, the Marmara Lake Basin has been the subject of extensive and intensive diachronic archaeological survey and excavation. See Roosevelt and Luke 2017; Luke et al. 2015; Roosevelt et al. 2018. To the best of our knowledge, full coverage archaeological surveys have yet to be completed in the areas around Cellad Lake (Metropolis), as well as the region around Yemişçigöl.

transparency, resembles that of a common pond, and seems alive with fish. Another circumstance marks it—flocks of swans and cygnets hover above the surface, and flights of various aquatic birds darken the air."

These narratives from the first half of the nineteenth century stand in stark comparison to the situation of the late nineteenth century. As with our other case studies, modernization practices coupled with major climatic fluctuations had direct consequences on the Lake Marmara wetlands. During the drought of 1862 lake waters retracted so significantly that they compromised seasonal fishing. Having just invested in the region, tax farmer Yorgi Kostantinidi could pay only ten percent of his first installment of fish taxes owed to the state (6,000 of 63,726 *guruş*); unable to pay more, and without leniency, he declared bankruptcy.[45] Going forward, only investors with extensive reserves were able to take on the risk associated with not only climatic changes but also the financial structure of concessions.

Increasingly, it was competitive economies that took a toll on already struggling wetlands. To mitigate their losses, those willing and able to take on the risk over-exploited natural resources. The Halime Hatun *waqf* had put into place and then enforced regulations regarding fishing techniques, seasonal access, and limits on the size and number of fish that could be harvested, as well as reeds. This provided sufficient checks and balances that sustained the ecosystem. As the *waqf*'s power waned in the nineteenth century and privatization increased, these regulations and policies were ignored; fishermen increasingly used previously forbidden harpoons as well as nets with smaller mesh.[46] The state also moved to capitalize on profits: annual installments for tax-farming increased to 212,500 *guruş* between 1882 and 1886. Financial stress was compounded during the regional drought between 1887 and 1889. During this period waters retracted significantly, even with seasonal rains and ever-flowing springs that supplemented lake waters. After this point, investors were unwilling (and unable in some cases) to stomach future risks. At the end of the four-year tax farming period, no new bidders turned out in 1887. Reasons moved beyond merely financial aspects. Investors lamented the deteriorating condition of the lake basin and surrounding wetlands. This lack of investment required local government mitigation both to manage the fishing and to collect the fish tax of 1887, a mere 46,580 *guruş*, significantly less than the annual 212,000+ *guruş* that had been levied.[47]

In 1888 a new local actor stepped forward: the Ottoman Public Debt Administration (OPDA; *Düyun-u Umumiye İdaresi*), which had been established in 1881 to secure the repayment of loans made to the Ottoman administration by various international bodies and European states. Also in 1888, a loan was taken from the Deutsche Bank,

45 PMO. ŞD. 398-57.
46 PMO. DH.MKT 766-1.
47 PMO. ŞD. 1390-4.

and thus the OPDA's control over taxes were extended to include the hunting taxes (of all sorts) and fishing from both the sea and lakes.[48] This gave the OPDA entry into the management of interior lakes and wetlands (Birdal 2010, 124). The poor condition of the Marmara Lake Basin in this period meant that it was not economically viable. Payment of the increased taxes became problematic, and things got worse. Extreme regional and local droughts in 1890 and 1893 were nearly catastrophic. Fish taxes collected by the OPDA between 1888 and the first three months of 1893 totaled only 95,863 *guruş*, a fraction of what was anticipated.[49]

The 1893 drought put in motion massive transformations for the Marmara Lake Basin. Farmers moved quickly to claim more than 7,425 *dönüms* of agricultural land along its northwestern shorelines. Their actions were endorsed by the OPDA, and in turn the sultan, with officials invoking Article 123 of the 1858 Land Code.[50] Their perspective was that management of this fertile landscape (as opposed to the waterscape) and the corresponding profits from agriculture would be far more reliable than from "unreliable" wetlands. What is more, by establishing precedent in the basin, efforts moved forward to drain and reclaim additional wetlands abutting the Gediz River basin along the eastern lakeshore and, notably, the Bağirgan *çiftlik*, owned by a member of the Karaosmanoğlu family. Authorities moved quickly. As waters evaporated and land was exposed, officials from the local branch of the imperial registry arrived to register eight parcels of land officially, which were then made available for sale or lease. The heirs of Halim Ağa, a local landowner, were among those with successful bids.

While reclamation schemes, state intervention, and economic interests in the Marmara Lake Basin follow patterns that are similar to what happened with Cellad Lake and Yemişçigöl, the final outcome in the Gölmarmara region was starkly different. The Izmir branch of the OPDA reacted very strongly against further encroachment by the Ottoman state. They placed priority on conservation of the basin's ecosystem, including not only the lake itself but also abutting ponds and wetlands. In fact, the OPDA took the case to court. At issue was interpretation of Article 123. The argument was that it could only be invoked if a lake and wetlands dried *completely*. Demonstrating the annual ebbs and flows of inundation and the *longue durée* of the lake and wetlands, the OPDA claimed the following:

> "Marmara(cık) lake is not like those lake-places where the lake totally vanishes. [On the contrary], at the moment the lake holds water that is about 12 hours to walk around and is of utmost importance and quality for the treasury."[51]

48 The sultanic *çiftliks* were exempted from the control of the OPDA, which can be seen as another impetus for the control of wetlands and lakes by the administration of sultanic *çiftliks*.

49 PMO. ŞD. 1390-4.

50 PMO. ŞD. 1390-4.

51 PMO. ŞD. 1390-4.

The OPDA argued that the ecological integrity of the lake and wetlands had deteriorated not only from climatic changes, but also, and significantly, from anthropogenic factors. Overfishing and new (forbidden) fishing technologies (nets, harpoons, etc.) had disastrous effects on local ecologies. In turn, the water quality had also deteriorated. Both were exacerbated by mismanagement of baseline infrastructure; canals that fed the lake and small ponds silted up with great impact on the diversity of local flora and fauna. The case demonstrated the damage caused by state seizures and villagers consistently evading established regulations that had been in place for at least three centuries. Illicit landgrabs and unregulated grazing along the shorelines had further harmed the ecological balance. In sum, the court case sought to demonstrate that the blatant disregard for the ecological balance collided with unusually warm conditions of the second half of the nineteenth century (*tekessür-ü hararet'ü-ş şems*)—which the State and local actors were leveraging to realize their economic and political interests throughout western Anatolia.

By placing the natural ecology at the center of their argument, the OPDA reformulated the narrative on wetland ecologies as naturally sustainable. They argued that, if well-managed, these zones were productive and impressive landscapes in their own right, not putrid, unproductive, malarial swamps only viable as future agricultural zones. Accordingly, seasonal desiccation was not sufficient evidence to justify private sale or lease, nor could opportunistic, seasonal "invasion" be permitted. Rather, shorelines were to be considered an inherent part of the wider lake basin (*gölün tetimmat ve müştemılatından bulunduğu*). The OPDA narrative invoked the early-modern holistic idea of wetland management. They fought against private sale and, in so doing, imposed formal conservation strategies. Prior quotas for fish size and numbers were strictly enforced. To restock Lake Marmara, spawns were brought from Europe, and the primary canal between a large natural spring north of the lake (Akpınar) that watered nearby fields and even small ponds, eventually emptying into the lake, was dredged and maintained. All in all, given its past condition and the possibilities for improvement, the lake was to be brought towards an ideal, managed state. The final response of Ottoman officials in Smyrna/Izmir to the OPDA decision was issued in 1897 and was apparently rather low profile. Ministers in Istanbul seem to have agreed without major issue that Article 123 could not be applied to Lake Marmara and contested neither the Smyrna office report nor the need to enforce established policies. This (re-)interpretation of Article 123 also set more stringent parameters for the protection of seasonal lakes and wetlands that retracted annually during the summer and early fall.

Yet, when drought struck again in 1898 large swathes of land were again exposed during the dry season, some 5,000 *dönüms* along the southern shorelines. The OPDA struggled to balance the ecological fluctuations and the fluid policies that had allowed *seasonally* exposed tracks of lakebed to be leased to local villagers for field agriculture for a fixed price.[52] In 1898 local villagers and external actors requested an open bidding process for *permanent* sale of these lands, recognizing their seasonal volatility (and thus significant investment risk). Ottoman administrators chose to negotiate directly with local and external stakeholders focused on the ecological integrity of the lake and wetlands, as well as the well-being of proximate communities. Assessment of lands for bidding was determined by a committee appointed by the Council of State (*Şura-yı Devlet*): two representatives from the local council, one representative from the office of land registry, the official engineer of the Saruhan district, and a representative of the OPDA. The details of these negotiations are not yet available in the archives. Given the identity of the negotiators, however, we suggest that negotiations crossed different scales, scrutinizing not only the economic conditions, but also the ecological integrity of the wetlands, technical/scientific obstacles and opportunities, and local interest. We do know the outcome of the 1898 situation: the lands were sold to the wife of the scribe of the Islamic Court of Appeals (*Meclis-i Tedkikat-ı Şer'iye*), Üryanizade Cemil Bey.[53]

Discussion

The decline in vibrant and sustainable wetlands had a significant impact on the natural cycles as well as the political systems and social relationships that had governed (albeit loosely) these zones for centuries. Precisely because these changes came gradually, there was not necessarily a single rupture point. In part, this was due to the ecological conditions of wetlands and the limitations of technical innovations. While redistribution of water and sediments was possible, it was often impossible to thwart annual inundations completely. At Cellad Lake and Akgöl, as well as Yemişcigöl, it took over a century, and even today the areas remain extremely wet; for Lake Marmara, reclamation efforts were abandoned. Indeed, the resilience of wetlands, even after periodic cycles of drought, reminds us of the fluid aspect of these fragile zones. What is more, drainage programs or construction (especially of infrastructure) often created peripheral and even more expansive wetlands. In fact, attempts to increase agricultural productivity by grafting onto the face of the landscape canals and diversion channels often *increased* hydrological instability. In this dynamic of human-induced landscape change, aquatic and terrestrial biota also shifted, reconfiguring and fundamentally altering the complex and rich biodiversity of the wetlands as well as the array of economic interests and political actors. Throughout this process, economies dominated by terrestrial, cultivated vegetation were increasingly controlled by state-supported actors. In turn, wetland zones became increasingly shallow and sick: without annual flooding cycles and the maintenance of canals, the process of cleansing and removing sediment led to stagnant zones; the once rich diversity of flora and fauna began to decline. Unsurprisingly, these zones became hotspots for mosquitos and thus malaria.

52 PMO. BEO 108781.
53 PMO. ŞD 57-11.

Rather than seek environmental balance for wetlands and a more natural approach to mitigating malaria, Ottoman authorities capitalized on the threat posed from malaria to pursue their own agendas for agricultural expansion and, in turn, greater control over the hinterlands of Izmir.

These case studies speak to the importance of detailed archival research for understanding local historiographies of climate change for understanding the "idea" of the environment. Here we have argued that western Anatolia adds to the growing body of Ottoman scholarship focused on how to conceptualize (and thus deploy) the idea of the environment and the need to balance data from local and regional scales, as argued by Deborah Coen (2018, 3 7). Our work demonstrates that there was not one Ottoman policy of wetland management, nor one concept of the environment. In fact, the sultan justified exploitation in large part to consolidate the Imperial power structure, which was showing severe signs of rupture by the late nineteenth century. We have illustrated that negotiations reached across multiple scales and in turn influenced how specific actors came to understand natural history and the importance of integrating policy and human manipulation of ecosystems. Seemingly small in spatial scale, these lakes and wetlands were connected to a much larger world of villagers, merchants, bureaucrats, administrators, entrepreneurs, and even imperial and trans-imperial institutions. Furthermore, when taken as parts of a puzzle, these wetlands were a significant part of western Anatolia. While each was unique, their micro-ecologies also reflect significant macro-climatic traits that in turn balanced patterns of migratory birds and animals, including transhumant lifeways.

At Tepeköy and Yemişçigöl human intervention altered the landscape and waterscape dramatically. In the case of the Marmara Lake Basin, competition among actors and challenges to Article 123 of the 1858 Land Code protected the lake and the wetland ecology from destruction. Knowledge of the natural history of the lake and the ecosystem and a past *waqf* structure committed to sustainable management both contributed to this outcome. In sum, the quest to control wetlands as well as move towards capitalist modes of production in western Anatolia pivoted on competition and conflict among various actors with different interests rather than on a monolithic group of entrepreneurs solely in pursuit of profits through agricultural production and the destruction of wetlands. It is the mosaic of actors and their varied interests and reactions that add value to understanding the scale of human agency, the historiographies of wetlands, and the climate changes of the late nineteenth and early twentieth centuries in western Anatolia.

Acknowledgments

Semih Çelik acknowledges the financial support provided by the Koç University Research Center for Anatolian Civilizations (ANAMED) during the 2019–2020 academic year for his research on the Lake Marmara wetlands. We thank Gygaia Projects for continued support of this research.

Works Cited

Aşkın, Anıl. 2017. "The Environment, Institutions, and Economy in the Nineteenth-Century Ottoman Empire: The Imperial Landed Estates of Mihaliç (1840-1850)." MA thesis, Boğaziçi University.

Birdal, Murat. 2010. *The Political Economy of Ottoman Public Debt: Insolvency and European Financial Control in the Late Nineteenth Century*. London: Bloomsbury Academic.

Christensen, Peter H. 2017. *Germany and the Ottoman Railways: Art, Empire, and Infrastructure*. New Haven: Yale University Press.

Coen, Deborah R. 2018. *Climate in Motion: Science, Empire, and the Problem of Scale*. Chicago: University of Chicago Press.

Cole, Camille L. 2016. "Precarious Empires: A Social and Environmental History of Steam Navigation on the Tigris." *Journal of Social History* 50, no. 1: 74-101.

Çelik, Semih. 2019. "Coping with Scarcity and Famines in Ottoman Anatolia c. 1650-1850." In *An Economic History of Famine Resilience*, edited by Jessica Dijkman and Bas van Leeuwen, 52-73. Oxfordshire: Routledge.

———. 2017. "Scarcity and Misery at the Time of 'Abundance beyond Imagination': Climate Change, Famines and Empire Building in Ottoman Anatolia (c. 1800-1850)." PhD diss., European University Institute.

Çelik, Semih, Christina Luke, and Christopher H. Roosevelt. In review. "Ottoman Lakes and Fluid Landscapes: Environing, Wetlands, and Conservation in the Marmara Lake Basin." *Environment and History* (in review).

Durbin, John P. 1847. *Observations in the East, Briefly in Egypt, Palestine, and Asia Minor*. Vol 2. New York: Harper&Brothers.

Derviş Paşa and Mehmed Hurşid Paşa. 1997. *Seyahatname-i Hudud*. Transliterated by Alaattin Eser. Istanbul: Simurg.

Efe, Ayla. 2016. "XIX. Yüzyılda Amik Gölü ve Dalyanları." İnsan *ve Toplum Bilimleri Araştırmaları Dergisi* 5: 933-54.

Fındıkoğlu, Fahri. 1962. "Türkiye'den Rusya ve Amerika'ya göç eden İslav muhacirleri." *Sosyoloji Konferansları* 4: 56-93.

Frangakis-Syrett, Elena. 1988. "Trade Between Ottoman Empire and Western Europe: The Case of Izmir in the Eighteenth Century." *New Perspectives on Turkey* 2: 1-18.

Frangakis-Syrett, Elena. 2009. "Banking in Izmir in the Early Twentieth Century." *Mediterranean Historical Review* 24, no. 2: 115-32.

Gamal-Eldin, Mohamed. 2019. "Cesspools, Mosquitoes and Fever. An Environmental History of Malaria Prevention in Ismailia and Port Said, 1869-1910." In *Seeds of Power: Explorations in Ottoman Environmental History*, edited by Onur İnal and Yavuz Köse, 184-207. Winwick: The White Horse Press.

Giraud, E. H. *Days Offs: With Gun, Rod, and Yacht around Izmir, Turkey*. London: William Clowes and Sons.

Goffman, Daniel, Edhem Eldem, and Bruce Masters. 1999. *The Ottoman City Between East and West. Aleppo, İzmir, and Istanbul*. Cambridge: Cambridge University Press.

Gratien, Chris. 2017. "The Ottoman Quagmire: Malaria, Swamps, and Settlement in the Late Ottoman Mediterranean." *International Journal of Middle Eastern Studies* 49: 583-604

Haykıran, Aysun Sarıbey. 2013. "Aydın Vilayeti'nde Çiftlikler (1839-1918)." PhD diss., Ege University.

Horden, Peregrine, and Nicholas Purcell. 2000. *The Corrupting Sea. A Study of Mediterranean History*. Oxford: Blackwell.

Husain, Faisal. 2014. "In the Bellies of the Marshes: Water and Power in the Countryside of Ottoman Baghdad." *Environmental History* 19: 638-64.

———. 2016. "Changes in the Euphrates River: Ecology and Politics in a Rural Ottoman Periphery, 1687-1702." *Journal of Interdisciplinary History* 47, no. 1: 1-25.

İnal, Onur. 2019. "Fruits of Empire: Figs, Raisins and Transformation of Western Anatolia in the Late 19th Century." *Environment and History* 25: 549-74.

İnalcık, Halil. 1994. "The Ottoman State: Economy and Society, 1300-1600." In *An Economic and Social History of the Ottoman Empire (1300-1914)*, edited by Halil İnalcık and Donald Quataert, 9-410. Cambridge: Cambridge University Press.

Karayaman, Mehmet. 2008. "Cellat Gölü'nden Sağlık Ovası'na." *Atatürk Araştırma Merkezi Dergisi* 24: 693-708.

Kaya, Alp Yücel. 2021. "Balkanlar ve Batı Anadolu'da İlk Birikimin Gelişim (1839-1914)." *Devrimci Marksizm* 45-46: 11-66.

———. 2019. "Les racines agraires d'un entrepreneuriat capitaliste: Les domaines fonciers de la famille Baltazzi à l'arrière-pays d'Izmir au xixe siècle." *Rives Mediterraneennes*, 59: 119-33.

Kokdaş, İrfan. 2018. "Manisa'da Çiftlik, Ağıl ve Mandıralardaki Tarımsal İşgücü Profili: Emek Hareketliliği ve Sosyo-Hukuki Dönüşümler, 1690-1870." *Meltem: İzmir Akdeniz Akademi Dergisi* 3: 42-55.

Kuru, Mehmet. 2017. "Locating an Ottoman Port-City in the Early Modern Mediterranean: Izmir 1580-1780." PhD diss., University of Toronto.

Küçükkalay, Mesud. 2008. "Imports to Smyrna between 1794 and 1802: New Statistics from the Ottoman Sources." *Journal of the Economic and Social History of the Orient* 51, no. 3: 487-512.

Luke, Christina, Christopher H. Roosevelt., Peter J Cobb, and Çiler Çilingiroğlu. 2015. "Composing Communities: Chalcolithic through Iron Age Survey Ceramics in the Marmara Lake Basin, Western Turkey." *Journal of Field Archaeology* 40, no. 4: 428-49.

McPhillips, Stephen, Astrid Meier, and Annika Rabo. 2019. "Ottoman Waterscapes: New Archaeological, Anthropological and Historical Studies of Hydraulic Landscapes in the Eastern Mediterranean, 1500 to the Present Day." *The Journal of the Council for British Research in the Levant* 2: 128–30.

Mitchell, Timothy. 2002. *Rule of Experts: Egypt, Techno-Politics, Modernity*. Oakland: University of California Press.

Nagata, Yuzo. 1995. *Studies on Social and Economic History of the Ottoman Empire*. Izmir: Akademi Kitabevi.

Özgün, Cihan. 2013. "Osmanlı Ağaç Kültüründe Yeni ve Egzotik bir tür: Ökaliptüs." Çağdaş *Türkiye Tarihi Araştırmaları Dergisi* 13, no. 26: 5–29.

Pehlivan, Zozan. 2020. "Niño and the Nomads: Global Climate, Local Environment, and the Crisis of Pastoralism in Late Ottoman Kurdistan." *Journal of the Economic and Social History of the Orient* 63, no. 3: 316–56.

Pianciola, Niccolo. 2019. "Cossacks and Sturgeons: Fisheries, Colonization, and Science around the Aral Sea (1873–1906)." *Journal of the Economic and Social History of the Orient* 62: 626–73.

Roosevelt, Christopher H., and Christina Luke. 2017. "The Story of a Forgotten Kingdom? Survey Archaeology and the Historical Geography of Central Western Anatolia in the Second Millennium BC." *European Journal of Archaeology* 20, no. 1: 120–47.

Roosevelt, Christopher H., Christina Luke, Sinan Ünlüsoy, Canan Çakırcılar, John M. Marston, Caitlin R. O'Grady, and Francesca Slim. 2018. "Exploring Space, Economy, and Interregional Interaction at a Second-Millennium BCE Citadel in Central Western Anatolia: 2014–2017 Research at Kaymakçı." *American Journal of Archaeology* 122, no. 4: 645–88.

Shopov, Aleksandar. 2019. "Cities of Rice: Risiculture and Environmental Change in the Early Modern Ottoman Balkans." *The Journal of the Council for British Research in the Levant* 2: 169–83.

Tabak, Faruk. 2008. *The Waning of the Mediterranean 1550–1870. A Geohistorical Approach*. Baltimore: The Johns Hopkins University Press.

Toğrol, Beğlan. 1987. "The Kuban Cossacks of Manyas Turkey and "The Cossacks" of Leo Tolstoy." *Psikoloji* Çalışmaları 15: 125–46.

Tunalı, Ayten Can. 2007. "Aydın Vilayeti Umumi Meclisi'nin 1909 Yılı Kararları." *Tarih Araştırmaları Dergisi* 42: 127–40.

Walsh, Robert, and Thomas Allom. 1838. *Constantinople and the Scenery of the Seven Churches of Asia Minor Illustrated*. 2 vols. London: Fisher, Son&Co.

Yücesoy, Özlem. 2018. "19. Yüzyıl Ortalarında Aydın Vilayeti'nde Çiftlikler ve Ortak Alanlar." MA thesis, Ege University.

EPILOGUE

Whither the Wind Blows: Environmental History and Archaeology in Anatolia

John Haldon
Christopher H. Roosevelt

Overview

The chapters in this volume reflect a range of important research trends in the fields of environmental and climate history, archaeology, and socio-political history in general, as well as the significance of Anatolia, in particular, as a field for interdisciplinary study. Because Anatolia supports a substantial microregional differentiation in climate, land use, and demographic history, exploration of relationships between historically attested extreme weather events, longer-term climate change, and socio-economic organization there presents a series of specific opportunities and challenges. The absence of continuous written records for many periods renders the uncertainties accompanying the integration of historical, archaeological, and climatic data particularly challenging, yet periods in which all three types of data overlap offer exceptionally favorable opportunities. Thanks to the introduction of new approaches and new sources, the study of environment and society in Anatolia promises to expand and deepen our understanding of how societies functioned, particularly under environmental stresses, and provides valuable examples for the study of other regions of the globe. The interaction of climate scientists and paleoenvironmentalists with historians and archaeologists promotes much better understandings of the methodological problems that specialists encounter, as well as more coherent, synthetically persuasive, and integrated interpretations of the various datasets that they deliver to each other.

The chapters of this volume exemplify these points very clearly. Rich case studies, ranging from prehistory through the Late Ottoman period, interrogate the linkages between different types of socio-environmental interaction from different angles, using diverse datasets, and at all scales, from decadal to centennial and millennial. They highlight socio-environmental asymmetries in which different degrees of socio-political complexity and population density preconditioned the potentials for inherent resilience under climate stress. They challenge simplistic accounts of societal transformation and, in particular, they underline the need to think about change and transformation as multi-causal phenomena, in which climate and environment could undoubtedly play roles, but in which those roles varied in their impacts according to both broader, global conjunctures, as well as the specifics of local societies, landscapes, and microclimatic phenomena. From this work, we understand that aridity does not necessarily result in negative societal change, while beneficial climate conditions may be favorable in the short term but at the same time support ultimately unsustainable economic pathways in the long term. Climate and the environment were major factors in the ways in which socio-economic systems evolved, but only very rarely were they sufficient to explain change by themselves. In particular, it has become clear that climate *impacts* cannot be determined by climate *conditions* alone: the texture and configuration of societies and the ways they may have developed within their particular environmental niches have to be taken fully into account.

By synthesizing the evidence derived from environmental and climate proxies along with that from historical and archaeological sources, the research presented in this volume illuminates these complex causal relationships and begins to disentangle the socio-environmental interconnections that generated the outcomes reflected in the data. This work thus also contributes towards identifying another current scientific concern, namely the factors that contribute to societal resilience and sustainability, as well as those that introduce vulnerability. At the same time, in bringing together the work of a diverse and interdisciplinary group of scholars—archaeologists, ecologists, paleoscientists, and historians—this volume contributes to the establishment of a common methodologically-innovative framework that can contribute to a better understanding of the past and thereby to broader scholarly and public science discourses.

A Few General Conclusions

A number of key conclusions are worth highlighting. First, and as already intimated above, it seems clear that climate impacted societies but also that different societal systems had (and have) different degrees of flexibility or resilience to impacts. These differences appear to depend on two key elements of any particular socio-environmental interaction: first, the severity and duration of the climatic or environmental challenge or stress and, although often remaining implicit in these chapters, the nature of societal/cultural habituation and response; and second, the pre-existing context in terms of societal complexity and the general environmental situation. We see plenty of evidence for adaptation, varying in nature by context and by preexisting conditions.

A second important general point that comes out of these chapters is the remarkable degree of resilience expressed by most societies over the longer term (something that also applies globally, although of course this volume focuses on Anatolia alone). As was stressed also in the symposium that led to this volume, it seems very clear that "collapses" that entailed societal breakdown at all levels—real systemic collapses in the fullest sense—were very rare. To the contrary, while very significant and often dramatic transformations can be found throughout the Anatolian past, close examination and careful juxtaposition of changes within their chronological and geographical contexts (where exactly and over what period of time did various levels of change in socio-economic, political, or cultural life take place?) suggest that most major historical transformations affected different parts of societies in different ways and with different outcomes, even if, when one compares the situation before and after, the earlier systems are no longer recognizable. What is interesting for Anatolia, but elsewhere also, of course, is assessing how such collapses and transformations appear in different types of data—historical, archaeological, and paleoenvironmental—and how the causal mechanisms actually worked.

Implicit in the authorship of most of the chapters of this volume is a third general key conclusion: addressing these sorts of questions and confronting the challenges they present requires a high level of collaboration. Only through such interdisciplinary collaboration can meaningful results be achieved that appropriately incorporate both earth and atmospheric science, as well as social science research.

Some Prospects for the Future

Taken as a whole, the chapters in this volume point to several additional areas in which ongoing and important discussions are likely to continue into the future. First are questions about the role and value of climate models and modeling, on the one hand, and of establishing clear interpretive frameworks within which to situate data and also from which to establish solid comparative bases for further discussion, on the other. With respect to climate modeling, the significant differences in spatial and temporal resolution, the representativeness of the forcings that are integrated, and the intensiveness of the required computational processing and resultant gargantuan output storage requirements remain important considerations. A key issue arising here is the extent to which a wide variety of inputs can be maintained while reducing temporal and increasing spatial resolutions in order to work more easily with data at historically and archaeologically meaningful scales (annual, decadal, generational, centennial).

Sampling data at appropriate temporal and spatial scales is also clearly essential beyond climate models and is what enables identifying and quantifying change within any interpretive framework. The interpretive frameworks invoked in this volume highlight not just the cyclical nature of socio-environmental systems (e.g., panarchy, resilience, and adaptive cycles), but importantly also their accretive or sedimentary natures, with each cycle building on the foundations of the previous. Thus the chapters collectively emphasize identifying antecedent states and "second natures." While these concepts have been considered important in analyses of archaeological materials for some time now—especially those from surveys, where the "palimpsest" or accretive nature of the surface record has long been recognized—applying them to a wider range of investigations into human-environment interactions may prove helpful moving forward.

Together with frameworks for analyzing and interpreting socio-environmental change, an essential methodological point on this topic is our basic ability to identify and quantify diachronic change itself. From using statistical coefficients to differentiate and highlight distinct patterning in land-use changes based on surface materials, to using pollen evidence to identify the effects of anthropic pressure on "initial states" of vegetation cover, the chapters of this volume demonstrate the value of and continuing need to develop innovative ways of comparing the synchronic snapshots of particular societal and environmental conditions provided by our datasets in order to create increasingly accurate understandings of the moving picture that is the record of past socio-environmental interaction. Developing methods of identifying and quantifying change runs parallel to continuing efforts to coordinate data from environmental and social systems that are recorded at differing scales and resolutions. At least one chapter in this volume offers an example of how to achieve this through a comparison of long-term climate and population data.

Other topics that will undoubtedly contribute to ongoing discussions include the transformative impacts of certain societal practices on the environment in light of discourse around the Anthropocene. The chapters in this volume bear clear evidence of enduring societal effects on the environment, including the coppicing and pollarding of oak woodlands already in the Anatolian Neolithic, as attested by anthracological analyses; a distinct proto-Beyşehir Occupation Phase resulting from socio-ecological, cultural, and economic developments evident in paleoenvironmental and archaeological records; changes to specifically Roman landscape signatures resulting from Turkish migrations and identifiable in palynological data; and economically driven desiccation of wetlands in Ottoman times revealed through archival research. The diachronic nature of this evidence should be useful to those engaged in the lively discussion about and critique of the use of the term Anthropocene, the identification of its onset, and its different and sometimes conflicting definitions according to different disciplinary perspectives.

Perhaps one of the most valuable outcomes of the volume remains the general discussion arising from most chapters about where the primary gaps in data or knowledge reside in current research on environment and society in Anatolia, in terms of both methodology and material evidence as well as geographical and chronological coverage. While such discussion needs to and will continue, this volume—following in the footsteps of the symposium from which it derives—takes important steps towards filling some of these gaps and identifying productive and functioning frameworks within which ongoing as well as future research can be integrated and within which all the different specialist fields that are essential to the enterprise of studying socio-environmental interactions can be accommodated. We hope that the authors of the chapters in this volume as well as their readers will continue to engage in this discussion to fill gaps in knowledge in diverse and innovative ways, and we look forward to seeing their results.

CONTRIBUTORS

Bülent Arıkan is an Associate Professor and Chair of the Department of Ecology and Evolution in Istanbul Technical University's Eurasia Institute of Earth Sciences, and the Director of its Applied Research Center for Climate Change. He is an anthropological archaeologist, and his research focuses on paleoenvironmental changes, social complexity, and human impacts on environments in the ancient Near East. Paleoclimate modeling, agent-based modeling of past land use, and landscape evolution are among the models he has been using. He ran several projects across Anatolia. He recently completed the project titled *From the Holocene to the Anthropocene: A New Look at the Ten-Thousand-Year Long History of Human, Environment, and Climate Relationships from the Perspective of Computational Sciences* (TÜBİTAK Grant# 215K130). This project combined paleoclimate modeling and multi-proxy-based paleoenvironmental reconstructions in central and southwestern Anatolia. The results of paleoclimatic changes for the 10,000-year-long period have been assessed to contextualize the social organization changes during prehistoric and protohistoric periods. Finally, two Early Bronze Age sites at different parts of the region were subjected to agent-based land-use modeling, revealing diverse land-use patterns. He is currently leading the Applied Research Center for Climate Change, and he is developing a field project.

<bulent.arikan@gmail.com>

Eleni Asouti is a Reader in Environmental Archaeology in the Department of Archaeology, Classics and Egyptology, University of Liverpool. Her research focuses on late Paleolithic and Neolithic human adaptations to climate and environmental change in southwest Asia and the eastern Mediterranean. She specializes in anthracology, the analysis of charred fuelwood and timber remains retrieved from archaeological sites, having worked with several projects in Anatolia. She also directs the Eastern Fertile Crescent (EFEC) prehistory project, excavating and surveying Epipaleolithic sites in the northwest Zagros, investigating the lifeways of pre-agricultural communities and paleoenvironments during the Pleistocene-Holocene transition.

<E.Asouti@liverpool.ac.uk>

Semih Çelik is a postdoctoral researcher and part time lecturer in the Department of History at Koç University. His research focuses on a wide range of topics in the field of Ottoman and Middle Eastern environmental history. He has published articles on human-animal relationships, famines, micro-ecologies, and climate change in Anatolia covering a period between the seventeenth and early twentieth century. Most recently, he has been given funding by The Scientific and Technological Research Council of Turkey (TUBITAK) for a project on the agricultural history of northwestern Anatolia.

<semihcelik@ku.edu.tr>

Mustafa Doğan concluded his Masters degree with a study of Mucur Obruk Lake, Central Anatolia (Kırşehir-Mucur), aiming to explain paleo-vegetational changes through the use of pollen analysis. His PhD research follows similar lines but plans to develop palynological and paleoecological approaches with a specific focus on charcoal and μ-XRF analysis. He is particularly interested in Quaternary paleoecological changes including vegetation, paleoclimate, and land-use and aims to understand long-term paleo-ecological changes in Anatolia over the last 20,000 years.

<mustafadogan@sdu.edu.tr>

Uğur Doğan is a Professor at Physical Geography in the Department of Geography, Ankara University. He has conducted research in various fields of geomorphology and Quaternary environments. His main research topics concern fluvial, karst, coastal, and tectonic geomorphologies and geoarchaeology. He partially focuses on Quaternary climate changes and its effects on Quaternary landscapes of Turkey.

<geoankara@gmail.com>

Warren J. Eastwood is a biogeographer and paleoecologist who researches past environmental change for the last 25,000 years or so in the eastern Mediterranean region, where he has worked for the past 30 years. His main specialism is elucidating natural versus human-induced vegetation change using pollen analysis (palynology) and works closely with archaeologists and historians. His research interests also include climate change and the impact of major volcanic eruptions and tephrochronology of volcanic ash layers preserved in lake and peat sediment archives. Warren is currently a member and Honorary Secretary of the British Institute at Ankara (BIAA).

<hon.secretary@biaa.ac.uk>

Hugh Elton is a Professor in Ancient Greek and Roman Studies at Trent University and served as permit holder for the Avkat Archaeological Project. He has been Director of the British Institute at Ankara and also directed the Göksu Archaeological Project in Turkey. His research focuses on the ways in which the Roman state worked in Late Antiquity. He has published numerous works examining the practicalities of ruling the empire, the impact that the Roman state had on Anatolia, especially on rural areas, and the effectiveness of the Roman army. His most recent book is *The Roman Empire in Late Antiquity: A Political and Military History* (Cambridge, 2018).

<hughelton@trentu.ca>

Josh England concluded his Master's degree at Trent University with a study of olive cultivation within the territory of the Roman city of Sagalassos, Pisidia, focusing on cultivation occurring at an elevation above 1000 masl. His current interests are in arboriculture in mountainous regions and agricultural decision-making processes. He is also a member of the Archaeological Institute of America.

<josh.robert.england@gmail.com>

Patrick Free obtained a BSc Hons in Physical Geography at the College of Life Sciences, University of Birmingham. As part of his studies, Patrick's undergraduate dissertation combined historical, archaeological, and paleoecological evidence to investigate climatic versus societal controls on high-elevation olive cultivation in southwestern Turkey. Patrick is currently pursuing a career in ecological restoration and is presently working on Rewilding Europe's Central Apennines, Italy project.

<patrickfree4@gmail.com>

Anneley Hadland (née McMillan) completed her PhD at the School of Geography, Earth and Environmental Sciences, University of Birmingham in 2013. Anneley's PhD utilized a novel suite of GIS models to investigate paleo vegetation changes in southwestern Turkey with an emphasis on the nature and character of high-elevation olive cultivation. Since completing her PhD, Anneley has become an experienced spatial technology consultant, working on a variety of Environmental, Emergency Management, and Defence applications, and is currently Head of Remote Sensing at Helyx SIS Ltd.

<a.hadland@helyx.co.uk>

John Haldon is Shelby Cullom Davis '30 Professor of European History Emeritus at Princeton University. He is currently Director of the Princeton *Climate Change and History Research Initiative* and Director of the *Program in Medieval Studies Environmental History Lab.* He is a member of the Archaeological Institute of America and the British Institute in Ankara, a corresponding member of the Austrian Academy of Sciences, and is also President of the International Association of Byzantine Studies. His research focuses on the history of the medieval eastern Roman (Byzantine) empire, in particular in the period from the seventh to the twelfth centuries; on state systems and structures across the European and Islamic worlds from late ancient to early modern times; on the impact of environmental stress on societal resilience in pre-modern social systems; and on the production, distribution, and consumption of resources in the late ancient and medieval world. He has published many books and articles addressing issues within these fields, most recently *A tale of two saints: the passions and miracles of Sts Theodore 'the recruit' and 'the general'* (Liverpool UP 2016); *The empire that would not die. The paradox of eastern Roman survival 640-740* (Harvard UP 2016); *Archaeology and Urban Settlement in Late Roman and Byzantine Anatolia: Euchaita-Avkat-Beyözü and its Environment* (ed. with Hugh Elton and James Newhard) (Cambridge UP 2018); and *The De Thematibus ('on the themes') of Constantine VII Porphyrogenitus. Translated with introductory chapters and detailed notes* (Liverpool UP 2021).

<jhaldon@princeton.edu>

Adam Izdebski is an environmental historian specializing in the history of the pre-industrial Mediterranean and Central Europe; associate professor at the Jagiellonian University in Krakow, Poland; and independent research group leader at the Max Planck Institute for the Science of Human History in Jena, Germany.

<izdebski@shh.mpg.de>

Ceren Kabukcu is a Leverhulme Early Career Fellow in the Department of Archaeology, Classics and Egyptology, University of Liverpool. Her research focuses on the evolution of plant food use, agricultural origins, and the vegetation history of southwest Asia. She specializes in the analysis of archaeological charred plant remains from late Paleolithic and Neolithic sites in Anatolia and the Zagros.

<C.Kabukcu@liverpool.ac.uk>

Ahmet Köse is a Lecturer in the Geographic Information Systems (GIS) program at Isparta University of Applied Sciences, Turkey. His PhD research synthesizes environmental data from Turkey using multivariate and quantitative methods in GIS and statistics-based work. Engaging with both present-day and paleoecological variables from the perspective of nature and human interactions, his research aims to reveal ecological regionalization in Turkey.

<ahmetkose1992@gmail.com>

Fikri Kulakoğlu is the Chair of the Department of Archaeology at Ankara University, Ankara, Turkey, where he teaches Ancient Near Eastern Archaeology. His current research interests include mainly Early Bronze Age, Assyrian Trade Colony Period, and Hittite Culture in Anatolia. He has participated in several major excavations in Anatolia, such as Samsat, Acemhöyük, Kaman-Kalehöyük, and Kültepe-Kanesh, where he assumed the directorship of excavations after Tahsin Özgüç. Since 2009, the second term of excavations at Kültepe have focused primarily on the mound, adopting new research strategies based on a multidisciplinary approach, cooperation, and being open to all disciplines.

<kulakoglu@yahoo.com>

Christina Luke is an Associate Professor in the Department of Archeology and History of Art at Koç University. Her research interests include field archaeology, as well as heritage and diplomacy in the eastern Mediterranean, Balkans, the Middle East, and North Africa, as well as Central America. Her current fieldwork pivots around the Marmara Lake Basin of Manisa, Turkey. In addition, she is engaged in archival research that explores the intersection of political ecologies and environing from the late seventeenth century through the Cold War. Her articles have appeared in many journals, including *Anatolian Studies, European Journal of Archaeology, History and Anthropology, International Journal of Cultural Policy, International Journal of Historical Archaeology,* and the *Journal of Social Archaeology*. Her books include *A Pearl in Peril: Heritage and Diplomacy in Turkey*, published by Oxford University Press. She also serves as editor of the *Journal of Field Archaeology*.

<christinaluke@ku.edu.tr>

Lee Mordechai is a Senior Lecturer at the Hebrew University of Jerusalem and Associate Director of Princeton University's *Climate Change and History Research Initiative*. His research examines premodern disasters in the eastern Mediterranean ranging from earthquakes in Late Antique Syria to flooding in Bronze Age Mesopotamia. His recent project focuses on the so-called Justinianic Plague in the Eastern Roman Empire. In all of the above, he adopts an interdisciplinary perspective, integrating methodologies from the digital humanities and the sciences, as well as frameworks from resilience and disaster studies. Lee has published his research in journals such as *Past & Present*, *American Historical Review*, and *Proceedings of the National Academy of Sciences*.

<lee.mordechai@mail.huji.ac.il>

James Newhard is Director of the Center for Historical Landscapes and Professor of Classics at the College of Charleston and served as the field director for the Avkat Archaeological Project. His fieldwork includes excavations and survey projects in Turkey, Greece, Albania, and the United States. His research focuses on the development of geospatial methods and informatics within landscape archaeology, the intersection between environmental and human histories, the emergence of complex state systems, and ancient economic processes.

<NewhardJ@cofc.edu>

Aziz Ören is an Assistant Professor in the Department of Geography at Süleyman Demirel University, Isparta, Turkey, with research interests in past environmental change in the eastern Mediterranean region. Having completed his Masters degree in 2012 and PhD in 2018, his main specialism is elucidating natural versus human-induced vegetation change using pollen analysis (palynology). His research interests also include analysis of lake and peat sediment archives. He works very closely alongside geologists, archaeologists, and historians. He is particularly interested in Quaternary environmental change including climate change, and human impact via pollen and spore analysis, non-siliceous fungal and algal microfossils, and charcoal in Late Quaternary and recent lake sediments and peat deposits.

<kuvaterner.paleo@gmail.com>

Charlotte Pearson is an Associate Professor at the University of Arizona's Laboratory of Tree-Ring Research, affiliated with the School of Anthropology and Department of Geosciences. She combines geoarchaeology and tree-ring research to study past human and environmental interactions. This includes dating archaeological wood samples using both dendrochronology and radiocarbon dating methods, synchronizing timelines based on event horizons in proxy records, and using chemical techniques on archaeological materials. Her current main research focus is on improving chronology for the period 2500–1100 BCE (and addressing debate over 'high' versus 'low' chronologies in the eastern Mediterranean) by creating high-precision datasets of annually resolved ^{14}C from calendar dated trees. She gained her PhD in Archaeological Science from the University of Reading, UK and pursued postdoctoral studies in dendroarchaeology at Cornell University as part of Peter Kuniholm's Aegean Dendrochronology Project, before joining the University of Arizona in 2013.

<clpearson@email.arizona.edu>

Neil Roberts is an emeritus professor of Geography at the University of Plymouth and visiting research fellow in Archaeology at Oxford University, UK. His research interests focus on understanding the long-term, two-way relationship between people, climate, and environment, including the analysis of lake sediment archives using diatom, stable isotope, pollen, and charcoal analyses; climate changes at the end of the last ice age and how they affected the emergence of Neolithic farming; and the human transformation of land cover in Europe and the Mediterranean during the Holocene. Following his doctorate at UCL and post-doctoral work at Oxford University, he was lecturer at Loughborough University (UK), where he wrote the first edition of his book *The Holocene*. In 2006, he was a member of the US National Academies committee on "Surface temperature reconstructions for the past 2000 years," established at the request of the US Congress. The following year he was visiting Blaustein Professor at Stanford University. He has also received an honorary doctorate from Ankara University. He has authored over a hundred refereed journal articles, has been an editor of *The Holocene* journal and *Quaternary Science Reviews*, and is currently editor-in-chief of the *Journal of Quaternary Science*.

<cnroberts@plymouth.ac.uk>

Christopher H. Roosevelt is a Professor in Koç University's Department of Archaeology and History of Art, Director of the Research Center for Anatolian Civilizations (ANAMED) and the Kaymakçı Archaeological Project, and Co-Director of Gygaia Projects. His work on Anatolian archaeology, landscape archaeology, digital archaeology, and related subjects has been published in books (*The Archaeology of Lydia from Gyges to Alexander* [Cambridge University Press, 2009]; *Gyges'ten Büyük İskender'e Lydia Arkeolojisi* [Koç University Press, 2017]), edited volumes (*Philanthropy in Anatolia through the Ages* [Istanbul: Vehbi Koç Foundation, 2020]; *Spatial Webs: Mapping Anatolian Pasts for Research and the Public*, [Koç University Press, 2021]), and many articles. He is a member of the Archaeological Institute of America and the British Institute in Ankara, a corresponding member of both the German Archaeological Institute and the Turkish Institute of Archaeology, and a fellow of the Society of Antiquaries, London.

<chroosevelt@ku.edu.tr>

Çetin Şenkul is an Associate Professor in the Department of Geography at Süleyman Demirel University, Isparta, Turkey. He is a Quaternary paleoecologist and a biogeographer with primary scientific interest in exploring microscopic biological remains (especially pollen) in all kinds of sediments all over Anatolia for the reconstruction of environmental settings, natural versus human-induced vegetation changes, and climate dynamics in the past. His research interests also include endemic plants and biogeographical modeling, enjoying establishing connections between extant and fossil taxa, and discovering their values as paleoecological indicators. In Quaternary paleoecology, he focuses on the Holocene dynamic of the forest-steppe ecotone working in Turkey in an archaeological and historical perspective.

<cetinsenkul@gmail.com>

Tevfik Emre Şerifoğlu is an archaeologist, art historian, and an assyriologist who is a research fellow at ANAMED and the current director of the Çadır Höyük excavations and the Taşeli-Karaman Archaeological Project. After completing his undergraduate education at Bilkent University, he continued his studies at the University of Cambridge, receiving his MPhil and PhD degrees in Assyriology/Ancient Near Eastern Archaeology. After teaching for several years at Çanakkale Onsekiz Mart University, he moved to eastern Turkey to start the Archaeology Department of Bitlis Eren University, and was the head of department until the end of 2018. He was a visiting researcher at the American Center of Oriental Research in Amman in 2011 (ARIT Mellink Fellowship), at the Oriental Institute of the University of Chicago in 2014 (Fulbright Fellowship), and at the Department of Archaeology and Ancient History of Uppsala University in 2019 (Swedish Institute Fellowship). He has also held a Newton Advanced Fellowship at the University of Leicester. He has previously co-directed several fieldwork projects including the Kilise Tepe Excavations (southern Turkey, 2009–13), the Cide Archaeological Project (survey, Black Sea coast, 2010–11), the Sirwan Regional Project (survey, northern Iraq, 2012–15), and the Lower Göksu Archaeological Salvage Survey Project (survey, southern Turkey, 2013–17).

<teserifoglu@gmail.com>

Madelynn von Baeyer is an archaeobotanist who uses botanical macrofossils to examine the reciprocal relationship between humans and plants. Particularly interesting to her is how environmental conditions and climate change impact agricultural systems and how changes in the human-environment relationship can foster cultural resilience in both ancient and modern populations. She works in the eastern Mediterranean and central Asia. She is currently a postdoctoral fellow at the Max Planck Institute for the Science of Human History on the FEDD: Fruits of Eurasia, Domestication and Dispersal team headed by Dr. Robert Spengler. Recently, she completed a Research Fellowship at the Harvard University Herbaria. Prior to that, she studied macrofossils from Çadır Höyük as part of her PhD project at the University of Connecticut.

<vonbaeyer@shh.mpg.de>

INDEX

T

U

V

W

Y

Z